4th International Conference on Computing and Wireless Communications Systems (ICCWCS 2022)

ITM Web of Conferences Volume 48 (2022)

Tangier, Morocco
21 - 23 June 2022

Editors:

Jamal Zbitou
Ahmed El Oualkadi
Mohamed Latrach
Larbi Setti

ISBN: 978-1-7138-6060-0

Printed from e-media with permission by:

Curran Associates, Inc.
57 Morehouse Lane
Red Hook, NY 12571

Some format issues inherent in the e-media version may also appear in this print version.

No changes have been made to the content of these proceedings. There may be changes to pagination and minor adjustments for aesthetics.

Printed with permission by Curran Associates, Inc. (2022)

For additional information, please contact EDP Sciences – Web of Conferences at the address below.

EDP Sciences – Web of Conferences
17, Avenue du Hoggar
Parc d'Activité de Courtabœuf
BP 112
F-91944 Les Ulis Cedex A
France

Phone: +33 (0) 1 69 18 75 75
Fax: +33 (0) 1 69 28 84 91

contact-edps@webofconferences.org

Additional copies of this publication are available from:

Curran Associates, Inc.
57 Morehouse Lane
Red Hook, NY 12571 USA
Phone: 845-758-0400
Fax: 845-758-2633
Email: curran@proceedings.com
Web: www.proceedings.com

TABLE OF CONTENTS

ICCWCS'22

Editorial Board

Organized by:

LabTIC

ENSA Tanger

جامعة عبد المالك السعدي
Université Abdelmalek Essaadi

Collaborations & Sponsoring

RMC

جامعة عبد المالك السعدي
Université Abdelmalek Essaadi

eseo GRANDE ECOLE D'INGENIEURS

Centre National pour la Recherche Scientifique et Technique

FACULTE DES SCIENCES ET TECHNIQUES SETTAT

جامعة الحسن الأول
UNIVERSITÉ HASSAN 1er

FACULTE DES SCIENCES ET TECHNIQUES TANGER

UQO UNIVERSITÉ DU QUÉBEC EN OUTAOUAIS

CIGMA

ENSA Tanger

ROYAUME DU MAROC
Ministère de l'Energie, des Mines et du Développement Durable

LabTIC

ENSA Fès

Imperial College London

Learning
UNIVERSITE HASSAN 1er

UNIVERSITÉ DE ROUEN NORMANDIE

Keynote Speakers

Prof. Pascal Dherbécourt, Institute of Technology in University of Rouen – Normandie, France

Prof. Ahmed Lakhssassi, Quebec University, Outaouais, Canada

Dr. Catalin Iulian Pruncu, Politecnico di Bari, Italy

Prof. Mostafa Hefnawi, Royal Military College (RMCC), Canada

Prof. Mohamed Latrach, ESEO Angers, France

Prof. Alexandre Jean René Serres, The Federal University of Campina Grande, Brazil

Prof. Abdulsalam Yassine, Associate Professor and Chair of the Software Engineering Department at Lakehead University, Thunder Bay, Ontario, Canada

Committee

Steering Committee

Ahmed Lakhssassi, University of Quebec in Outaouais, Canada
Mostafa Hefnawi, Royal Military College (Rmcc), Canada
Mohamed Latrach, Eseo-letr Angers, France
Rachid Mandry, Fsts, Hassan 1st University, Morocco
Abdellah Lakhouili, Fsts, Hassan 1st University, Morocco
Mohammed Aboulfatah, Fsts, Hassan 1st University, Morocco
Ahmed Errkik, Fsts, Hassan 1st University, Morocco
Larbi El Abdellaoui, Fsts, Hassan 1st University, Morocco
Abdelmounim Elhassane, Fsts, Hassan 1st University, Morocco
Jamal Zbitou, ENSA of Tangier, UAE, Morocco
Bahaj Mohamed, Fst of Settat, Hassan 1st University, Morocco

Organizing Committee

Jamal Zbitou, ENSA of Tangier, UAE, Morocco
Ahmed EL Oualkadi, ENSA of Tangier, UAE, Morocco
Ahmed Errkik, Fsts, Hassan 1st University, Morocco
Abdelali Tajmouati, Fsts, Hassan 1st University, Morocco
Mostafa Hefnawi, Royal Military College (Rmcc), Canada
Ahmed Lakhssassi, University of Quebec In Outaouais, Canada
Mohamed Latrach, ESEO, France
Bahaj Mohamed, Fst of Settat, Hassan 1st University, Morocco
Aytouna Fouad, ENSA of Tetuan, UAE, Morocco

Local Committee

Samadi Hassan, ENSA of Tangier, UAE, Morocco
Chahboun Asaad, ENSA of Tangier, UAE, Morocco
Chahboun Nouha, ENSA of Tangier, UAE, Morocco
Chraibi Lotfi, ENSA of Tangier, UAE, Morocco
El haddad Mohammed, ENSA of Tangier, UAE, Morocco
El mamouni Anass, ENSA of Tangier, UAE, Morocco
Ezzine Abdelhak, ENSA of Tangier, UAE, Morocco
Ghailani Mohamed, ENSA of Tangier, UAE, Morocco
Laaziz Yassin, ENSA of Tangier, UAE, Morocco
Moussaoui Mohamed, ENSA of Tangier, UAE, Morocco
Mezroui Soufiane, ENSA of Tangier, UAE, Morocco
Houda Bel Mokadem, ENSA of Tangier, UAE, Morocco
Cherkaoui Abdeljabbar, ENSA of Tangier, UAE, Morocco

Technical Program Committee

Semma Alami Fsts, Hassan 1st University, Morocco
Jamil Akhtar, Manipal University Jaipur (Muj), India
Nijas C M, Cochin University of Science and Technology, India
Luay H. Tahat, Gulf University for Science and Technology, Kuwait
Larbi El Abdellaoui, Fsts, Hassan 1st University, Morocco
Aboelmagd Noureldin, Royal Military College of Canada
Mostafa Baghouri Ensam Casablanca Morocco
Ahmed Rhallabi, Pcm Imn Nantes, France
Ahmed Lakhssassi, University of Quebec, Canada
Abdelali Tajmouati, Fsts, Hassan 1st University, Morocco
Fumie Costen, University of Manchester, United Kingdom
Rachid EL Alaiji Ensa of Tangier, Uae, Morocco
Saad Chakkor Ensa of Tangier, Labtic, University of Abdelmalek Essaâdi Morocco
Hossam Hassanein, Queen's University, Kingston, Ontario, Canada
Laaziz Yassin, Ensa of Tangier, Uae, Morocco
Asmaa Zugari Fs Tetuan Uae, Morocco
Ahmed El Oualkadi , ENSA of Tangier, UAE, Morocco

Samira Khoulji, Ensa of Tetuan, Uae, Morocco
Mourad Nedil, University Of Quebec, Abitibi-Témiscamingue (Uqat), Canada
Mostafa Hefnawi, Royal Military College, Canada
Ahmed Errkik, Fsts, Hassan 1st University, Morocco
Rouijaa Hicham Fsts Uhi Morocco
Mounir Arioua, Ensa of Tetuan, Uae, Morocco.
Mohamed Latrach, Eseo-Ietr Angers, France
Abdelmounim Elhassane, Fsts, Hassan 1st University, Morocco
Mohammed El Gibari, Ietr – University of Nantes, France
Naoufal Raissouni, Ensa of Tetuan, Morocco
Mohammed El Haj Tirari, Insea of Rabat, Morocco
Olivier Latry, University of Rouen Normandy, France
Ahmed Sulyman, Embry-Riddle Aeronautical University, Usa
Tayeb A. Denidni, Inrs University, Montréal (Quebec), Canada
Elkhider Sidi Ahmed, University of Nantes, France
Adnane Latif, Ensa of Marrakech, Morocco
Eric Joubert , University Of Rouen Normandy, France
Abdelhamid Zouhair, FST of Tangier, Morocco
Alexandru Vulpe, Politehnica University of Bucharest, Romania
Mohamed Moussaoui, Ensa of Tangier, Uae, Morocco
Catalin Iulian Pruncu, Politecnico Di Bari, Italy
Najib El Ouanjli Fsts, Hassan 1st University, Morocco
Hassan Ammor, Emi Rabat, Morocco
Ankur Singh Bist, Kiet Ghaziabad, India
Vladescu Alina, National Institute of Optoelectronics, Romania
Otman Chakkor, Ensa of Tetuan, Morocco
Pascal Dherbécourt, University of Rouen Normandie, France
Sorin Vlase, University of Transilvania, Romania
Ridouane Er-Rebyiy Emsi Marrakech Morocco
Mehdi Rahmati, Rutgers University, The State University of New Jersey, Usa
Hichame Maanane, Thales Group, France
Boutaina Hdioud, Ensias of Rabat, Morocco
Cosmin Danisor, Politehnica University of Bucharest, Romania
Otman Oulhaj, Fp of Larache, Abdelmalek Essaâdi University, Morocco
Rafik Lasri Fp Of Larache, Uae Tetuan Morocco
Rachid Mandry Fsts Uh1 Morocco
Mohamed Massoudi, University of Rouen Normandy, France
Ana Dragulinescu, Politehnica University of Bucharest, Romania
Cherkaoui Abdeljabbar Ensa of Tangier, Uae, Morocco
Alexandre Jean René Serres, Federal University of Campina Grande, Brazil
Robson Hebraico Cipriano Maniçoba, State University of Southwest Bahia, Brazil
Aziz Mahboub Fst of Tangier Uae Morocco
Elharoussi Mustapha Fsts Uh1 Morocco
Nasiri Badr Fsts Uh1 Morocco
Bennani Dosse Saad Ensa of Fez Morocco
Abdelhak Ezzine, Ensa of Tangier, Uae, Morocco
Kulwant Singh, Manipal University Jaipur, India
Ghailani Mohamed, Ensa of Tangier, Uae, Morocco
Anass Elmamouni Ensa of Tangier, Uae, Morocco
Abdellfattah Ba-Razzouk Fsts Uh1 Morocco
El Allami Mhammed Crmef Settat Morocco
Lahcen Moumoun Fsts Uh1 Morocco
Youssef Ben Youssef Ensa Berrechid Uh1 Morocco
Philippe Eudeline, Thales Group, France
Aytouna Fouad, Ensa of Tetuan, Uae, Morocco
Marius Vochin, Politehnica University of Bucharest, Romania
Divya Rishi Shrivastava, Manipal University Jaipur, India
Noha Chahboun, Ensat Uae, Morocco
Sarfaraz Nawaz, Swami Kesvanand Institute of Technology, Jaipur, India
Issam Zahraoui Limie Laboratory Isga Casablanca, Morocco
Said Harchi Limie Laboratory Isga Casablanca, Morocco

Preface

The Fourth Edition of the International Conference on Computing and Wireless Communication Systems (ICCWCS 2022) has been held in ENSA of Tangier, Abdelmalek Essaadi University, Tetuan, Morocco, during June 21-23, 2022. This event is organized by LabTIC (ENSA of Tangier, Abdelmalek Essaadi University) in collaboration with Association "MRSI". The ICCWCS'22 conference is intended to provide an ideal platform to discuss, to share experiences and to open opportunities for collaboration between the different existing working groups in the areas of engineering and information technology. It is also an excellent opportunity for young researchers to start in the work of publication and dissemination of their work and contributions of research.

The conference tracks were: Track 1 - Antennas & Propagation; Track 2 - Passive Components, Circuits and Subsystems; Track 3 - Active devices; Track 4 – Renewable Energy and Track 5 - Computer Science, Intelligent Systems and Information Technologies. In addition to the presentations brought by the authors, there were seven invited Keynotes Speakers.

It was also a great pleasure to work with such an excellent organizing committee team for their hard work in organizing and supporting the conference. In particular, both the Local and Technical Program Committees. Many thanks to all the participants, Keynotes Speakers, organizing and TPC committees and sponsors. We strongly believe that Fourth Edition of ICCWCS conference provides a good forum for all researcher, developers. We also expect that the future Editions of ICCWCS conference will be as successful, as indicated by the contributions presented in this volume.

Proceedings editors:

Prof. Jamal ZBITOU. LabTIC, ENSAT, Abdelmalek Essaadi University, Tetuan, Morocco.
Prof. Ahmed EL OUALKADI. LabTIC, ENSAT, Abdelmalek Essaadi University, Tetuan, Morocco.
Prof. Mohamed LATRACH. ESEO Angers, France.
Prof. Larbi SETTI. FPL, Abdelmalek Essaadi University, Tetuan, Morocco.

Statement of Peer review

In submitting conference proceedings to *Web of Conferences*, the editors of the proceedings certify to the Publisher that

1. They adhere to its **Policy on Publishing Integrity** in order to safeguard good scientific practice in publishing.
2. All articles have been subjected to peer review administered by the proceedings editors.
3. Reviews have been conducted by expert referees, who have been requested to provide unbiased and constructive comments aimed, whenever possible, at improving the work.
4. Proceedings editors have taken all reasonable steps to ensure the quality of the materials they publish and their decision to accept or reject a paper for publication has been based only on the merits of the work and the relevance to the journal.

Title, date and place of the conference

4th EDITION OF THE INTERNATIONAL CONFERENCE ON COMPUTING AND WIRELESS COMMUNICATION SYSTEMS.
21-23 June 2022 ENSA of Tangier, Morocco

Proceedings editor(s):

Prof. Jamal ZBITOU. LabTIC, ENSAT, Abdelmalek Essaadi University, Tetuan, Morocco.
Prof. Ahmed EL OUALKADI. LabTIC, ENSAT, Abdelmalek Essaadi University, Tetuan, Morocco.
Prof. Mohamed LATRACH. ESEO Angers, France.
Prof. Larbi SETTI. FPL, Abdelmalek Essaadi University, Tetuan, Morocco.

Date and editor's signature

04/08/2022

J. ZBITOU
A. El OUALKADI
M. LATRACH
L. SETTI

Real-Time Multiple Input Multiple Output (MIMO) Radar Using Software Defined Radio

Robert Gilpin[1,*], *Joey* Bray[1], and *Mostafa* Hefnawi[1]

[1] *Royal Military College of Canada, Electrical Engineering Department, Kingston, Ontario, Canada*

Abstract. In this paper, commercially-available software-defined radios (SDRs) are used to build a 64-channel, reconfigurable Active Electronically Scanned Arrays (AESA) radar operating in C-band (NATO G-band). The SDRs are used to design and implement a 3-dimensional multi-input and multi-output (MIMO) radar. The flexibility of the SDRs has been harnessed to evaluate the performance of a linear frequency modulated continuous wave (LFMCW) MIMO radar using three different methods of achieving the orthogonality, namely Time Division Multiplexing (TDM), Frequency Division Multiplexing (FDM), and Code Division Multiplexing (CDM). In addition, the radar's parameters are user-selectable and can be rapidly changed such that the radar can be used in different environments without requiring changes to the hardware. Measurements indicate that the radar is capable of detecting and localizing multiple targets in all 3-dimensions, including bearing, range, and Doppler. The MIMO radar operates in real-time, with a refresh rate of only 3 seconds. Experimental results are generated for the TDM mode of operation with further research reporting on the CDM and FDM modes of operation.

1 Introduction

A modern military radar provides situational awareness to the operator and the larger combat system in real time. The radar must be able to sense the entire environment to detect and track multiple targets. The faster a target can be detected and tracked, the more time the operator has to interpret the target and eliminate a threat, if required. As weapons, particularly missiles, continually increase in speed and improve in accuracy, an operator or anti-missile combat system needs as much time as possible to counter the threat posed [1].

Radars have traditionally consisted of customized equipment that cannot be easily reconfigured or modified. Although improvements to the radiofrequency front-end, such as the adoption of Active Electronically Scanned Arrays (AESA) have provided flexibility to the radar's beamforming, sector scans are often performed using conventional techniques in which a transmit beam is slowly steered through a sector to illuminate targets in a sequential manner, leaving the radar blind to threats when the transmit beam is pointed in other directions. Recent research in the field of communications has revealed a new scanning mode, known as the multiple input, multiple output (MIMO) mode. MIMO can be used by an AESA to illuminate the entire field of view (FOV) of the radar simultaneously, where beamforming is done in post-processing, thus providing uninterrupted situational awareness. However, the MIMO radar mode is more challenging to implement because of the added requirements of orthogonality between its channels, and because of the drastic increase in the required digital signal processing.

In this research, commercially available software-defined radios (SDRs) are used to build a 64-channel, reconfigurable AESA radar operating in C-band (NATO G-band). The SDRs are used to design and implement a 3-dimensional MIMO radar. For the first time, the flexibility of the SDRs have been harnessed to evaluate the performance of three different methods of achieving the orthogonality required for MIMO using a linear frequency modulated continuous waveform (FMCW). In addition, the radar's parameters are user-selectable and can be rapidly changed such that the radar can be used in different environments without requiring changes to the hardware.

This article provides a brief background section on radar basics and the concept of MIMO radar. Following the background, the MIMO radar design will be described. A selection of results is presented to emphasize the importance of MIMO radar research followed by potential areas for further research and a conclusion.

2 Background

2.1 Radar Basics

RAdio Detection And Ranging (RADAR) systems detect targets or objects by sending out high-frequency Electromagnetic (EM) waves that are reflected off the object and that propagate back to the radar to determine range, radial velocity and angle. A diagram showing the elements of a traditional, single source generic radar is shown in Fig. 1. The transmitter generates the radar waveform which

*e-mail: Robert.Gilpin@forces.gc.ca

 https://doi.org/10.1051/itmconf/20224801001

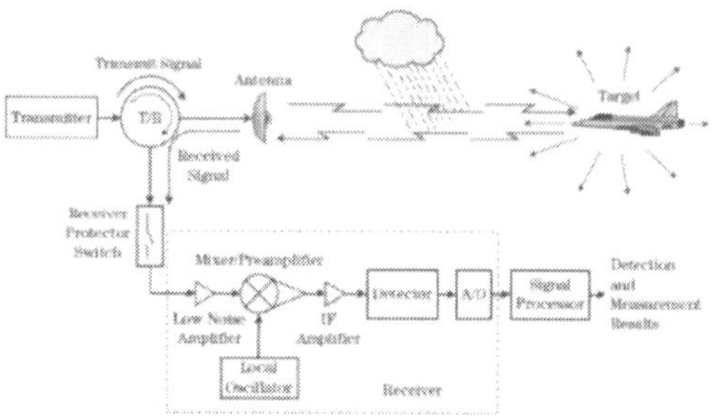

Figure 1. Basic elements of a radar including a target and atmospheric factors [2]

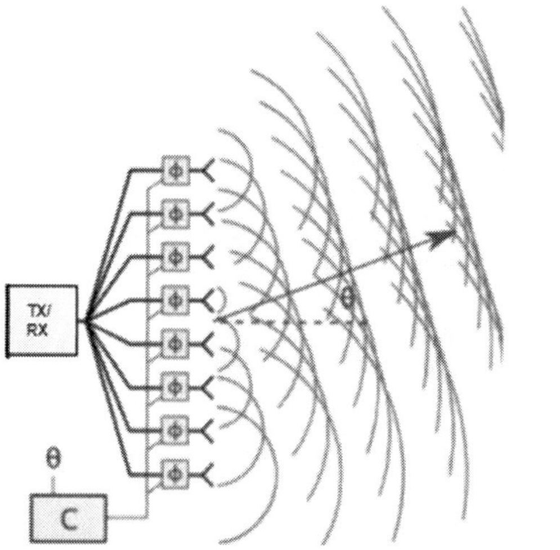

Figure 2. PESA radar architecture showing a single transmitter (TX), phase shifters (ϕ) being controlled by a computer system (C) with a desired transmit angle (θ). The red hemispherical lines show the direction of EM radiation [4]

is sent to the antenna (through a duplexer in some cases) and into free space. The EM wave travels to the target, is reflected and returns to the antenna. The signal from the antenna enters the receiver and is processed to determine the target range, velocity and azimuth [2].

2.2 MIMO Radar

The radar that has gained popularity due to its ability to scan a large FOV is the Phased Array Radar (PAR) [3]. To understand MIMO radar a brief description of PARs is required. First, a Passive Electronically Scanned Array (PESA) will be described followed by an Active Electronically Scanned Array (AESA) and finally a MIMO radar will be described.

Fig. 2 shows a traditional PAR known as a PESA. In the PESA, a single transmitter generates a high-powered signal that is transmitted through individual antenna elements. Each transmit element has an associated computer-

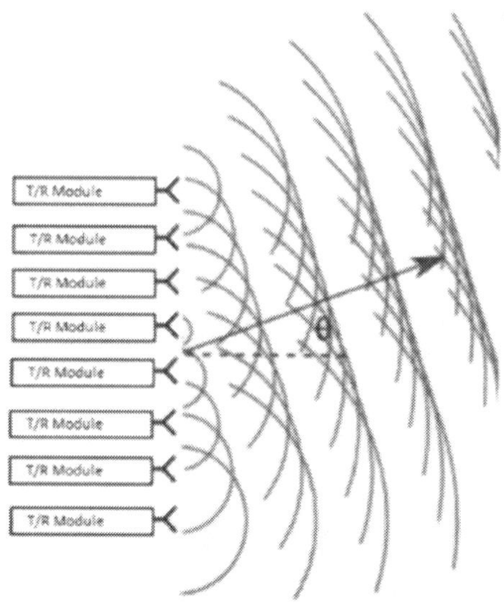

Figure 3. AESA radar architecture showing multiple T/R transmitting at a desired angle (θ). The red hemispherical lines show the direction of EM radiation [4]

controlled phase shifter (ϕ) that creates a directional beam by adding a phase shift to the signal from each element. The narrow beam is sequentially steered through the FOV by changing the phase at each element.

Fig. 3 displays the second generation of phased array radar, the AESA radar. This is the current technology used in modern radars like the AN/SPY family of PARs. Each element of the AESA is a Transmit Receive (TR) module which gives access to element control of the array. The AESA can create multiple beams using sub arrays which decreases the scan time of radar and allows the AESA to perform multiple functions at once, i.e. track while scan with multiple beams. Multiple beams allow for faster scanning however the radar requires time to scan the entire FOV [3].

Figure 4 shows an extension of the AESA known as MIMO radar. Rather than using sub arrays within the radar to form multiple beams like the AESA, each TR element acts independently, transmitting an orthogonal waveform. The transmit beams are wide, 180 degrees typically, and can illuminate the entire FOV with one sequences of pulses since there is no directivity on transmit, this is sometimes referred to as scan on receive [5].

To take advantage of MIMO technology the receiver must be able to separate the transmitted orthogonal signals on reception. This is done using Time Division Multiplexing (TDM), Frequency Division Multiplexing (FDM), or Code Division Multiplexing (CDM) [6]. For this paper only TDM will be discussed with FDM and CDM measurements happening in future publications.

In TDM each TR module transmits at a specified time interval ensuring the waveforms do not overlap. A figure representing TDM is shown in Fig. 5 with a FMCW chirp

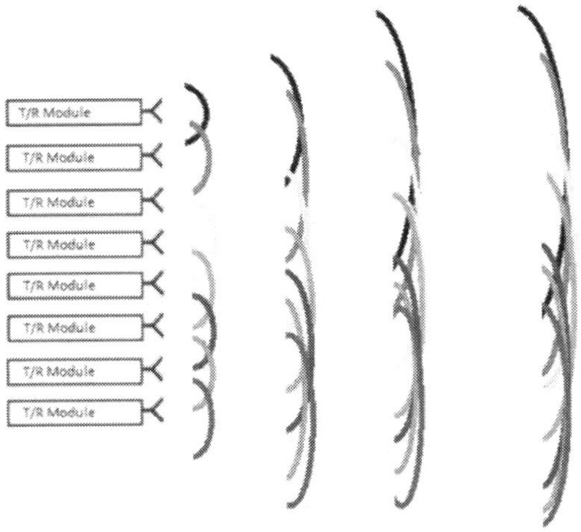

Figure 4. MIMO Radar (AESA - MIMO Mode) where each colour represents an orthogonal waveform transmitted from a TR Module

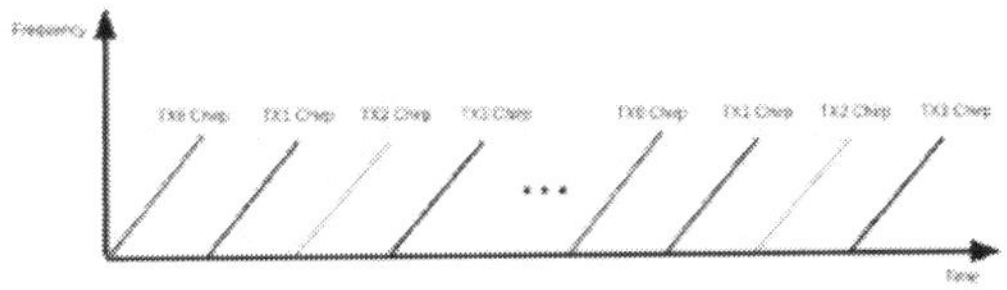

Figure 5. FMCW Signal using TDM, the TX indicates which TR module is transmitting

as the base waveform with 4 TR modules. Although easy to implement with perfect orthogonality between transmit channels, TDM is slow because each transmit element is given a place in time. This is not the case in FDM and CDM as those forms of multiplexing all signals are sent simultaneously. TDM servers as an excellent proof of concept before moving onto the mover complex multiplexing strategies.

3 MIMO Radar Design

With an understanding of PARs and MIMO radar theory the next section discuses the MIMO radar design. It describes the five phase process used to design and build an AESA radar capable of MIMO modes of operation.The design is broken down into five phases shown in Fig. 6.

In phase 0, the radar performance parameters are determined for a short-range surveillance AESA-MIMO radar using basic radar theory found in [2] or any other radar text book. To be able to test the radar in a lab environment two sets of requirements are created, a lab and a surveillance mode, shown in Table 1.

To obtain the required performance described in Table 1, Phase 1 of the design uses radar theory to determine the required specifications of the RF front and DSP algorithm

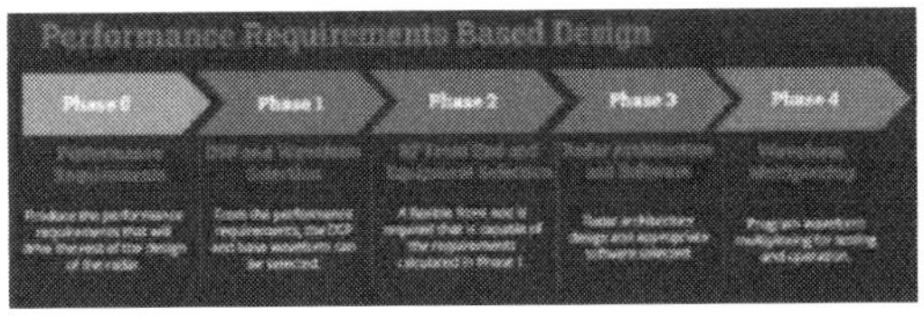

Figure 6. MIMO Radar Design Phases

Parameter	Symbol	Unit	Surveillance Mode	Lab Mode
Maximum Range	R_{max}	m	150	50
Range Resolution	R_{res}	m	0.75	0.75
Maximum Velocity	v_{max}	m/s	10	1.5
Minimum Velocity	v_{min}	m/s	-10	-1.5
Velocity Resolution	v_{res}	m/s	0.01	0.01
Angular Resolution	θ_{res}		2	2
Scanning Ability	N/A	N/A	Azimuth Only	Azimuth Only
MIMO Modes	n/a	n/a	CDM, TDM, FDM	CDM,TDM,FDM

Table 1. Radar Performance Requirements

Parameter	Surveilance Mode	Lab Mode
AESA Antenna Size (TX/TR)	8x8	8x8
Operating Frequency (f_0)	5 GHz	5 GHz
Bandwidth (B)	200 MHz	200 MHz
ADC Sampling Frequency ($f_{samp,min}$)	204 MHz	204 MHz
Coherent Operation Capable	Yes	Yes
Range Application	Medium = 150 m	Short = 50 m
Range Resolution	0.75 m	0.75 m

Table 2. MIMO Mode AESA Radar Minimum Performance Parameters

which are displayed in Table 2. Based on the DSP parameters and the short-range application, the FMCW waveform is selected as the base waveform.

Once the radar parameters from Phase 1 are calculated, Phase 2 of the design selects the RF and processing equipment for the radar. The cornerstone of the radar is the Ettus Research N-series (N321/N320) SDR [7]. Each radio has 2 transmit and 2 receive channels, each capable of 200 MHz of instantaneous bandwidth. To obtain the desired beamwidth of 2 degrees four SDRs were purchased to provide 8 transmit and 8 receive RF channels producing a 64 channel MIMO radar. Fig. 7 shows the 4 SDRs with associated cables and equipment. A 148-core host processor running Linus Mint and GNU Radio performs the required control and signal processing of the MIMO radar.

To construct the antenna array, the Tagolas 10 GHz ultra wideband antenna is selected. An antenna analysis was conducted in the anechoic chamber at the Royal Military College to ensure the individual antenna elements would meet the requirements of the MIMO radar antenna array.

Following the equipment selection, Phase 3 is the design of the MIMO radar architecture and software development in C and C++. A high-level diagram of the architecture is shown in Fig. 8. Moving right to left in Fig. 8, an 8 transmit by 8 receive antenna array is constructed using the Tagolas antenna elements. The next block to the left shows the MIMO radar hardware rack, consisting of

 https://doi.org/10.1051/itmconf/20224801001

Figure 7. Four Ettus Research SDRs and a Clock Destitution Unit of the MIMO Radar

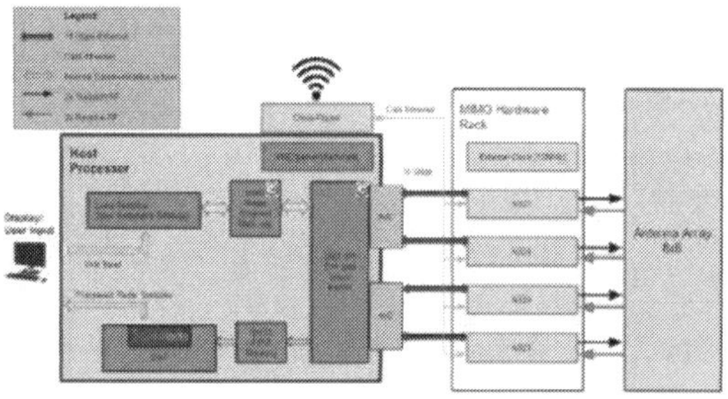

Figure 8. High level schematic of the MIMO radar, hardware components are grey and software components in blue

Figure 9. Photo of the final MIMO Radar during testing

the 4 Ettus research SDRs, and the clock distribution unit. The MIMO Radar Hardware rack is connected to the host processor. The host processor is running the main radar program and the DSP program used to process the radar data in real-time for display. Fig. 9 is a photograph of the MIMO radar.

4 Results

The performance of the multi-mode MIMO radar in terms of its ability to detect multiple targets in range, bearing, azimuth, and velocity, was tested in 2 different environments. The first environment was the lab environment. The second was the RMC Athletic Dome which provides a wide-open space to test the various parameters of the radar using humans as targets. The MIMO radar operates at a very low output power which ensured the safety of the personnel involved in all stages of testing. Several videos showing the radar in full operation can be found on YouTube at https://www.youtube.com/watch?v=eu5Wo0JYB3M&list=PL_lkRDKeAnhpYMzRbNTOyXM1O-xJ4aRwL or search YouTube for "Robert Gilpin MIMO Radar".

The first test environment was the lab, it provided an area to prove the operation of the MIMO radar during development and early radar operation. A target table was used to prove all three dimensions of operation: range, bearing, and velocity. In the lab it was clear that the MIMO Radar was fully operational in TDM mode. To gain better measurements in a less cluttered environment the Radar was moved to the RMC Athletic Dome. There were two main tests conducted in the RMC Dome: single target testing and multi-target testing.

4.1 Single Target Test

The single target consisted of a human starting right in front of the antenna array at 0 and then walking at a slow pace until the human was no longer visible on the MIMO radar display. The maximum range of the display is reached due to the SNR of the target dropping below the noise floor. The range, velocity, and bearing can be estimated in this single target test.

The TDM technique is used as a baseline to compare future radar performance techniques including CDM and FDM. Fig. 10 shows the target walking at $R = 5$ m with a velocity of $v = -1.2$ m/s at approximately -10 (right of boresite). In Fig. 10, the zero Doppler bin has been blanked in the range vs. velocity graph. However, the zero Doppler bin was not blanked in the range vs. bearing display. The zero Doppler bin was not blanked in range vs. bearing because that functionality is not possible with the DSP algorithm employed in GNU Radio. The maximum range observed was 21.75 m.

4.2 Multiple Target Identification

A radar must be able to detect several targets in the FOV simultaneously. The second test performed in the Dome was a 2 target in which test two humans start at zero range, walk at a normal pace, split apart after several steps, and then returning back to the starting point. This test is used to demonstrate the MIMO radar is able to detect multiple targets and separate them in bearing, velocity, and range.

The results from the TDM can be found in Fig. 12. The figure shows the point at which the two humans walking toward the radar are separated by a null in the bearing display.

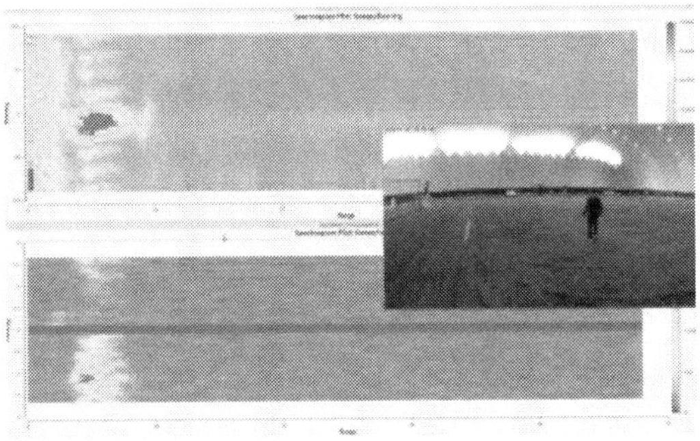

Figure 10. TDM MIMO mode real-time display of range vs. bearing on top, range vs. velocity on the bottom and an inserted photograph of the target and environment

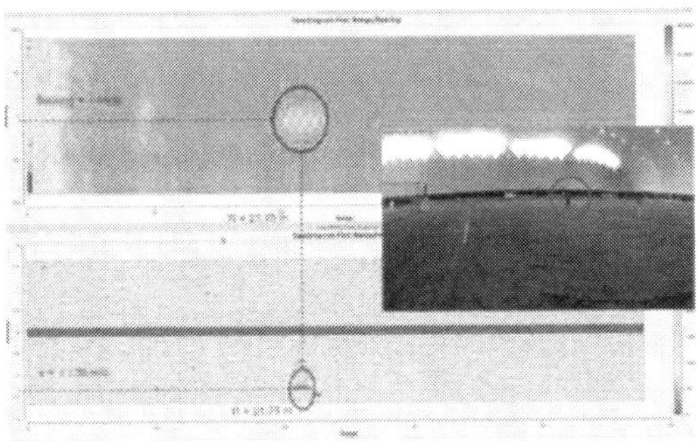

Figure 11. TDM MIMO radar real-time display showing the estimated maximum range of 21.75 m at a velocity of $v = -1.35$ m/s (opening)

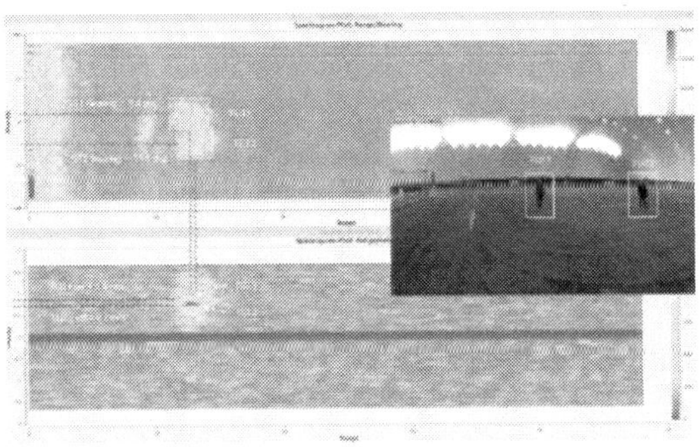

Figure 12. TDM MIMO radar real-time display, multi-target test, where TGT stands for target. TGT1 at a bearing of $\theta_{TGT1} = 5.8$, velocity of $v_{TGT1} = 0.9$ m/s and a range of $R_{TGT1} = 12.4$ m. TGT2 at a bearing of $\theta_{TGT2} = -14.5$, $v_{TGT2} = 0.7$ m/s and a range $R_{TGT1} = 12.6$ m

Both the single and multi-target tests are designed to prove functionality of the radar and not provide the full resolution and performance of the radar. Further research will implement a background subtraction algorithm and static testing will be conducted to provide a fully specified radar resolution cell.

5 Conclusion

This research aimed to do three things, all of which were demonstrated. The first was to create a AESA radar using SDRs capable of MIMO operation. Ettus Research SDRs were used to build a fully functional AESA Radar capable of MIMO operation using all three multiplexing techniques. The second aim was to ensure the signal processing remained real-time, which was demonstrated in all three test environments. The final aim of the research was to compare TDM, FDM and CDM modes which was completed in a multi-target environment.

This research has provided a basis for more work to continue in MIMO radar development in the radar community. It has demonstrated that modern SDRs can be used to build an AESA radar to allow for further development of multiplexing strategies for use in MIMO radar. Finally, this research has shown the advantages MIMO operation could bring to future radars for use in a military or civilian application.

References

[1] C. Payne, *Principles of Naval Weapon Systems*, U.S. Naval Institute blue & gold professional library (Naval Institute Press, 2010), ISBN 9781591146674

[2] M. Richards, J. Scheer, W. Melvin, J. Scheer, *Principles of Modern Radar: Advanced Techniques, Volume 2*, EBSCO ebook academic collection (Institution of Engineering and Technology, 2012), ISBN 9781891121531

[3] M. Skolnik, *Radar Handbook, Third Edition*, Electronics electrical engineering (McGraw-Hill Education, 2008), ISBN 9780071485470

[4] *Difference between AESA radar and PESA radar*, `https://www.rfwireless-world.com/Terminology/AESA-radar-vs-PESA-radar.html`

[5] J. Bergin, S. McNeil, L. Fomundam, P.A. Zulch, *MIMO phased-array for SMTI radar*, in *2008 IEEE Aerospace Conference* (2008), pp. 1–7

[6] *Analog | Embedded Processing | Semiconductor Company | TI.com*, `https://www.ti.com/`

[7] *National Instruments Ettus Research USRP*, `https://www.ettus.com/`

A compact dual-band patch coupler design for modern wireless applications

Khadija Abouhssous[1*], *Asmaa* Zugari[2], and *Alia* Zakriti[1]

[1] Laboratory of Science and Advanced Technology, Department of Civil and Industrial Sciences and Technologies, National school of Applied Sciences, Abdelmalek Essadi University, Tetouan, Morocco
[2] Information and Telecommunication System Laboratory, FS, Abdelmalek Essadi University, Tetouan, Morocco

Abstract. This paper presents a compact rectangular patch hybrid coupler design which can be operated at dual resonant frequencies and is suitable for modern wireless applications. The dual-band functionality while maintaining practical compactness is achieved by introducing a series of parallel rectangular slots. The proposed design is optimized by the CST studio suite software and etched on the FR4 substrate. The good agreement between the simulated and measured frequency responses indicate that the presented coupler is capable of operating from 2.44 GHz to 3.1 GHz and from 5 GHz to 5.8 GHz, with fractional bandwidth of 24.44% and 14.8%, respectively, and a size reduction of 56.28%. The robust simulation results indicate that the proposed design is suitable for microwave integrated circuits (ICs) that require small size and dual functionality.

1 Introduction

The rapid development of wireless communications systems requires the creation of new systems that can operate in multiple frequency bands and support multiple standards. The multi-band RF components, consisting of single circuits and operating at multi-specified frequencies, are compact solutions for modern wireless communication systems. Therefore, the directional coupler [1], being an important passive device, must be redesigned to meet these requirements. These passive devices are widely used in various applications such as phase shifters [2], mixers, power amplifiers and bridging of many microwave circuits [3]. There are three main topologies for hybrid couplers. These are the branch line [4], coupled line [5] and patch line [6] structure. Each of these topologies has its specific benefits and drawbacks.

Various approaches have been suggested to design hybrid dual-band couplers. In reference [7] the authors have replaced the conventional coupler branches by a composite metamaterial transmission line on the right and left sides for dual-band functionality. The losses of the local components and interconnecting holes in the left-hand TLs are important. In [8], the cross-coupled branch coupler achieves the dual frequency band and improved frequency bandwidth by using the additional cross-coupled branches. The use of arms with connection lines was highly appreciated to ensure the implementation of a dual-band functionality and consequently, increase the overall surface of the couplers [9]. In Ref [10][5], dual-band performance is realized with the extension of ports and, as a consequence, increasing the coupler's total size. however, However, miniaturization is of particular interest for wireless equipment. Within this context, several works are oriented to realize dual-frequency structures while maintaining a compact size. For example, in [11], a dual-band inline coupler with reduced size is proposed by the authors, with a topology composed by two π-shaped TLs. In reference [12], the dual-band operation and miniature area are complemented by the use of non-uniform microstrip transmission lines. In addition, in reference [13], a dual-band miniature coupler with a power division of 3 dB is presented. This compact design is obtained through the folding of sections of the transmission line within a circuit. And in [14], the dual-band performance is achieved by implementing artificial transmission lines (TLs) based on open ring resonators (OSRRs).

This work presents the design, the simulation and the fabrication of a compact dual-band hybrid coupler with rectangular patch loaded by parallel rectangular slots as shown in Figure 1. Our proposed design has various benefits compared to conventional dual-band systems. First, the patch design minimizes the undesirable effects caused by junction discontinuity in the BLC design as the operating frequency rises. Second, the patch offers a structure which is simple, easily and inexpensively manufactured. Third, the rectangular slot-loaded hybrid patch coupler is able to offer a wider bandwidth than the branched line coupler-based designs. Finally, the slot-loaded rectangular patch hybrid coupler has a

* Corresponding author: abouhssoukhadija@gmail.com

convenient phase response at the output ports and can provide a quadrature phase difference in both bands. In this paper, a conventional single-band rectangular patch hybrid coupler is presented firstly, and then a compact-sized dual-band rectangular patch hybrid coupler topology is defined by introducing parallel rectangular slots.

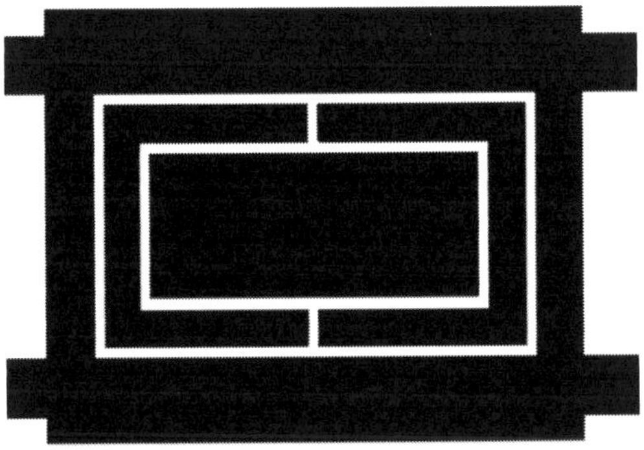

Fig.1. Design of the proposed dual-band coupler with rectangular patch

2 Conventional patch coupler

The design procedure starts by using the standard transmission line model equations as a guideline for the design of the conventional microstrip patch coupler, as shown in Figure 2. Through the adjustment of the center of the frequency band to 3.5 GHz and the properties of the material used, namely FR-4, the conventional patch coupler dimensions are achieved which are given in Table 1.

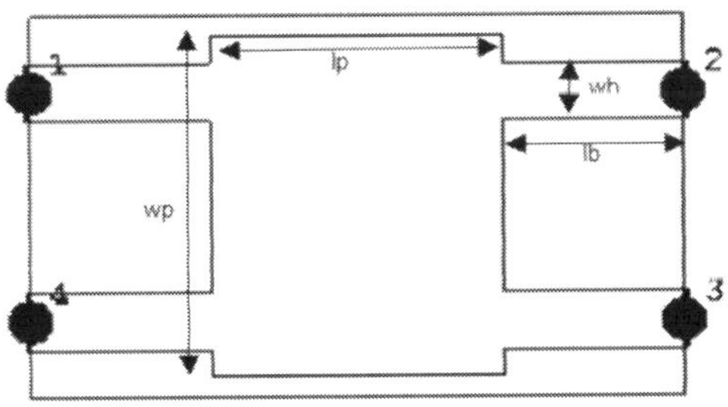

Fig. 2. Conventional patch coupler design

Table 1. Dimensions of the conventional patch coupler

Parameters	Value (mm)
wp	25.4
lp	21
wb	4.35
lb	13

The area occupied by the traditional patch coupler design is (25.4mm*47mm). The Figure 3 and Figure 4 show the simulated frequency responses of the design, where the return loss$|S_{11}|$and isolation$|S_{41}|$ are 24.7 dB and 56.5 Db at the resonant frequency of 3.5 GHz respectively. Furthermore, the coupling factor $|S_{31}|$ and insertion loss $|S_{21}|$ are equal to 3.7 dB and 4.4 dB, respectively, with a phase difference of the output ports of 88.1°.

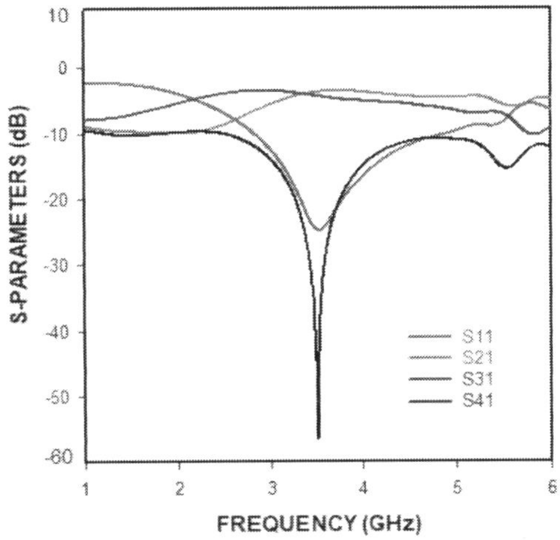

Fig.3. Conventional patch coupler frequency response

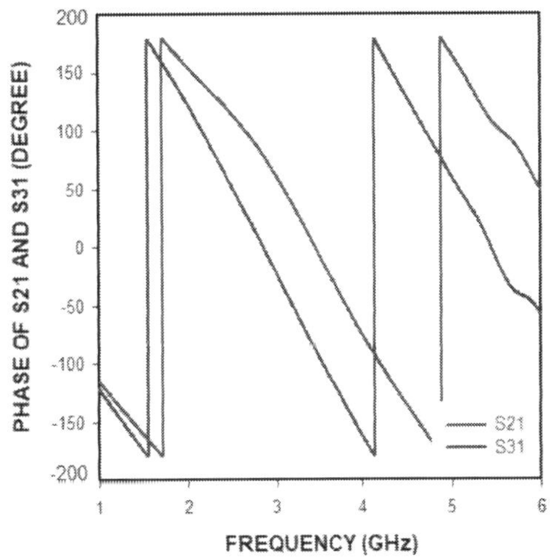

Fig.4. Two output ports phase of the conventional patch coupler

3 Compact dual-band patch coupler

To create dual-band functionality while simultaneously overcoming the bulky patch coupler size, a series of parallel rectangular slots have been etched on the patch, which introduces more resonances and achieves dual-band functionality characteristics. Figure 5 shows the dual-band hybrid coupler with rectangular patch, is simulated in an optimized way using CST studio suite software. The parameter dimensions used for this design are given in Table 2, of which the overall area of the proposed structure is (29mm*18mm), so it occupies 43.72% compared to the conventional structure, therefore it makes a saving of area by 56.28%.

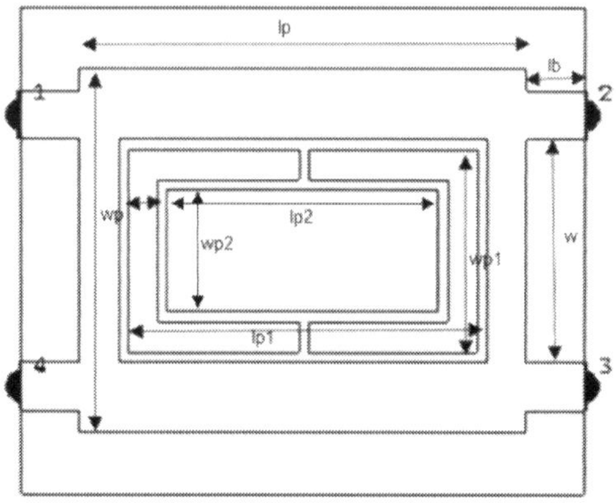

Fig.5. Proposed dual-band patch coupler design

Table 2. Dimensions of the proposed dual-band patch coupler

Parameters	Value (mm)
lp	23
wp	18
lp_1	19
wp_1	11
lp_2	16.5
wp_2	8.5
w	11
wb	2.4
lb	3
g	0.5

As shown in Figure 6, the proposed patch coupler has a good frequency response in two frequency bands. The return loss $|S_{11}|$, and the isolation $|S_{41}|$ are 46.6 dB, 25.4 dB at 2.6 GHz, and 22 dB, 31.6 dB at 5.3 GHz, respectively. And the simulated $|S_{21}|$, $|S_{31}|$ are 3.7 dB, 3.2 dB at 2.6 GHz, and 4.1 dB, 3.9 dB at 5.3 GHz, respectively. The fractional bandwidths with $|S_{11}| < 10$ dB and $|S_{41}| < 10$ dB are 36.15%, 33.42% at 2.6 GHz, and 16.79%, 20.94% at 5.3 GHz, respectively.

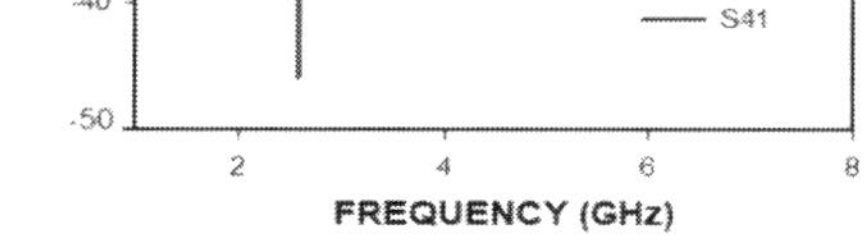

Fig.6. Proposed dual-band patch coupler frequency response

Figure 7 shows the simulated phase shifts between the two output ports. The phase differences are 88.4° and 92.7° respectively at the two reasoning frequencies.

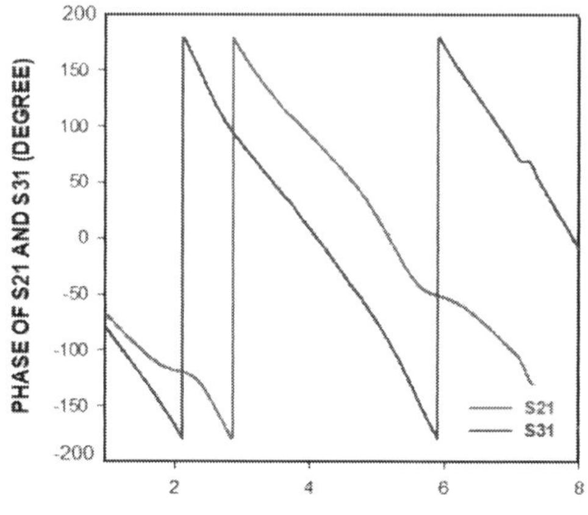

Fig.7. Phase of two output ports of the proposed dual-band patch coupler

Figure 8 shows the construction of the optimized hybrid quadrature coupler prototype, Where the dimensions of the miniature coupler are 29 mm*18mm, representing 56.28% of reduction compared to the conventional coupler. In Figure 9, the proposed design's frequency responses are shown, in which the manufactured construction demonstrates a strong match with the results of the simulation, so that the proposed coupler has two bandwidths ranging from 2.44 GHz to 3.1GHz and 5 GHz to 5.8 GHz respectively while keeping high performance. The phase difference measured between the two output ports $S_{21}$and $S_{31}$is equal to 88% for the 1st band centered at 2.72 GHz and 88.2% for the 2nd band centered at 5.4 GHz, indicating that the envisaged layout is adequate for connecting the signals diagonally for the two reasonable frequencies.

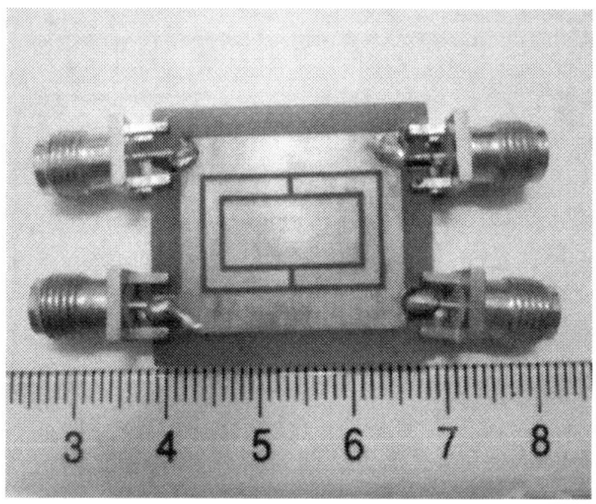

Fig. 8. Proposed dual-band patch coupler photograph

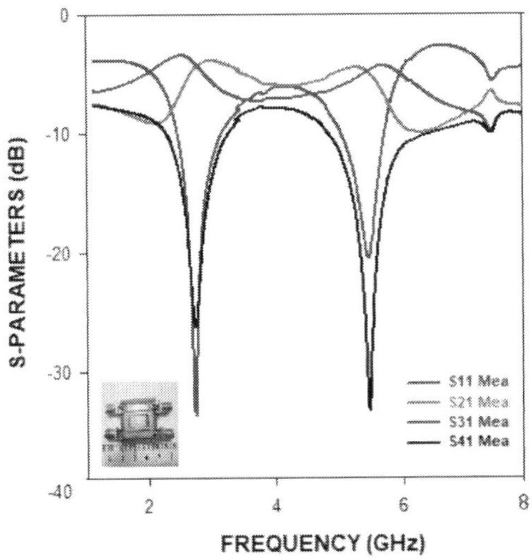

Fig.9. Frequency response of the manufactured dual-band patch coupler

4 Result and discussion

In this chapter, we compare the traditional coupler and the proposed design. The results of this comparison are shown in Table 3, which demonstrates that the proposed configuration provides a 56.28% size reduction over the conventional design. Simultaneously, the proposed structure provides dual-band behaviour while maintaining high performance. The high agreement of the proposed dual-band patch coupler's simulated and measured performance can be seen in Figures 10 and 11, and demonstrate the success of this structure in meeting the need for multiple standards and dual-band operation.

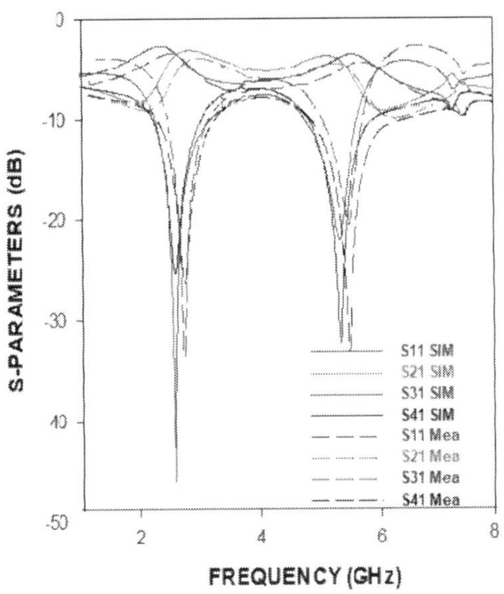

Fig.10. Proposed dual-band patch coupler simulation and measurement results

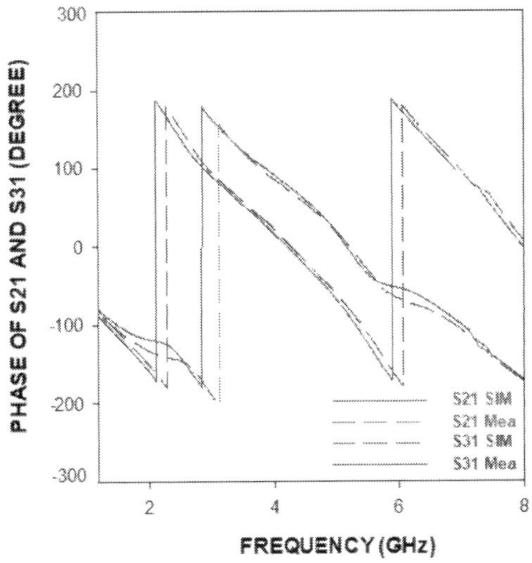

Fig.11. The proposed dual-band patch coupler simulation and measurement phase

Table 3. Comparison between the conventional and the proposed model of the dual-band patch coupler.

Parameters	Conventional Result	Simulation Result		Measurement Result	
Frequency (GHz)	3.5	2.6	5.3	2.72	5.4
$S_{11}(dB)$	-24.7	-46.6	22	-33.29	-20.13
$S_{21}(dB)$	-4.4	-3.7	4.1	-3.8	-4.5
$S_{31}(dB)$	-3.7	-3.2	3.9	-4	-4.5
$S_{41}(dB)$	-56.5	-25.4	31.6	-26.28	-33
Phase Difference (°)	88.1	88.4	92.7	88	88.2
FBW (%)	62	36.15	16.79	24.44	14.8
Area (mm^2)	25.4*47	29*18			
Percent reduction (%)	56.28%				

With the aim of properly evaluating the results of the envisaged coupler. A comparative study of the dual-band coupler presented here versus some others reported earlier in frequency response, phase shift and physical dimensions is given in Table 4. The encouraging results of the dual-band patch coupler indicate that it is a good choice for wireless communications systems, which require systems that can cover several bands and support several standards.

Table 4. Comparison between proposed coupler with previous works

Ref	Freq. (GHz)	$S_{11}(dB)$	$S_{21}(dB)$	$S_{31}(dB)$	$S_{41}(dB)$	D.P(°)	Size (mm^2)
[10]	1	-36.7	-3.3	-3.1	-32.9	90	33.66×32.34
	2	-25.3	-3.1	-3.7	-29.9	89.89	
[11]	0.87	-26	-3.3	-3.67	-34	89.3	31×31
	1.79	-21.6	-3.09	-3.9	19.9	91.4	
[12]	0.9	-18.3	-3.1	-3.25	-18.33	90	65.34×65.34
	1.8	-19.1	-3.08	3.14	-18.7	90	
This work	**2.6/2.72**	**-46.6/-33.29**	**-3.7/-3.8**	**-3.2/-4**	**-25.4/-26.28**	**88.4/88**	**29×18**
	5.3/5.4	**-22/-20.13**	**-3.9/-4.5**	**-4.1/-4.5**	**-31.68/-33**	**92.7/88.2**	

5 Conclusion

A miniature dual-band hybrid patch coupler has been presented. Parallel slots are used to achieve dual-band operation while maintaining a compact size and lower complexity. The simulation and optimization of the proposed hybrid patch coupler is performed by the CST studio simulation software and then etched on the FR-4 substrate. The results measured show the ability of the proposed structure to operate from 2.44 GHz to 3.1 GHz and from 5 GHz to 5.8 GHz, with a size of (29mm × 18mm). The scattering parameter and output port phase difference results have excellent performance and high agreement between measurement and simulation results, showing that this coupler prototype is well suited to modern wireless implementations that require dual-band operation and a compact size.

References

1. Zaidi, Aijaz M., Mirza Tariq Beg, Binod Kumar Kanaujia, Karun Rawat, Sachin Kumar, Karumudi Rambabu, Satya P. Singh, et Aimé Lay-Ekuakille. « A Dual-Band Rat-Race Coupler for High Band Ratio Wireless Applications ». IEEE Transactions on Instrumentation and Measurement 70 (2021)
2. Liu, Hongmei, Xuejiao Wang, Tielin Zhang, Shao-Jun Fang, et and Zhongbao Wang. « Design of Full-360° Reflection-Type Phase Shifter Using Trans-Directional Coupler with Multi-Resonance Loads ». Progress In Electromagnetics Research Letters 101 (2021)
3. Shao, Qiang, Fu-Chang Chen, Yi Wang, et Qing-Xin Chu. « Design of 4 × 4 and 8 × 8 Filtering Butler Matrices Utilizing Combined 90° and 180° Couplers ». IEEE Transactions on Microwave Theory and Techniques 69, no 8 (août 2021)
4. Roshani, Sobhan, et Saeed Roshani. « A Compact Coupler Design Using Meandered Line Compact Microstrip Resonant Cell (MLCMRC) and Bended Lines ». Wireless Networks 27, no 1 (1 janvier 2021)
5. Islam, Rakibul, Asif I. Omi, Mohammad A. Maktoomi, Christine Zakzewski, et Praveen Kumar Sekhar. « A New Coupled-line Based Dual-band Branch-line Coupler With Port-Extensions ». Progress In Electromagnetics Research M 105 (2021)
6. Liu, Jia, Yongqiang Wang, et Kaixue Ma. « A Compact Folded Patch Coupler Based on SISL Platform ». IEEE MTT-S International Microwave Workshop Series on Advanced Materials and Processes for RF and THz Applications (IMWS-AMP), 323-25, 2021.
7. Lin, Xian Qi, Ruo Peng Liu, Xin Mi Yang, Ji Xin Chen, Xiao Xing Yin, Qiang Cheng, et Tie Jun Cui. « Arbitrarily dual-band components using simplified structures of conventional CRLH TLs ». IEEE Transactions on Microwave Theory and Techniques 54, no 7 (juillet 2006)
8. Collado, C., A. Grau, et F. De Flaviis. « Dual-band planar quadrature hybrid with enhanced bandwidth response ». IEEE Transactions on Microwave Theory and Techniques 54, no 1 (janvier 2006)
9. B. Li, W. Wu. « Compact Dual-Band Branch-Line Coupler with Power Dividing Ratio» Journal of Electromagnet: Waves and Applications;25(4):607–15(2011)
10. Kim, Hyunchul, Byungje Lee, et Myun-Joo Park. « Dual-Band Branch-Line Coupler With Port Extensions ». IEEE Transactions on Microwave Theory and Techniques 58, no 3 (mars 2010)
11. Gai, Chao, Yong-Chang Jiao, et Yu-Long Zhao. « Compact Dual-Band Branch-Line Coupler With Dual Transmission Lines ». IEEE Microwave and Wireless Components Letters 26, no 5 (mai 2016)
12. Taravati, S., et M. Khalaj-Amirhosseini. « Compact Dual-band Stubless Branch-line Coupler ». Journal of Electromagnetic Waves and Applications 26, no 10 (1 juillet 2012)
13. A. Bekasiewicz and S. Koziel. « Miniaturised Dual-band Branch-line Coupler» Electronics Letters - Wiley Online Library ; Vol. 51 No. 10 pp. 769–77(2015
14. Ghaffarian, Mohammad Saeid, Gholamreza Moradi, et Pedram Mousavi. « Dual band/ Dual Mode Branch-Line/Rat-Race Coupler Using Artificial Transmission Line ». In 2019 27th Iranian Conference on Electrical Engineering (ICEE), 1622-26, (2019).

Wideband meta-material-inspired reconfigurable antenna using Infinity Split Ring Resonator

Zohra Zerrouk[1*], *Larbi Setti*[2]

[1] Physics Department, faculty of sciences, Abdelmalek Essaadi University, Tetouan, Morocco
[2] LSTA, FPL, Abdelmalek Essaadi University, Tetouan, Morocco

Abstract. In this article, a reconfigurable antenna inspired by the broadband meta-material using the Infinity slotted ring resonator (I-SRR) is designed and computer-based for wireless communication. Antenna takes an infinitely split ring resonator as a radiator powered by a coplanare wave guide (CPW). The procedure for extracting parameters of the meta-material property of the I-SRR is discussed in detail, by which the existence of negative permeability and the new resonance frequencies are verified. The proposed antenna has advantages of simple design, miniaturisation that can be used for wireless mobile communication system.

Index terms – frequency reconfigurable, meta-material, negative permeability, split ring resonator (SRR), wideband antenna, coplanar waveguide (CPW)

1 introduction

Over the past few years, there has been a remarkable development in the field of communication, notably in wireless communications. There is an increasing need for innovative materials to improve the performance of antennas, which are components that transmit and receive electrical signals. Since the first demonstration on the negative of electric permittivity and magnetic permeability by V.G.Veselago in 1968 [1], and its manufacture by Smith and his team [2], Meta-data have a strong interest in the area of antennas due to their properties, that are not found in nature. Synthesized meta-materials are classified into three categories: the first one is the meta-material with negative permittivity, the second one is the meta-material with negative permeability, and the last one is the meta-material with both are negative, the permittivity and the permeability [3].

Bandwidth is a very important parameter for transmitting data via the antenna. The multi-band antenna covers millimeter waves for 5G technology was presented in [4-5]. In [6-7], a compact coplanar waveguide (CPW)-fed meta-material inspired multi-band is designed to cover WLAN, C-band, UMTS and WiMax. A multi-band reconfigurable antenna inspired hexagonal meta-material was presented in [8]-9]. A wideband antenna of a monopole type for frequency reconfiguration using varactor diode was illustrated in [10]. Complementary Split Ring Resonator is a kind of meta-materials that have a negative epsilon. In [11] the four rectangular Split Ring Resonator was printed in ground plane for improve gain value with a rectangular patch. In [12], the Infinity Split Ring Resonator was loaded in ground plan improve gain and efficiency.

In this work, we focus on developing the shape of meta-materials and using it as a primary resonator to perform broadband reconfiguration. The simulator used is CST MICROWAVE STUDIO

2 Extraction of negative permeability

The electromagnetic response of that material is determined by electric permittivity and magnetic permeability. In conventional materials, these parameters are positive, but in artificial materials, the two are negative or one of them is negative. The most commonly used techniques to extract negative permittivity or negative permeability are the Nicholson-Ross-Weir (NRW) [13] and Robust [14] methods. These techniques are based on parameters S: render in the coefficient of reflection S11 and the coefficient of transmission S21.

The permeability expression is given by the following equation

$$\mu = z * n \qquad (1)$$

Where n is the refractive index and z is the wave impedance.

According the NWR and Robust methods, the expression of n and z are given by the following equation:

* Corresponding author: zohra.zerrouk@gmail.com

$$z = \left(\frac{(1+S_{11})^2 - S_{21}^2}{(1-S_{11})^2 - S_{21}^2}\right)^{1/2} \quad (2)$$

$$n = \frac{1}{kd}\cos^{-1}\left[\frac{1-S_{11}^2+S_{21}^2}{2S_{21}}\right] \quad (3)$$

Fig.1. Infinity-SRR unit cell

The proposed infinity split ring resonator is located between two waveguide ports on the right and left of the z-axis. The wave is excited from the positive Z axis (port1) to the negative Z axis (ports2) to calculate the parameter S11 (reflected from port 1) and the parameter S21 (transmission to port 2 due to 1**)**.

Fig.1. shows the photographs of the improved infinity SRR cell. Scattering parameters are essential to extraction of negative permeability. The S11 and S21 parameters are extracted from CST Microwave Studio. The simulated results presented in Fig.4. are consistent.

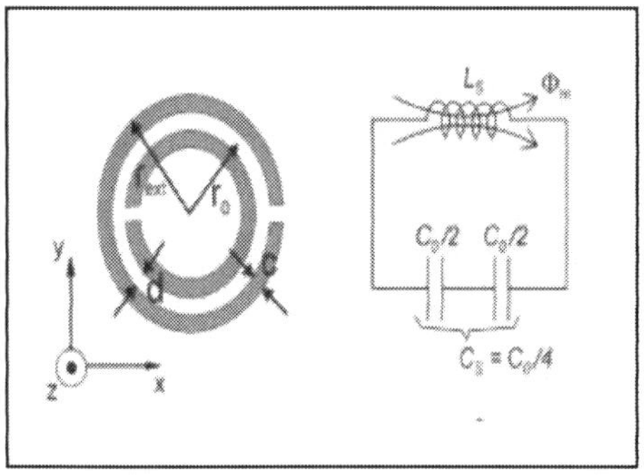

Fig.2. circular split ring resonator and its equivalent circuit [15]

The combination of two circular split ring resonators (Fig.2) results an infinitely shaped split ring resonator whose resonance frequency is given by the following formula[16],[17] :

$$f = \frac{1}{2\pi\sqrt{L_T C_{eq}}} \quad (4)$$

Where

$$C_{eq} = \left(2a_{avg} - \frac{g}{2}\right)\frac{\sqrt{\varepsilon_e}}{c_0 Z_0} + \frac{\varepsilon_0 c h}{2g} \quad (5)$$

And

$$L_T = 0.00508l(2.303\log_{10}\frac{4l}{d} - 2.451) \quad (6)$$

C_{eq} is the equivalent capacitance and L_T is the total inductance

The development of the structure leads to the final structure presented in **Fig.1** and its electrical equivalent is given in **Fig.3**

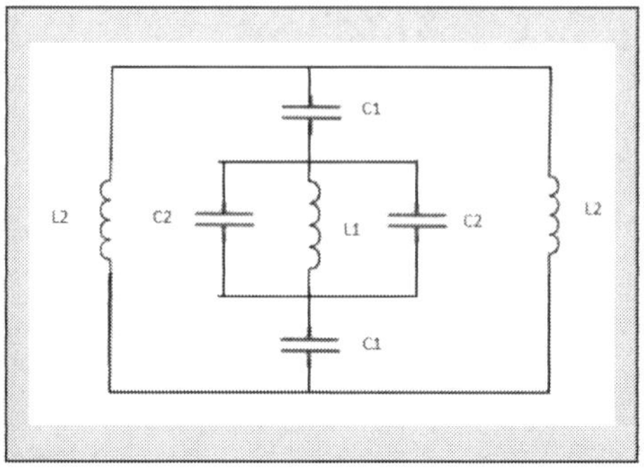

Fig.3.equivalent circuit of Infinity-SRR

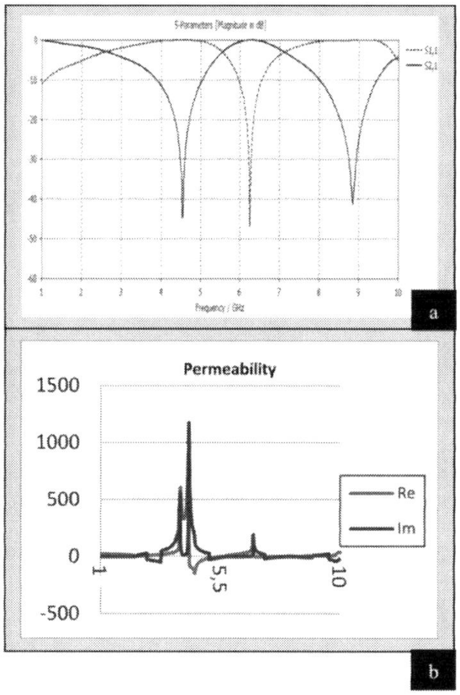

Fig.4. (a) S11&S21 parameters vs frequency and (b) real and imaginary permeability plot vs frequency

3 Antenna design

3.1 antenna configuration

The suggested antenna has infinity shape which CPW-ground planes as presented in Fig.1. Its design is imprinted on a 32 x 35.56 x 0.381 mm3 low –cost

Rogers RO3003 substrate ($\varepsilon_r = 3,\ tan\delta = 0.001$). This proposed antenna has infinity split ring radiating element to create wideband reconfigurable for wireless communication. The radiating area consists of pairs of infinite metal rings, which are one inside another. Top and side views of the antenna are presented in Fig.5.

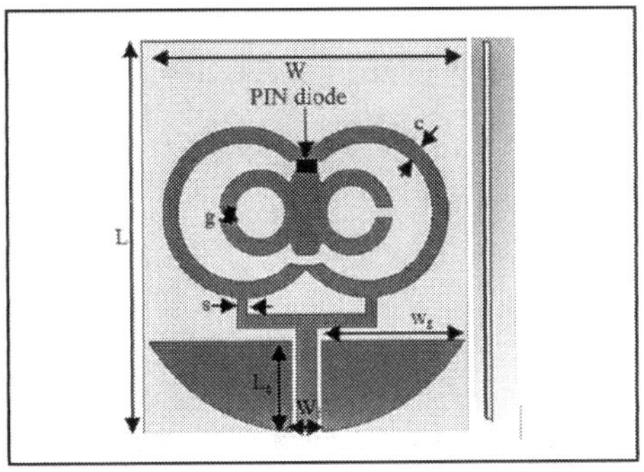

Fig.5. Geometry of proposed reconfigurable antenna

The parasitic inner infinity is connected to the outer infinity ring using one PIN diode in the higher gap between them. After numerous optimization runs using the CST simulator that have based on the finite integration technique (FIT). It is the most used solver due to ease in simulations. Final geometry of the proposed infinity meta-material is illustrated in Fig.5. The optimised dimensions obtained for the reconfigurable antenna inspired by the proposed meta-material are summarized in Table 1.

Table 1. Design Parameters of the proposed antenna

parameter	Dimension (mm)
L	36.56
W	32
Wf	3
Lf	9.78
Lg	8.49
Wg	14
g	1
c	1.5

3.2 Equivalent circuit model of PIN diode:

PIN diode is one of the electronic switching components that have been widely used to reconfigure antennas thanks to their speed of switching and its simple integration within antennas. PIN diode used in this work is MPP4203 from Microsemi.[18]

Fig.6. shows the corresponding circuit for a PIN diode. ON state is replaced by a resistor of 3 Ω, and OFF state is replaced by a resistor of 25 KΩ simultaneously with a capacitor of 0.08 pF.

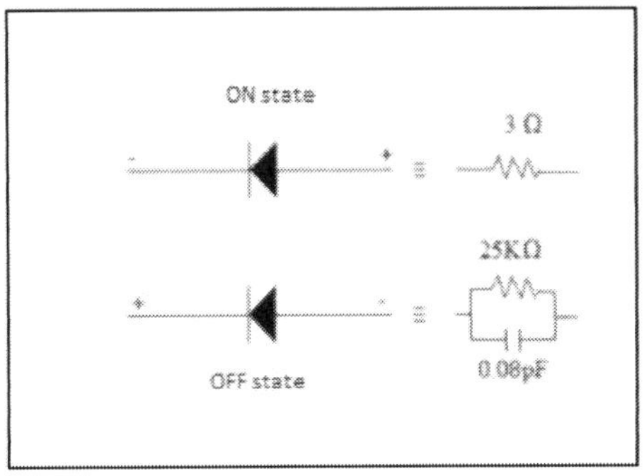

Fig.6. Equivalent circuit of PIN diode

3.3 Parametric Study

3.3.1 the effect of substrate

Parameter S11 (Reflection coefficient) is simulated for different substrate types Three substrates chosen for this study are FR-4 (dielectric permittivity constant of 43, dielectric loss tangent of 0.025), Rogers RT5880 (dielectric permittivity constant of 2.2, tangent dielectric loss of 0.009), and Rogers RO3003 (dielectric permittivity constant of 3, dielectric loss tangent of 0.001). The results are plotted in Fig.7. The impedance bandwidth of the proposed antenna is observed to remain stable and frequency resonances are offset. The Rogers RO3003 substrate gives the good results.

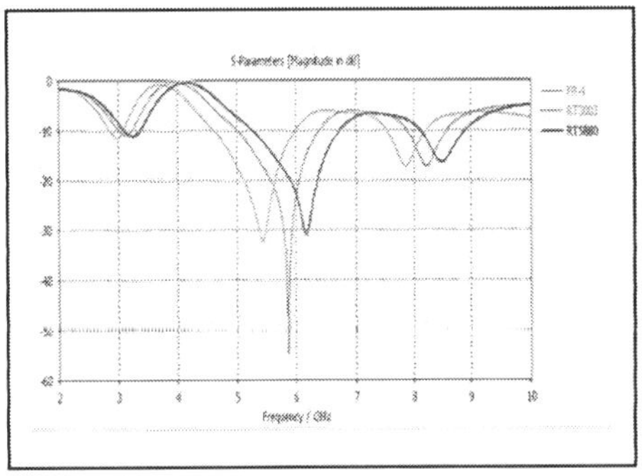

Fig.7. S11 for different types of substrate

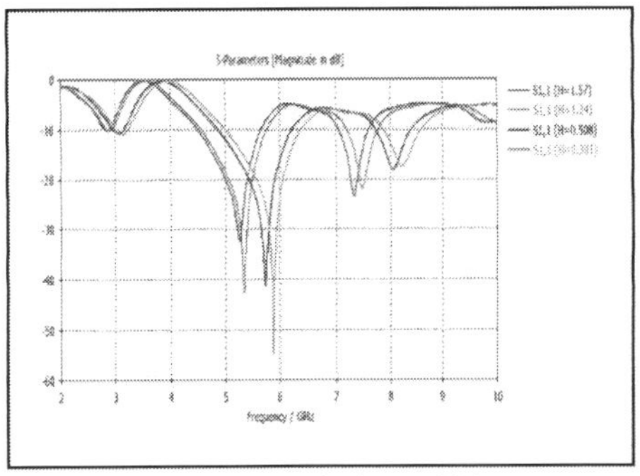

Fig.8. S11 for four values of substrate thickness

3.3.2 the effect of substrate thickness

Fig.8. presents the simulation reflection coefficient for different values of substrate thickness h. It has been observed that the impedance bandwidth of recommended antenna stay unchanged and that frequency resonances are significantly affected. When H=0.381 mm, the antenna fits better.

3.3.3 the effect of width of the rings

The width of the rings has a direct effect on inductance and capacity. The increase in this width decreases the value of the inductance and the value of the capacity. Therefore, rings with low thickness have a small frequency. The optimal thickness value is c=1.5mm in order to reach the desired frequency. Fig.7. shows to obtained results with the other dimensions remains fix (Table1).

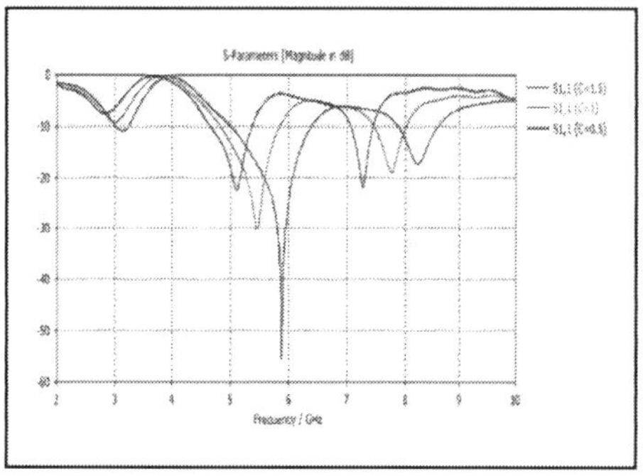

Fig.9. S11 for different values of width of rings c

4 results and discussion

4.1 reflection coefficient

The simulation reflection coefficient S11 results for the both configurations are presented in Fig.10. PIN diode ON configuration resonates at 4.832 GHz in the simulation with -10 dB impedance bandwidth of 2.268 GHz, and OFF configuration resonates at 5.84 GHz and 7.8 GHz in the simulation with -10 dB impedance bandwidth of 1.65 GHz and 687 MHz, respectively.

The results obtained for the coefficient of reflection presented in Table 2 demonstrate that the proposed antenna is suitable for CPW.

There is an advantage to the reconfigurable antenna, it is bandwidth improvement. In this work the bandwidth gets larger. The antenna covers the 2.5 GHz range.

Another parameter is very interesting than S11 parameter; it is the Voltage Standing Wave Ratio (VSWR) which determines the reflected power. Depending on the results presented in Table 2, for the both configurations, the value of VSWR is optimal, which means that most of the power is provided to the antenna.

Table 2. Diodes configurations, S11and VSWR values

	Fr(GHz)	S11 (dB)	BW(%)	VSWR
Configuration 1(ON)	4.832	-26.28	44.68	1.1
Configuration 2 (OFF)	5.84	- 40,87	28.3	1.01
	7.8	- 28.1	8.79	1.08

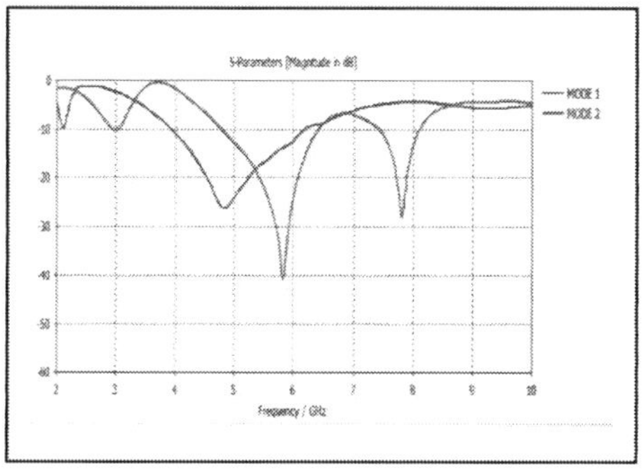

Fig10. Simulated S11 value of the proposed antenna

4.2 Gain and efficiency

The gain of the proposed reconfigurable antenna for the observed operating frequencies are simulated namely the CPW monopole antenna with PIN diode ON/OFF as dipected in Fig.10. and the Fig.11. shows the simulated gain over frequency.

The peak gain for CPW monopole antenna inspired meta-material goes up to 3 dB at 4.832 GHz with diode ON, and 3.58 dB at 5.84 GHz and 4.53 dB at 7.8 GHz with diode OFF.

In this work, the CPW monopole reconfigurable antenna inspired infinity SRR is wideband antenna. It covers the frequencies defined under the C-band (4-8GHz). The most important appears is Industrial, Scientific and Medical (ISM) band at 5.84 GHz.

One of the major parameters is the efficiency of radiation. It describes how such an antenna transmits and receives RF signals. It has defined as the relation between the total power radiated by the antenna and the total power input received from generator. Radiation efficiency value simulated is 92.97% at 4.83 for the configuration 1, and for configuration 2 its value for the bi-band are 94.32% and 98.27% at 5.84 GHz and 7.8 GHz respectively. The gain and radiation efficiency are presented in Table 3.

The simulated surface current distribution for the both configuration at 4.832 GHz, 5.84 GHz, and 7.8 GHz are shown in Fig. 13 respectively.

The recommended antenna is successful in terms of gain and efficiency. Finally, the proposed reconfigurable antenna settings are compared to other references in Table 4.

Table 3. Simulated Gain and Efficiency

	Fr(GHz)	GAIN (dB)	EFFICIENCY (%)
Configuration 1	4.832	3	92.97
Configuration 2	5.84	3.58	94.32
	7.8	4.53	98.37

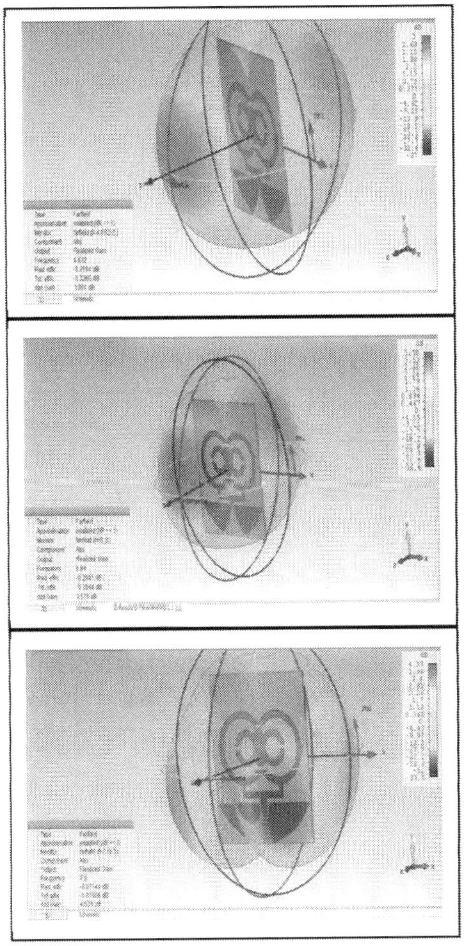

Fig.11. the simulated gain for the three bands and proposed antenna position

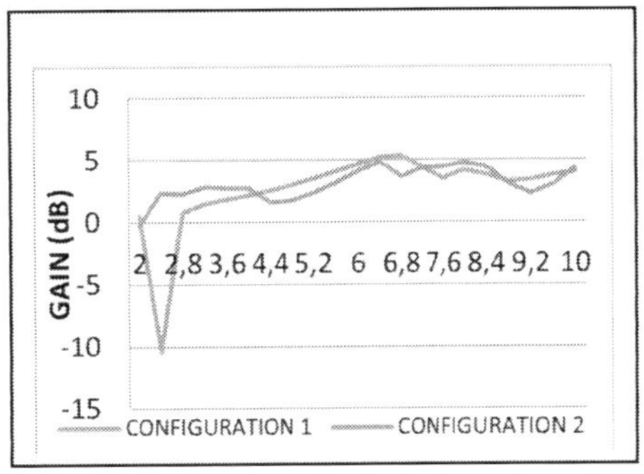

Fig.12. gain vs frequency for the both configurations

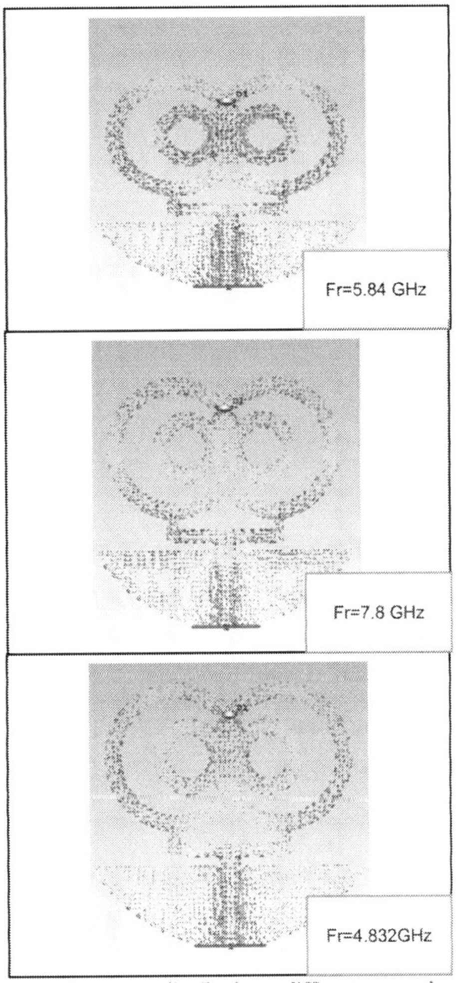

Fig.13. the current distribution at different resonating frequencies.

Table 4. Comparison of the proposed antenna with other works

Ref.	Patch	Dimension(mm^2)	EFFICIENCY (%)	Gain (dB)	Meta-proporty
[6]	Penta-rings	40 x 40	Not virified	1.5-6.5	yes
[7]	Hexagonal	27.38 x 29	55-81	0.634-3.75	no
[8]	hexagonal	26.5 x 30	Not virified	2.6 – 3.8	no
This work	**I-SRR**	**36.56 x 32**	**92.97 – 98.27**	**3-4.58**	yes

5 Conclusion

In this article, a new broadband meta-material based on the reconfigurable antenna using Infinity Split Ring Resonator (I-SRR) was introduced. The proposed antenna has developed for C-band wireless communications. A PIN diode was used as an ON and OFF switch. The proposed antenna covers three bandes, one of them is wideband range that appears in configuration 1. One of the benefits of reconfiguring is to increase the bandwidth of this antenna. The antenna resonates effectively, with gains and efficiencies presenting better values. The proposed antenna introduces a new structure usable in wireless networks.

References

1. V.G.Veselago , *the electrodynamics of substrates with simultaneously negative value of ε and μ*, soviet physics uspekhi **volume 10**, number 4, pp. **509-514**.
2. D.R.Smith, W.J.Padilla, D.C.Vier, and S.Schultz, *Composite Medium with Simultaneously Negative Permeability and Permittivity*, pp. **1-4**,(2000)
3. C.Caloz, T.Itoh, *Electromagnetic metamaterials: transmission line theory and crowave applications*, John Wiley&Sons; (2005).
4. M.Anouar, L.Setti, *Multiband Microstrip Patch Antenna for 5G Wireless Mobile Phone*, International Journal of Science and Research , IJSR, **Volume 9 Issue 10**, (October 2020)
5. J.O.Abolade, D.B.O.Konditi, *A shorting-pin based compact meander multiband printed monopole antenna*, Heliyon 7 (2021) e08390
6. N. Thamil Selvi, P. Thiruvalar Selvan, S.P.K. Babu, R. Pandeeswari, *Multiband metamaterial-inspired antenna using split ring resonator*, computers and Electrical Engineering **84**(2020) 106613
7. C.Zhu, T.Li, K.Li, Z.Jian su, X.Wang, H.Qing Zhai, L.Li, and C.Hong Liang , *Electrically Small Metamaterial-Inspired Tri-band Antenna with Meta-mode*, DOI: 10.1109 /(LAWP. 2015). 2421356.IEEE Antennas and Wireless Propagation Letters.
8. Z.Zerrouk, S.Larbi, *Multiband Reconfigurable antenna for Wireless Communications Systems Using metamaterials*, 16th 16th International Conference on Cyber Warfare and Security , ICCWC; DOI.10.3419/IWS.21.108: **436-440** (2021)
9. V.Sharma, S.Meena and R.K.Saraswat, *Split-Ring Resonator for WLAN/WIMAX/X Standard*, http://doi.org/10.1007/978-981-15-5566-4_4: **35-41**
10. A.Benbarka, L.Setti, A.Tribak, H.Nachouane, and H.Tizyi , *Frequency Tunable Filtenna Using Defected Ground Structure Filter in the Sub-6GHz for cognitive Radio Applications*, Progress In Electromagnetics Research C, **Vol.118**,**213-229**, (2022)
11. R.Pandeeswari and S.Raghavan , *Microstrip Antenna with complementary split ring resonator loaded ground plane for gain enhancement*, Microwave and Optical Technology Letters/ **Vol.57**, No.2 (February 2015), **DOI** 10.1002/mop
12. Z.Zerrouk, S.Larbi, *Frequency Reconfigurable Anetnna Using Infinity Complementary Split Ring Resonator for Wireless Communication*, ICECET. **DOI:** 10.1109/ICECET52533.2021.9698543 (2022)
13. E.J.Rothwell, J.L.Frasch, S.M.Ellison, P.Chahal, and R.O.Ouedraogo, *Analysis of the Nicolson-Ross-Weir Method for Characterizing the Electromagnetic Properties of Engineered Materials*, Progress In Electromagnetics Research, **Vol. 157** , **31-47**, (2016)
14. X.Chen, T.M.Grzegorczyk, B.I. Wu, J. Pacheco, Jr., and J. A. Kong, *Robust method to retrieve the constitutive effective parameters of metamaterials*, PHYSICAL REVIEW E **70**, 016608 (2004)
15. N.A.Reddy, S.Raghavan, *Split ring resonator and its evolved structures over the past decade*, International Conference on Emerging Trends in Computing, Communication and Nanotechnology ICECCN (2013)
16. N.Pekçokguler, G. Dundar, H. Torun, A.Yalçinkaya, *A novel equivalent circuit model for split ring resonator with an application of low phase noise reference oscillator*, INTEGRATION the VLSI journal **61** (2018) **160-166**
17. F.E.Terman, *Radio Engineers' Handbook*, Mcgraw-Hill, New York, (1943)
18. F.Rahmani, N.Amar Touhami, A.Belbachir Kchairi, N.Taher, *Circular Planar Antenna with Reconfigurable Radiation Pattern using PIN Diodes*, 13th International Conference Interdisciplinarity in Engineering, INTER-ENG 2019, Procedia Manufacturing **46** (2020) **760–765.** DOI: 10.1016/j.promfg.2020.04.001

Design of a Circular Patch Antenna with a reflector for GPR applications

Ouboudrar Ismail, Lagmich Youssef, Oulhaj Otman and Amina Aghanim

Laboratory of Sciences and Advanced Technologies (LSTA), Polydisciplinary Faculty of Larache (FPL), Abdelmalek Essaadi University, Tetouan, Morocco.

Abstract. In this paper, a structure of a patch antenna operating at 0.5 GHz is proposed for ground-penetrating radar (GPR) applications, with an optimized performance at low frequencies. The structure was designed using FR4 substrate, with a 4.4 permittivity value, and a tangent loss equivalent to 0.021. The prototype design is backed by an optimized rectangular reflector, which has improved the antenna gain and directivity. The performance of the antenna is evaluated in terms of the return loss, gain, and the radiation pattern for two cases, with and without a reflector. When using the reflector, a good radiation pattern is obtained, a wide operational bandwidth ranging from 0.4GHz and 1.25 GHz is achieved, and an increased gain from 4.3 to 7.43dB was realized.
Keywords: Antenna, Patch, GPR, reflector

1 Introduction

Microstrip antennas are one of the most popular antennas in the wireless communication market. We can find applications of microstrip antenna in various fields of technology like in mobile communication systems[1], satellite communications[2], spacecraft, radars[3], wireless phones and wireless computer networks[4]

This type of antenna received a considerable attention starting in 1970 in a paper written by R.Munson (1974)[5], although the idea of a microstrip antenna can be traced back to G.A.Deschamps 1953 [6]. A microstrip Patch Antenna, in its simplest form consists of a radiating patch on one side of a dielectric substrate, and ground plane on the other side of the substrate.[7]

The major advantage which makes the antenna so popular is their low profile, lightweight, easy fabrication, and low-cost solution and can be integrated with other microstrip components in monolithic applications like RFIC or MMIC [8] However they have also some drawbacks such as the low frequency bandwidth, the radiation efficiency deterioration as frequency and antenna array size increases due to an augmentation in the feeding network losses and limited to relatively low power levels.[7]

Few words about feeding methods since the microstrip antenna is some sort of a resonator we need to excite the fields in this generator in this resonator and there is various ways to feed the antenna. Some of the most common feeding techniques are Microstrip-Line feed, Microstrip Inset Line Feed, Co-axial Line Feed, Aperture Coupled Line Feed, Proximity Coupling Feed.[9]-[7]

The Ground-penetrating radar (GPR) technique consists of employing radio waves and detecting the reflected signal to show the denoted structure [10]. Given the advantages of superior agility, high data density, and short survey time in large non-accessible areas of interest, it has been widely investigated and applied to glaciology, archeology, avalanche rescue, mineral exploration, and landmine detection.[10]

In this paper, a design patch circular antenna operating in the frequency of 0.5 GHz, with a reflector to compensate the gain is presented. The organization of this work is the following: The second section elaborates the principle of working and the different types of ground penetrating radar systems, while the third section was dedicated to the conception of the circular patch antenna. The fourth section consists of illustrating and discussing the simulated results of the structure with and without the reflector using CST environment and finally a conclusion with an opening to some short incoming perspectives.

2 Ground penetration radar

2.1 Working principle

Much previous research has concerned ground-penetrating Radar (GPR), which refers to an electromagnetic (EM) technique mapping conditions below the ground level and the building materials [11]. This system aims to transmit a radio wave signal, into the probed material and receive the reflected pulses after contacting and detecting the discontinuity [12]. It should be mentioned that the discontinuity could be at the level of the boundary or even at the interface level of the material with different dielectrics or even at the level of a subsurface object contact such as delimitation. The range of the transmitted wave signal goes from the range of some megahertz to some gigahertz [13]. The fig.1 shows clearly this process.

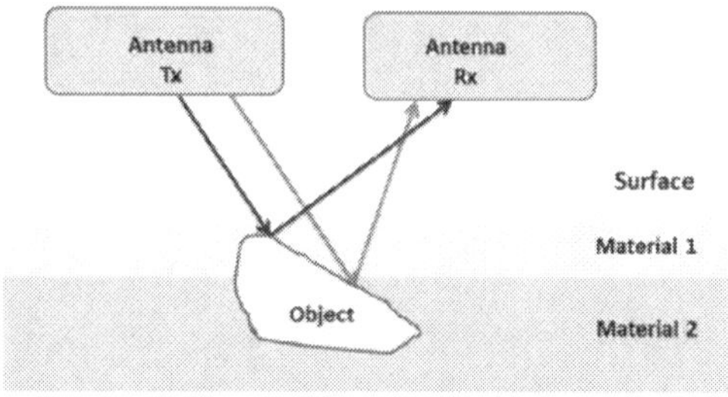

Fig 1. Working principle of GPR

2.2 Types of GPR architecture

In the literature, three commonly used types of radar architecture are studied in the GPR system,a mono-static GPR where the transmitting and the receiving radar pulse are accomplished by the same antenna. Also, a second type named bi-static GPR, where the transmitting and receiving process are established by an antenna each one. A single antenna performs the third type, called multi- static, here the transmitting signal while multiple receiving antennas in an array configuration are required for the receiving part[14].

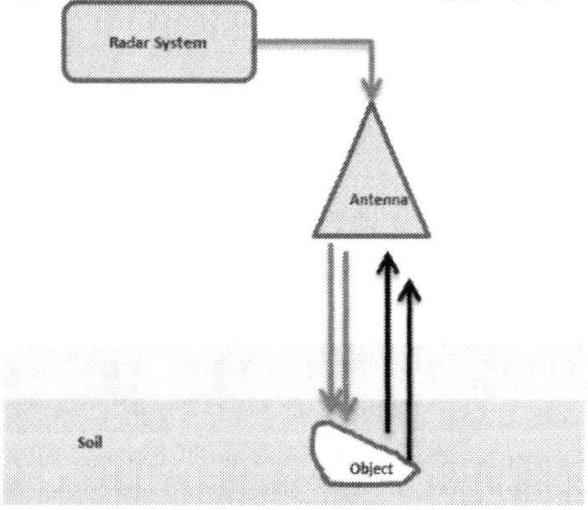

Fig 2. Mono-static architecture radar

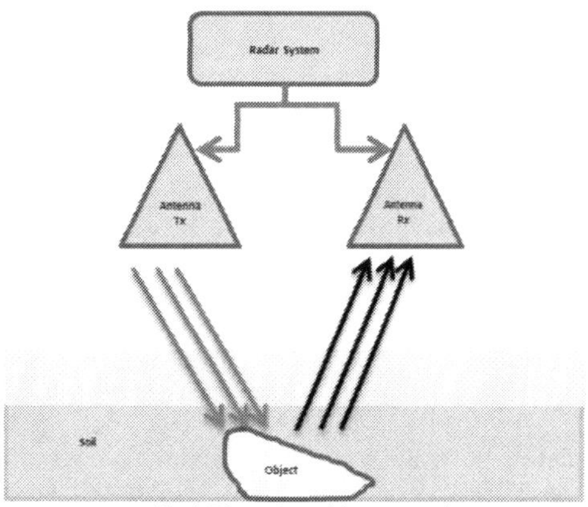

Fig 3. Bi-static architecture radar

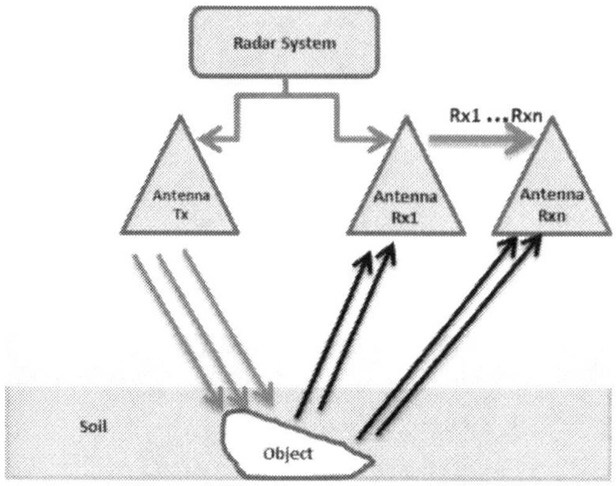

Fig 4. Multi-static architecture radar

3 Patch antenna Design

The antenna under consideration is generally used in GPR systems. This structure has been designed using a FR4 substrate [15], the choice was performed based on the low coast and easiness of fabrication of the concerned material. The dimensions of this substrate are h_s=1.57 mm, the dielectric constant is ε_r=4.3 and the tangent of losses is equivalent to 0.021 [15].

The topology of the patch antenna was designed under CST environment, where the dimensions of the substrate are 200 mm x 250 mm. The micro-strip line fed of 50Ω is characterized with dimensions of 17 mm x 35 mm. It should be noted that the ground plane length is the same as the feeding line length [16].

In order to obtain a wide operational bandwidth, the modification and reduction of the ground plane was performed based on two parameters [17][19], which are W_g and L_g respectively referring to the width and the length of the ground plane. The carried out modification was performed at the punctual contact between the feeding line and the patch antenna. The radius circle size was established using (1) [5] and gave the value r=81 mm. Both figures Fig.3 show the dimensions of the concerned structure.

$$r = \frac{F}{\sqrt{\left\{1 + \frac{2h}{\pi \epsilon_r F}\left[\ln\left(\frac{\pi F}{2h}\right) + 1.7726\right]\right\}}} \quad (1)$$

Where $F = \frac{8.791 \times 10^9}{f_r \sqrt{\epsilon_r}}$ (2)

- h : The thickness of the substrate in cm.
- f_r : The resonant frequency.
- ε_r: The dielectric constant of the substrate.

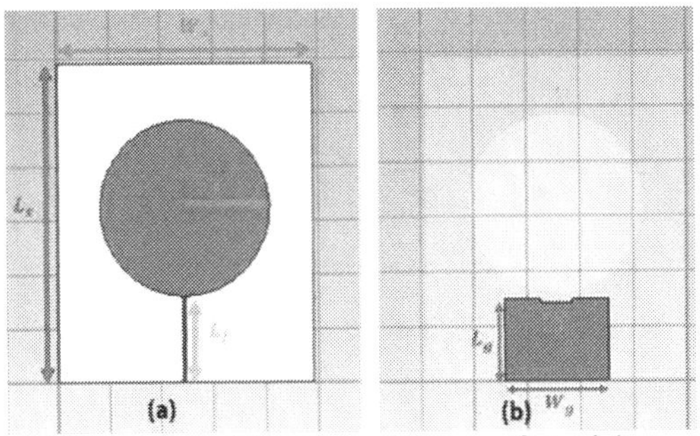

Fig 5. (a)Upper face of the antenna (b) Lower face of the antenna

In order to achieve a directional radiation pattern, we have used a structure in form of a reflector connected to the antenna with four plastic cylinders, their length is d=130mm, in the beginning the value of this length was d=10mm then we has optimized this parameters, until we got an optimized result with the value of d=130mm. The geometry is showed in Fig.5.

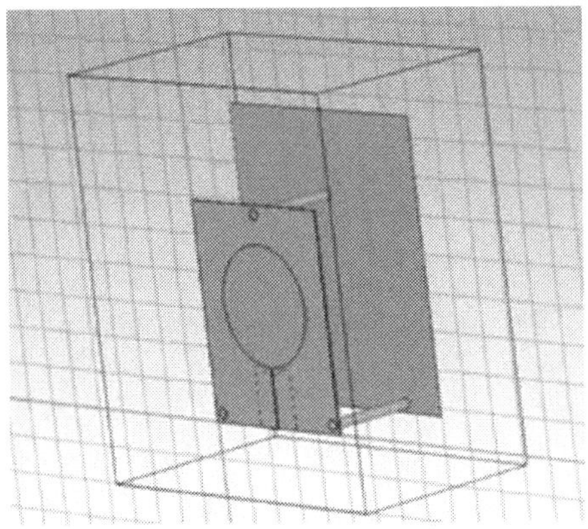

Fig 6. Antenna and reflector system

The overall optimized dimensions of the proposed antenna are showed in Table 1.

Table 1. Geometry Parameters

Parameters	Size (mm)
Substrate Width (W_s)	245
Substrate Length (L_s)	295
Ground plane Length(L_g)	80
Ground plane Width(W_g)	70
Feed line Width (W_f)	1.9
Feed line Length(L_f)	80
Reflector with a spacing(d)	130
Width of slot (W_c)	30

4 Results and Discussing

In this section, a discussion of the obtained results is presented in terms of the return loss, the bandwidth (MHz), the directivity (dBi) and the gain (dB) for the considered structure with and without reflector in order to establish the directional radiation pat-tern.

4.1 Reflection Coefficient

The carried out results of the reflexion coefficient of our structure is shown in fig.5 where it depicts $\mathbf{S_{11}}$ of the antenna with and without reflector at an operating frequency of 0.5 GHz. The blue plot of the reflexion coefficient of the structure without reflector has a return loss of 39.90 dB, and a bandwidth at -10 dB ranging from 0.45 GHz to 1.32 GHz. While we can clearly see that the $\mathbf{S_{11}}$ results has been optimized when the reflector was added. The red plot has showed a return loss of 41.1 dB and a bandwidth at -10 dB ranging from 0.44 GHz to 1.25 GHz.

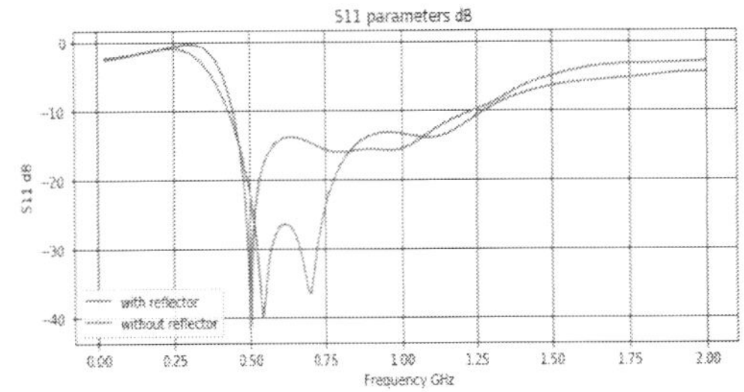

Fig 7. Simulated S11 with and without reflector

4.2 Radiation Pattern and Gain

The simulated 3D far-field radiation pattern for the patch antenna has been elaborated in this section. The comparison radiation pattern of the proposed antenna at 0.5 GHz with and without reflector has been presented in Fig.8. The performance of the proposed antenna is compared with preexisting antennas, and analysis is presented in Table 2.

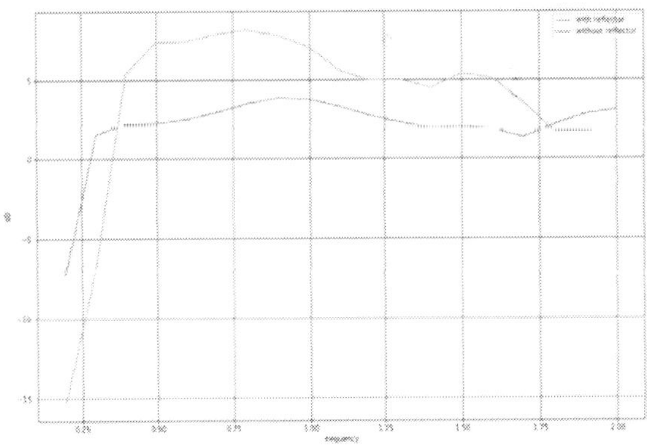

Fig 8. Gain of the structure with and without reflector

Table 2. Compared results

Reference	[18]	[14]	Proposed design
Antenna dimensions (cm2)	18x22	10x10	24.5x29.5
Frequency (GHz)	0.6- 4.6	1.1-5.5	0.45-1.32
Peak gain (dB)	5.01	4.3	7.43
Reflector type	Planner	Planner	Planner

 https://doi.org/10.1051/itmconf/20224801004

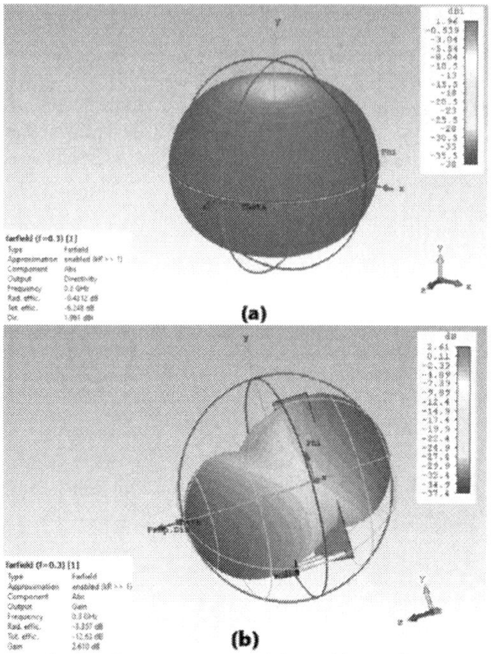

Fig 9.(a) Radiation pattern at 0.3 GHz without reflector.(b) Radiation pattern at 0.3 GHz with reflector.

Adding the reflector to the structure adopter we cloud have se we that the directivity and gain had increased in the three different previous frequencies.

For example Fig.9 depicts comparative performances illustrating of the gain, where it has increased by an amount of 1.07 dB going from 1.53 to 2.6 also Fig.9 shows the appearance of the directivity

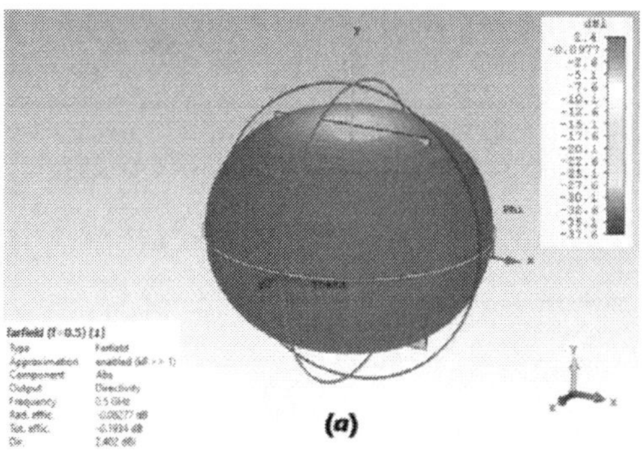

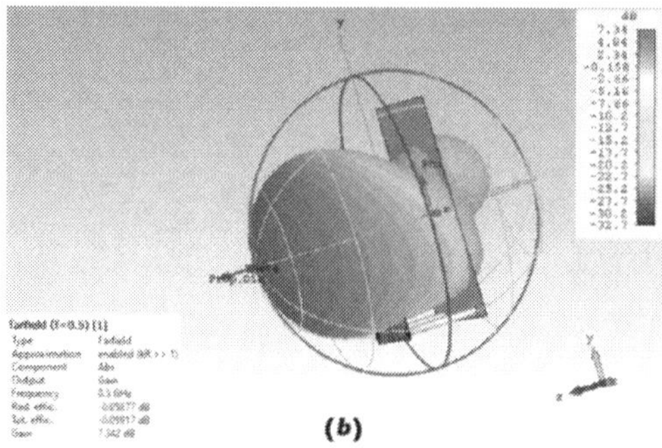

Fig10.(a) Radiation pattern at 0.5 GHz without reflector. (b) Radiation pattern at 0.5 GHz with reflector.

The antenna has radiated omnidirectional with gain 2.8 dB. When the reflector was positioned at the back of the antenna, the gain has increased to 7.43 dB and the radiation pattern is more directives. In this regard, it is clear that the fact of including the reflector at the back of our antenna has highly improved the performances in terms of gain as well as the radiation pattern.

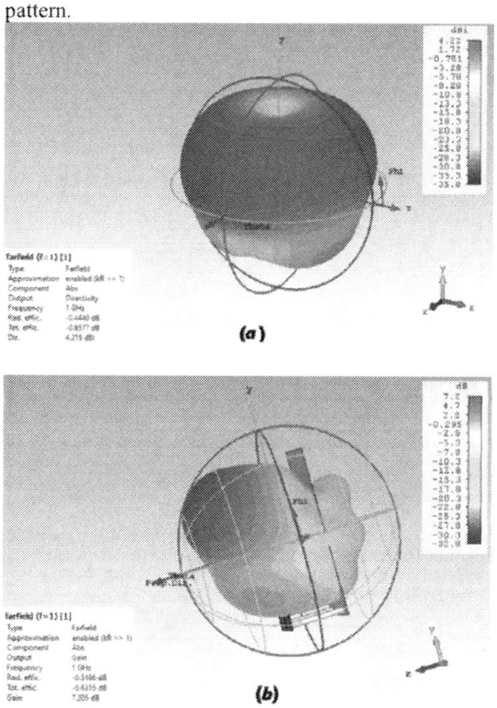

Fig11.(a) Radiation pattern at 1GHz without reflector.(b) Radiation pattern at 1 GHz with reflector.

At the third frequency we can clearly say the performances of the antenna has decreased in term of directivity as well as change in the gain going from 4.21dB to 7.20 dB .

Conclusion

In this work, we had presented and proved the efficiency of a reflector in increasing and improving the directivity of the radiation. This was effectuated by comparing the structure with and without reflector in terms of the following properties: the return loss, the gain, the radiation pattern of patch circular antenna operating in the frequency 0.5 GHz, with a large bandwidth ranging from 0.43 to 1.25 GHz. By this, we can clearly say that this type of antenna is suitable for GPR applications. In this regard, we are expecting to develop our structure and bring an antenna array to see if the reflector will prove again its efficiency in improving the radiation directivity in terms of gain and polarization.

References

1. N. O. Parchin, H. J. Basherlou, and R. A. Abd-Alhameed, 7 (n.d.)
2. T. Iida and J. N. Pelton, in Handbook of Satellite Applications, edited by J. N. Pelton, S. Madry, and S. Camacho-Lara (Springer New York, New York, NY, 2013), pp. 397–426
3. E. T. Rahardjo, D. P. Yusuf, Basari, and F. Y. Zulkifli, in 2015 Asia-Pacific Microwave Conference (APMC) (IEEE, Nanjing, China, 2015), pp. 1–3
4. R. Karli and H. Ammor, 4 (2012)
5. E. Strumwasser, J. A. Short, and R. J. Stegen, in (1952)
6. (n.d.)
7. C. A. Balanis, Antenna Theory: Analysis and Design, 2nd ed (Wiley, New York, 1997)
8. V. H. R. K. V.Harsha Ram Keerthi, IOSR-JECE 5, 49 (2013)
9. A. Bansal and R. Gupta, Int. j. Inf. Tecnol. 12, 149 (2020)
10. D. J. Daniels, Ground Penetrating Radar (2004)
11. M. N. A. Karim, M. F. A. Malek, M. F. Jamlos, L. Y. Seng, and N. Saudin, in 2013 IEEE International RF and Microwave Conference (RFM) (IEEE, Penang, Malaysia, 2013), pp. 253–257
12. M. Cherniakov, M. Salous, V. Kostylev, and R. Abdullah, in European Radar Conference, 2005. EURAD 2005. (IEEE, La Defense, Paris, France, 2005), pp. 165–168
13. N. J. Willis, Bistatic Radar (SciTech Publishing, Edison, NJ, 2005)
14. M. N. A. Karim, M. F. A. Malek, M. F. Jamlos, A. Z. Abdullah, N. S. Md Noorpi, N. M. Mokhtar, and M. A. Jusoh, in Theory and Applications of Applied Electromagnetics, edited by H. A. Sulaiman, M. A. Othman, M. Z. A. Abd. Aziz, and M. F. Abd Malek (Springer International Publishing, Cham, 2015), pp. 59–67
15. J. R. Aguilar, M. Beadle, P. T. Thompson, and M. W. Shelley, 6 (n.d.)
16. N. Kumar and N. Sharma, IJECE 6, 23 (2019)
17. O. Oulhaj, N. Touhami, M. Aghoutane, and A. Tazón, 11, 32 (2016)
18. A. Raza, W. Lin, M. K. Ishfaq, M. Inam, F. Masud, and M. H. Dahri, International Journal of Antennas and Propagation 2021, 1 (2021).
19. Bembarka A, Setti L, Tribak A, Nachouane H, Tizyi H (2022) Frequency tunable filtenna using defected ground structure filter in the sub-6 ghz for cognitive radio applications. PIER C 118:213–229.

RFID System for Respiratory Rate Monitoring

Jéssyca I. L. Araujo[1*], *Alexandre J. R. Serres*[1], and *Danilo F. S. Santos*[1]

[1]Post-Graduation Program in Electrical Engineering, Federal University of Campina Grande, Campina Grande, Brazil

Abstract. In this work, an RFID system was developed to monitor the respiratory rate. This monitoring was based on variations in the Received Signal Strength Indicator (RSSI) by the reader due to respiratory movements. In some breath measurement configurations, the maximum read range was greater than 1 m, which is the distance considered in the measurement configurations described in some scientific papers. The RSSI data, obtained by reading the RFID tag, were processed using the Daubechies wavelet of order 3 (db3) and considering three levels of signal decomposition. Less noisy breathing signals were obtained in the measurements with the participant seated and the RFID tag with the polyethylene foam on the abdomen. The respiratory frequencies acquired from the variations of the RSSI parameter were validated by comparison with the results obtained using a vital signals monitor.

1 Introduction

Vital signals are indicators of the health status of patients, providing relevant information about the body's functions. The monitoring of these signals allows the identification of health problems and a faster intervention by professionals in the area, in order to stabilize the patient's clinical condition, and to evaluate the results obtained after the intervention [1].

Respiratory rate is the least registered vital signal, despite being an important indicator of serious diseases, such as prognosis of cardiac arrest, abdominal pathology or sepsis [2]. In addition, with respiratory rate monitoring, sudden infant death syndrome (SIDS) [3], chronic obstructive pulmonary disease (COPD), obstructive sleep apnea syndrome (OSAS), Rett syndrome and asthma attacks can be detected [4]. Thus, it is possible to verify the importance of monitoring this vital signal, especially in critical cases, in which emergency care is crucial.

Some equipment used to monitor respiratory function is uncomfortable, requires devices attached to the body, and can even be invasive [5-6]. Among the methods for this monitoring, there are those that evaluate the temperature of the exhaled air, since it has been proven that the increase in the temperature of the exhaled air is related to the inflammation of the airways and to the increase in vascularization that is characteristic of the lung cancer [4, 7]. Some of these methods involve the use of masks or cannulas, non-contact infrared technology that requires expensive computer infrastructure and tools, or wireless systems with the sensor located in the region below the nose [7]. A wireless system has also been developed whose monitoring is based on the humidity of the exhaled air [4]. Other forms of monitoring use devices with sensors, such as a narrow chest strap, and some wireless technologies require custom Doppler radios, which are not readily available on the market [4, 6]. Therefore, it is observed that some of these methods can cause discomfort for the patient, are of high cost or are not easily accessible.

The use of technologies that provide this monitoring continuously associated with patient comfort is of great interest. Therefore, Radio Frequency Identification (RFID) technology is an alternative to existing technologies, since it allows this continuous monitoring through wireless communication, in a non-invasive way, and the use of flexible substrates can provide comfort to the individual. In addition, passive RFID tags do not require an internal battery for their operation, which contributes to both reducing weight and reducing costs and increasing their useful life.

In this work, the RFID UHF system that was developed for monitoring the respiratory rate is presented, and it is possible to determine the value of this frequency from the variation data of the Received Signal Strength Indicator (RSSI). Section 2 presents the tag design with the simulation results obtained after optimization of the tag presented in [8], as well as the made tag. In Section 3, the measurement settings are described, the respiratory signals obtained from reading the tags, that is, the RSSI variation signals due to respiratory movements, as well as the respiratory frequency results obtained from these variations of the RSSI and with the vital signals monitor. Finally, the conclusions are drawn in Section 4.

* Jéssyca Iasmyn Lucena Araujo: jessyca.araujo@ee.ufcg.edu.br

2 Passive and flexible RFID tag design

2.1 Simulations

The RFID tag consists of an improvement of that presented in [8], with alteration of the L3 dimension of the port to facilitate the soldering of the chip and adjustments in other dimensions in order to achieve a resonance frequency close to 915 MHz with a lower value of the coefficient of reflection (S_{11}). Its simulation was also carried out using the ANSYS® Electronics Desktop™ software.

The tag can be seen in Fig. 1 (a) with its respective layers. The dimensions of its constituent parts are shown in Table 1 and indicated in Fig. 1 (b) and Fig. 1 (c). The distance between the CSRRs (Complementary Split Ring Resonators) cells is 1.1 mm and between the folded dipole antenna and the CSRR structure is 1 mm. Its total dimension is 63.9 x 107 mm².

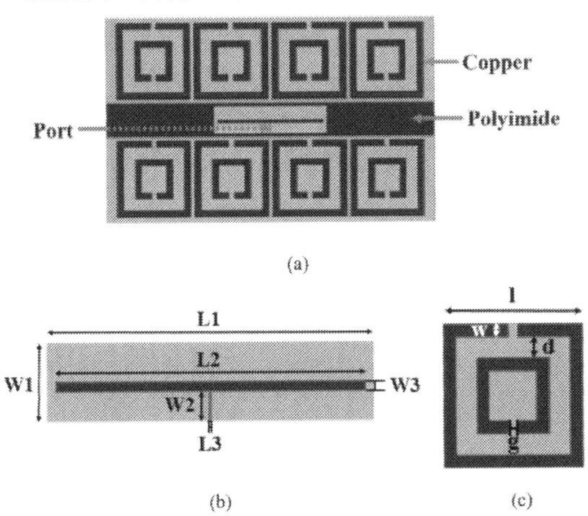

Fig. 1. Simulated RFID tag: (a) tag layers, (b) dimensions of the folded dipole antenna, (c) dimensions of the CSRRs cells.

Table 1. Dimensions of the RFID tag in [mm].

L1	L2	L3	W1	W2	W3
36.3	34.4	0.2	8.7	1.1	3.3
	l	w	d	g	
	24.2	2.2	3.6	1.6	

With this tag, it was possible to obtain a reflection coefficient of -49.99 dB at the operating frequency (915 MHz), as shown in Fig. 2. In Fig. 3, the antenna input impedance at this frequency is observed, with Za = (11.45 + j199.2) Ω. In this way, it is noticed a very good conjugate matching with the chip impedance (Zc = 12.7-j199 Ω for the NXP SL3S4021 chip at 915 MHz [9]).

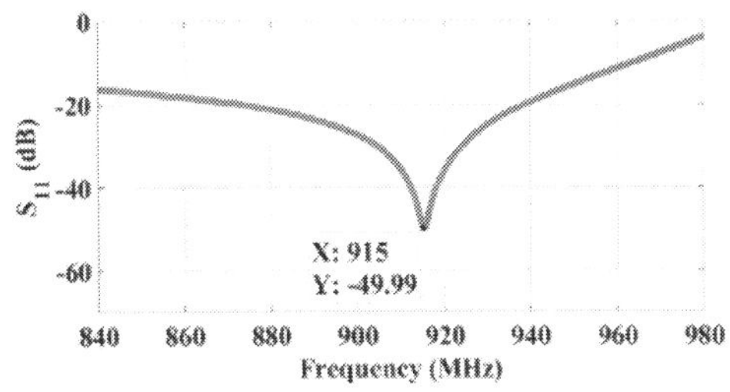

Fig. 2. Reflection coefficient (S_{11}) of the simulated tag.

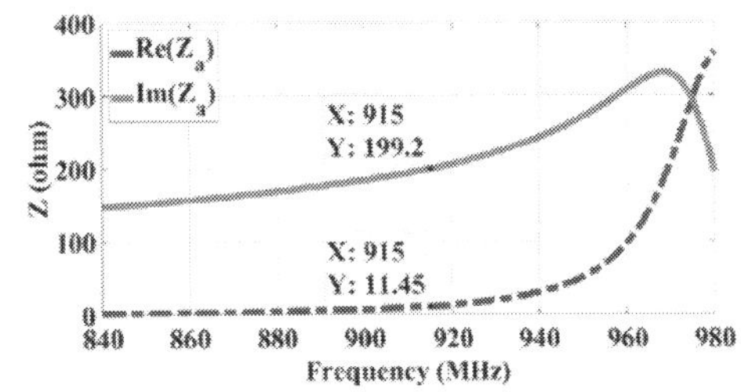

Fig. 3. Input impedance of the simulated tag.

Fig. 4 and Fig. 5 show the radiation patterns of the tag. It is possible to check the maximum positive gain value (5.47 dBi). In addition to this parameter, a maximum directivity value of 6.28 dB and a radiation efficiency of 82.98% were obtained.

Fig. 4. 3D radiation pattern of the simulated tag.

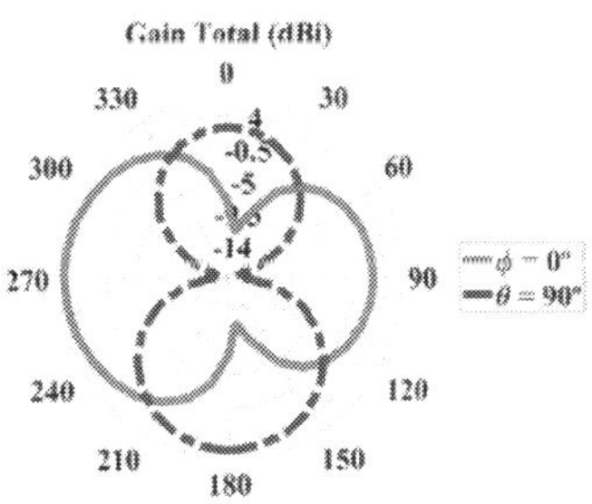

Fig. 5. 2D radiation patterns of the simulated tag.

The average Specific Absorption Rate (SAR) on the tag substrate was also simulated, as can be seen in Fig. 6, in order to certify that the limits specified by the International Commission on Non-Ionizing Radiation Protection (ICNIRP) were being met. As the limits established by ICNIRP for public and occupational SAR for the trunk are 2 W/kg and 10 W/kg [10], respectively, it is noted that the value obtained meets these requirements.

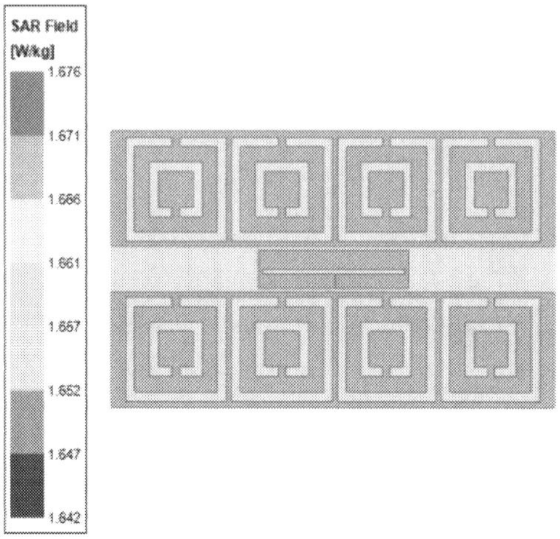

Fig. 6. SAR behavior on the RFID tag substrate at the frequency of 915 MHz.

The maximum read range for this tag, calculated using the Friis equation [11], is 24.60 meters at 915 MHz.

2.2 Fabrication

The RFID tag was made using the corrosion process of the copper face of the polyimide laminate using an iron perchloride solution. For this, the adhesive of the tag layout was pasted on this conductive layer, in order to prevent corrosion of the copper located under it. Subsequently, the substrate was cut and part of the adhesive to remove the chip was removed. To avoid cutting the copper located on the structure's margins by the manual manufacturing process, a margin of the substrate was left and the total dimension of the structure was 64 x 108 mm^2. The made tag is shown in Fig. 7.

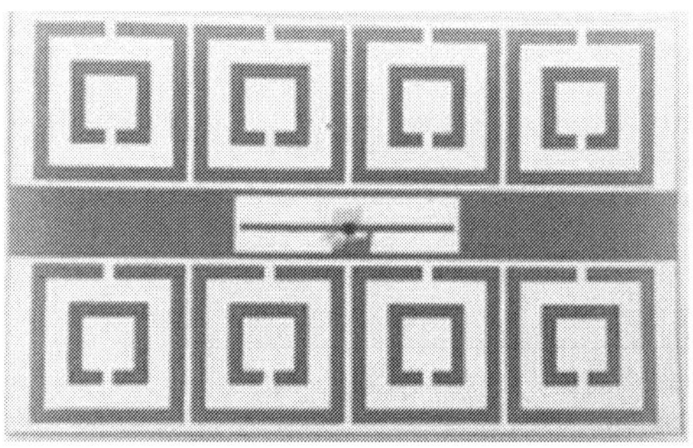

Fig. 7. Manufactured RFID tag.

3 Results and analysis of measurements

3.1 Experiments Setting

The configuration of the tests to obtain the respiratory rate is shown in Fig. 8. In these tests, the RSSI measurements from the reading of the RFID tag on the abdomen and the measurements with the Lifemed LIFEtouch.10 vital signals monitor were performed simultaneously for later validation of the respiratory rate results obtained by the RSSI variations. The 5-lead ECG cable was used on the multiparametric monitor and the electrodes were glued to the body in the positions indicated in Fig. 8 and also on the legs.

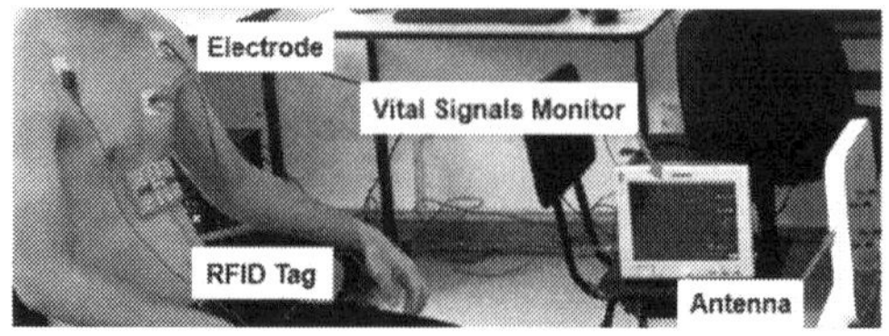

Fig. 8. Measurement setting.

As it was not possible to read the tag placed directly on the abdomen due to the influence of the human body on its performance, a 11 mm layer of polyethylene foam was inserted between the RFID tag and the body, which reduces the influence of the body on the performance of the tag. The foam was adhered to the polyimide substrate and to the human body by means of a thin double-sided tape, which does not cause damage to the skin.

In addition, the position of the tag on the abdomen was varied between the positions shown in Fig. 9, in order to observe in which position it is possible to obtain a greater range read of the tag and a better signal of breathing. Position 1 corresponds to the top edge of the tag placed just below the sternum and position 2 with it placed approximately four fingers below the sternum. The position of the individual was also varied, making measurements with him sitting and standing.

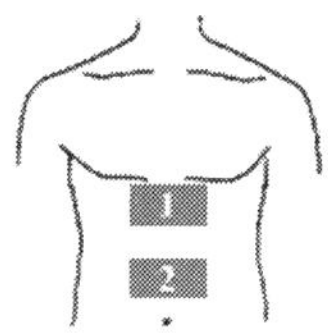

Fig. 9. RFID tag positions on the human body (adapted from [1]).

In those tests, the Intermec IF2 network reader was used and the Laird PAL90209H antenna was connected to it. This antenna has a gain of 9 dBic and operates in the frequency range of 902-928 MHz, which includes the operating frequency of the 915 MHz tag.

To carry out those measurements, the project was submitted to the Research Ethics Committee of the Hospital Universitário Alcides Carneiro (CEP-HUAC) for evaluation and was approved in accordance with the opinion number 3.742.182 and Presentation Certificate for Ethical Appreciation (CAAE) number 21811419.3.0000.5182.

3.2 Results and Analysis

The maximum read ranges after varying the distance between the tag and the antenna connected to the RFID reader are presented in Table 2. According to this table, it can be noted that it was possible to achieve greater range with the tag in position 1 and the participant standing. It is also noticed that all distances were greater than 20 cm, respecting the requirement of a minimum distance of 20 cm between the transmitting antenna and the human body of the Federal Communications Commission (FCC) [14].

Table 2. RFID tag maximum read range.

Position Tag	Participant's Position	Maximum Read Range (m)
1	Seated	1,09
2	Seated	0,865
1	Standing	1,23
2	Standing	1,05

The difference between the maximum read ranges obtained in the measurements and the value obtained from the simulation data (24.6 m), may be, among other factors, due to flaws in the manual manufacturing process and the influence of the human body on the performance of the RFID tag.

In some configurations, however, the read range was greater than 1 m, which is the distance considered in some works found in the literature, as can be seen in Table 3.

Table 3. Comparison between some works that use RFID technology to monitor breathing.

Reference	Year	Substrate	Read Range (m)
[12]	2016	Mix of wool and lycra and 55 mm polyethylene foam	1.00
[13]	2017	Fabric	1.00
[6]	2017	-	6.00
[4]	2017	Polyimide and polyurethane film	0.60
[3]	2018	-	1.00
[7]	2018	Bio-silicone membrane	0.50
This work	2020	Polyimide and 11 mm polyethylene foam	1.23

Because the signals received by the reader from reading RFID tags are quite noisy, it was necessary to process these signals to filter out noise. For that, the MATLAB software toolbox dedicated to the Wavelet Transform was used, which presents, among its applications, noise attenuation. Daubechies wavelet of order 3 (db3) was applied considering three levels of signal decomposition, which allowed the visualization of RSSI variations with respiratory movements and, consequently, the calculation of respiratory frequency from the number of signal peaks.

The measurements were carried out over a period of 2 minutes, however, as the respiratory rate corresponds to the number of respiratory cycles in 1 minute, only the data from the time interval in which it was possible to achieve better results was considered.

In Fig. 10, it is possible to observe the signals obtained when considering the seated participant. For the tag in positions 1 and 2, a respiratory rate of 20 rpm (breaths per minute) was obtained. The RSSI values varied between -60.3 dBm and -63 dBm for the tag in position 1 and between -59.5 dBm and -62 dBm for the tag in position 2.

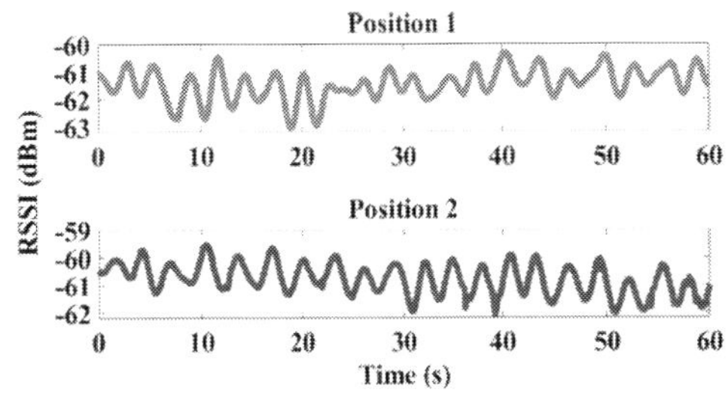

Fig. 10. Breathing signals obtained from reading the RFID tag with the 11 mm polyethylene foam and the participant seated.

The results for the standing participant, on the other hand, can be seen in Fig. 11. Note that the respiratory rate is approximately 20 rpm for the tag in position 1 and 19 rpm with it in position 2. Furthermore, in the measurement with the tag in position 1 on the abdomen, a variation of the RSSI was obtained between -58.3 dBm and -62.7 dBm, and, with it in position 2, this variation was between -64 dBm and -65.4 dBm.

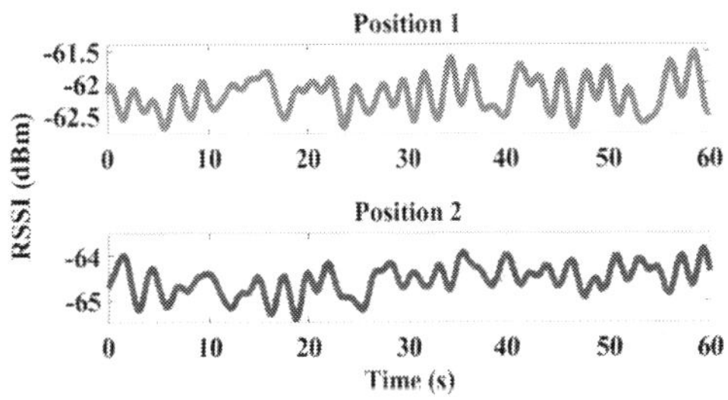

Fig. 11. Breathing signals obtained from reading the RFID tag with the 11 mm polyethylene foam and the participant standing.

According to the signals in Fig. 10 and Fig. 11, it is clear that, after processing, a less noisy result was obtained with the tag in position 2 and the participant being seated. However, in the other configurations it is also possible to verify variations of the RSSI with the respiratory movements.

In order to validate the values of the respiratory frequencies obtained from the variations of the RSSI, a comparison was made with the results obtained using the multiparametric monitor. Table 4 shows the values obtained from the RSSI variations and the frequency range in which the respiratory rate obtained through the signal monitor varied during the measurement. It is verified that the frequency obtained from the RSSI data are within the range of variation of the frequency obtained using the multiparametric monitor, thus validating the results of these measurements.

Table 4. Respiratory rate values obtained from RSSI variations and through the vital signals monitor.

Position Tag on the Abdomen	Participant's Position	Respiratory Rate Measured from Changes in RSSI (rpm)	Range of Respiratory Rate Variation with the Vital Signals Monitor (rpm)
1	Seated	20	15-22
2	Seated	20	16-22
1	Standing	20	20-22
2	Standing	19	17-21

4 Conclusion

In this work, a UHF RFID system for monitoring respiratory rate was developed, using a passive and flexible RFID tag. In the results of simulations of the tag, a maximum positive gain value of 5.47 dBi was observed and it was found that the average simulated SAR complies with the limits established by ICNIRP, ensuring its use on the body. Regarding the monitoring of respiratory rate, the possibility of this monitoring was verified through the variation of RSSI data obtained from the reading of the tags by the RFID reader. A better result was achieved in the measurement with the participant in the sitting position and the tag with the polyethylene foam on the abdomen. The respiratory rate data obtained from the RSSI variation were validated by comparison with the results obtained by the vital signals monitor. Regarding the maximum read range, it was noted that it was possible to obtain a range greater than 1 m in some configurations, this distance used in the measurements presented in some papers in the literature.

The authors would like to thank COPELE/UFCG and CNPq for their institutional and financial support, the measurements volunteer, and Simões who works at Embedded/DEE/UFCG for soldering the RFID tag chip, without which this paper would not be possible.

References

1. P. A. Potter, A. G. Perry, Fundamentos de Enfermagem, **8** (Elsevier, Rio de Janeiro, 2013)
2. M. A. Cretikos, R. Bellomo, K. Hillman, J. Chen, S. Finfer, A. Flabouris, Respiratory rate: the neglected vital sign, Med. J. Aust., **188**, 11 (2008).
3. R. Zhao, D. Wang, Q. Zhang, H. Chen, A. Huang, *CRH: A Contactless Respiration and Heartbeat Monitoring System with COTS RFID Tags*, in 2018 15th Annual IEEE International Conference on Sensing, Communication, and Networking (SECON), Hong Kong (2018)
4. M. C. Caccami, C. Miozzi, M. Y. S. Mulla, C. Di Natale, G. Marrocco, *An epidermal graphene oxide-based RFID sensor for the wireless analysis of human breath*, in 2017 IEEE International Conference on RFID Technology & Application (RFID-TA), Warsaw (2017)
5. W. Mongan et al., *A Multi-Disciplinary Framework for Continuous Biomedical Monitoring Using Low-Power Passive RFID-Based Wireless Wearable Sensors*, in 2016 IEEE International Conference on Smart Computing (SMARTCOMP), St. Louis (2016)
6. Y. Hou, Y. Wang, Y. Zheng, *TagBreathe: Monitor Breathing with Commodity RFID Systems*, in 2017 IEEE 37th International Conference on Distributed Computing Systems (ICDCS), Atlanta (2017)
7. C. Occhiuzzi, C. Caccami, S. Amendola, G. Marrocco, *Breath-monitoring by means of Epidermal Temperature RFID Sensors*, in 2018 3rd International Conference on Smart and Sustainable Technologies (SpliTech), Split (2018)
8. J. I. L. Araujo et al., *Flexible UHF RFID Tag Gain Enhancement using Metamaterial Structure*, in 2019 4th International Symposium on Instrumentation Systems, Circuits and Transducers (INSCIT), São Paulo, Brazil (2019)
9. NXP, SL3S4011_4021: UCODE I²C [Online]. Available: https://www.nxp.com/products/identification-and-security/smart-label-and-tagics/ucode/ucode-ic:SL3S4011_4021 (Accessed 24 Apr. 2020)
10. International Commission on Non-Ionizing Radiation Protection, Guidelines for limiting exposure to time-varying electric, magnetic, and electromagnetic fields (up to 300 GHz), Health Physics **74**, 4 (1998)
11. I. Aznabet, M. Ennasar, O. El Mrabet, S. Tedjini, M. Khalladi, *Meander-line UHF RFID tag antenna loaded with split ring resonator*, in 2016 5th International Conference on Multimedia Computing and Systems (ICMCS), Marrakech (2016)
12. D. Patron et al., *On the Use of Knitted Antennas and Inductively Coupled RFID Tags for Wearable Applications*, IEEE Trans. Biomed. Circuits and Syst. **10**, 6 (2016)
13. S. A. Vora et al., *On implementing an unconventional infant vital signs monitor with passive RFID tags*, in 2017 IEEE International Conference on RFID (RFID), Phoenix (2017)
14. Federal Communications Commission. FCC review of RF exposure policies [Online]. Available: https://www.fcc.gov/document/fcc-review-rf-exposure-policies (Accessed 2 Jan. 2020)

Characterization of Low-Cost UWB MIMO antenna model for K/Ka/Q/U/V bands Applications

Abdellatif Berkat[1], *Zeyneb Berkat*[2], *Noureddine Boukli Hacene*[3], *Mohamed Latrach*[4]

[1] Dept. of Telecommunication, Faculty of Electrical Engineering, Djillali Liabès University, Sidi Bel Abbes, Algeria
[2] Dept. of Electronics, Faculty of technology, Saad Dehleb University, Blida1, Algeria
[3] Dept. of Telecommunication, Faculty of technology, Abou-Bakr Belkaid University, Tlemcen, Algeria
[4] RF-EMC research group, IETR, ESEO, Angers, France
Email: berkat.abdellatif@gmail.com; berkat.z.telecom @gmail.com; bouklin@yahoo.com; Mohamed.LATRACH@eseo.fr

Abstract. A dual-element and a four elements low-cost multiple-input-multiple-output (MIMO) antenna are introduced in this paper. It achieves a bandwidth (BW) of 25 GHz (22-45 GHz) and 35 GHz (22- 60 GHz) respectively, and the VSWR is less than 2 for both configurations. The proposed antennas average isolation of 20 dB over the entire operating band, and a peak realized gain of 7.6 dBi for the dual element antenna and 7.59 dBi for the four-element antenna, which are suitable for ultra-wideband at K/Ka/Q/U/V bands Applications.

Index Terms—dual element MIMO antenna, four element MIMO antenna, ultra-wideband antenna, patch antenna.

1 Introduction

Ultra-wideband (UWB) technology has attracted much attention in the last years, Due to its outstanding advantages of low power and high data rate, is regarded as a good choice in several wireless communications, such as imaging systems, ranging, and satellite communication. Multiple-input-multiple-output (MIMO) technology is another promising technology [1], which can play an essential role in wireless communication systems to meet the requirements of omnidirectional characteristics, more bandwidth and higher data rate [2], UWB-MIMO can make full use of their advantages and provide more big data rate [1]. On the other hand, some techniques of development for UWB antennas have been introduced in [3]. A UWB crescent-shaped monopole antenna for a 5G millimeter-wave test frequency band has been proposed in [4]. Ref. [5] applied a Genetic algorithm to design an Ultra-wideband.

Others work to design a compact MIMO antenna. Ref [6] used a 6-elements MIMO system with EBG structure in the form of a square spiral cell. a compact microstrip four-port dual MIMO antenna for the Sub-6G band is discussed in [7].

Another challenge to combine a UWB technology and MIMO antenna which can be applied to wireless systems, for example, ref [8] based on split-ring resonator (SRR) and complementary split-ring resonator (CSRR). A parasitic unit decoupling was used to prove UWB-MIMO in ref.[9]

With a small profile, low-cost structure, two configurations of UWB-MIMO antenna which the first model of dual-element cover 22GHz-45 GHz UWB and the second of four elements covers 22 GHz- 60 GHz UWB are studied in this article.

Simulation is achieved using the Finite Integration Technique (FIT) based computer simulation technology microwave studio (CST MWS) .

2 The primary structure antenna

Fig. 1 (a) and (b) show the geometry of the primary proposed antenna on CST Microwave and HFSS respectively. The proposed model consists of a new configuration radiating patch inspired from the L-shaped antenna cited in [10]. The UWB is achieved by inverting the L-shaped radiating element and modifying the 50 Ω coaxial cable position.

The S11 result depicted in Fig. 1 operates from 22 GHz to 45 GHz that appeared in K, Ka and Q bands technology The major advantage of this structure antenna is resumed in its small size, low-cost profile, and good results.

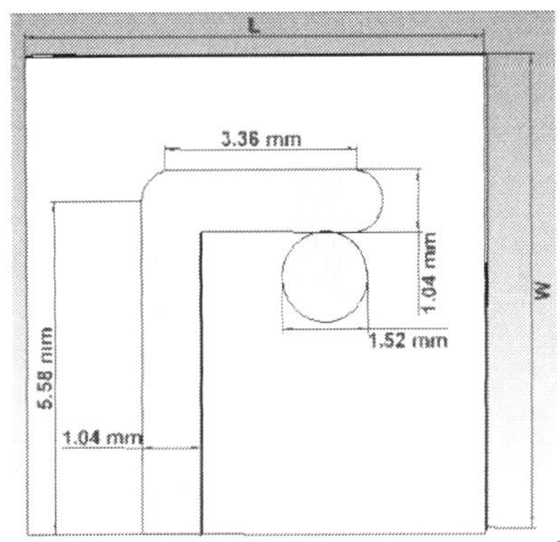

(a)

https://doi.org/10.1051/itmconf/20224801006

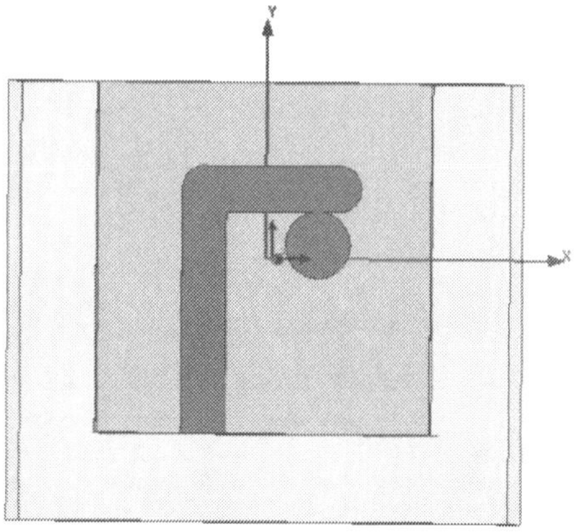

(b)

Fig1. Primary structure antenna, (a) on CST Microwave, (b) on HFSS.

Fig. 2 (a) shows the variation effect of the length (L) of the proposed structure on the reflection coefficient results. The simulated S11 result obtained with L=W=10mm operates from 22 GHz to 45 GHz that appeared in K, Ka, and Q bands applications. On the other hand, fig.2 (b) shows the comparison between the simulation on CST Microwave and the other on HFSS, the both obtained S11 results obtained with L=10mm indicates good conformity in the desired frequency bands. Thus, the mismatch in Fig.2 (b) curves is related to the high losses of the used substrate at high frequency bands.

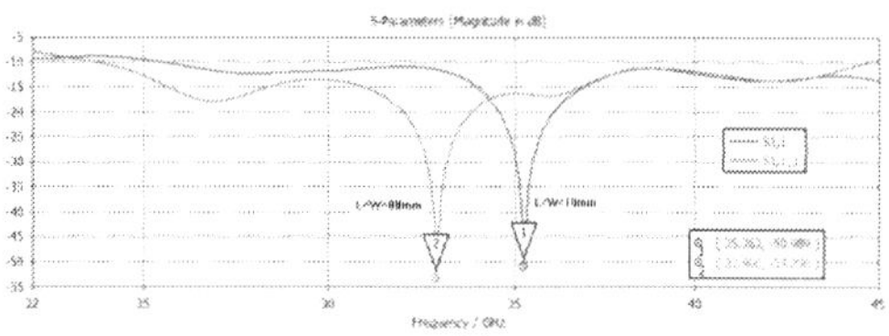

(a)

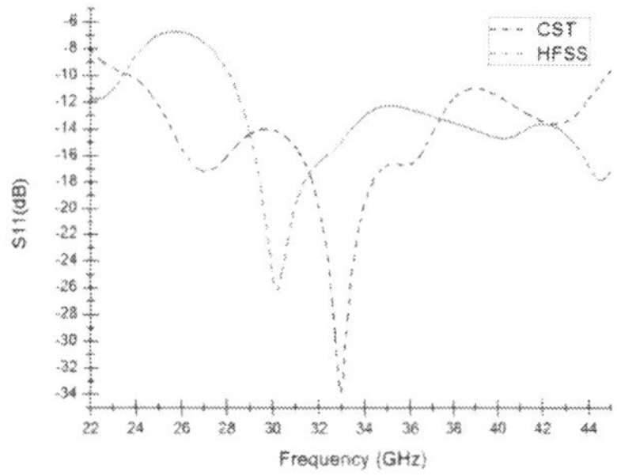

(b)

Fig2. (a) The S11 result of the primary structure, (a) comparison on CST with two different dimension of antenna, (b) comparison result, CST Microwave and HFSS.

To determine the bandwidth of an antenna, two relationships should be mentioned: a return loss of -10 dB corresponds to a VSWR less than 2. In fig .2 and fig.3 we can inspect those last conditions.

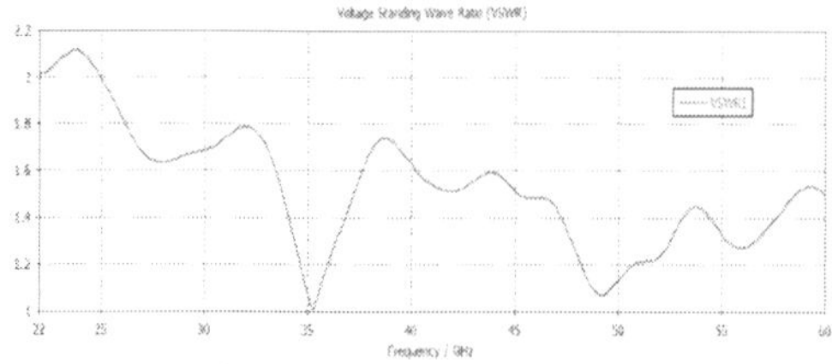

Fig3. VSWR of the proposed antenna

Fig. 4 shows the current distribution on the surface of the radiating element at 33 GHz. The maximum current concentrated in the center of the patch can be produced due to the inclined shape of the patch depending on the port excitation.

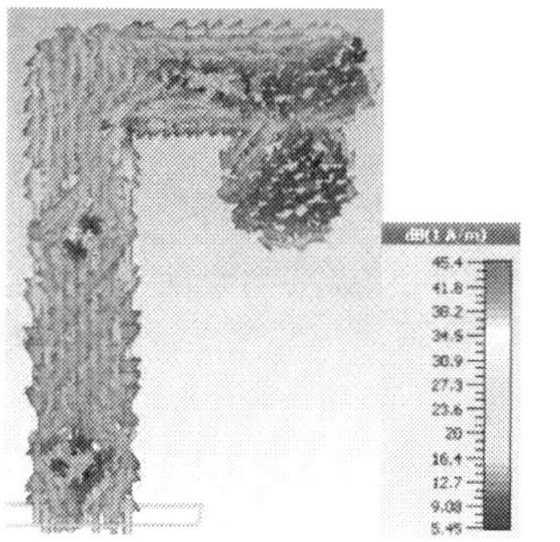

Fig4. The current distribution on the surface of the primary element antenna at 33 GHz.

3 UWB-MIMO structure

3.1 First configuration of dual elements MIMO antenna

In this suggestion, and for the first 1X2 MIMO configuration, the substrate size is increased up to 16x08 mm, and the two radiating elements are studied in two configurations as seen in Fig.5 (a) and (b), respectively.

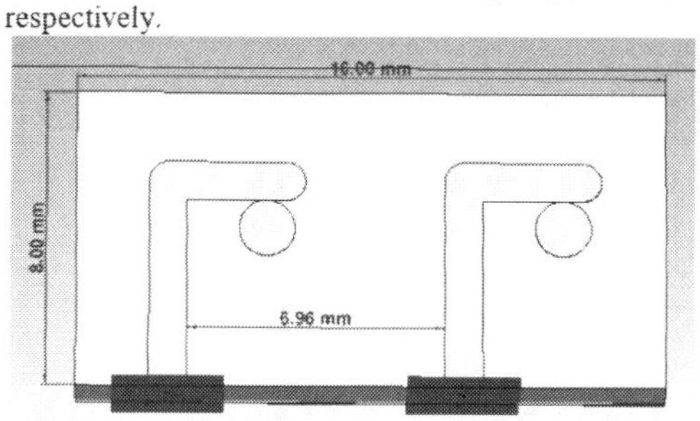

(a)

https://doi.org/10.1051/itmconf/20224801006

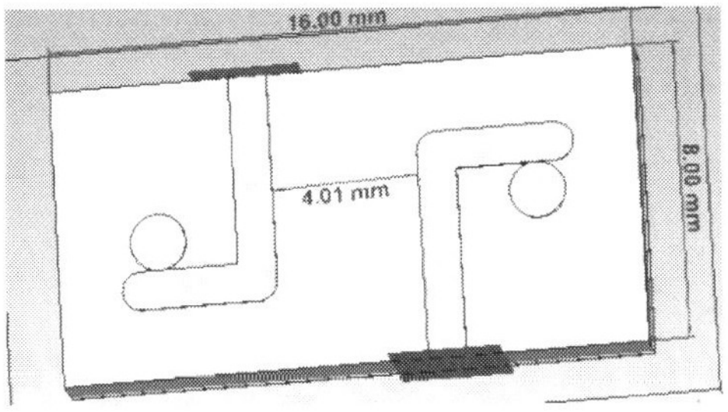

(b)

Fig5. The proposed MIMO 1X2 antenna, (a) of the first MIMO antenna model, and (b) the second MIMO antenna model.

The simulated Sij of the MIMO configurations shown in fig. 5 is depicted in Fig.6 (a) and (b) respectively. Both S11 results of the proposed configurations cover the band assigned for UWB applications that operates from 25 GHz to 45 GHz. On the other hand, the isolation between the radiating elements is greater than 20 dB over the entire operating frequency, as depicted in fig.6. (b). The proposed dual-radiating element can be used for the K, Ka and Q technologies.

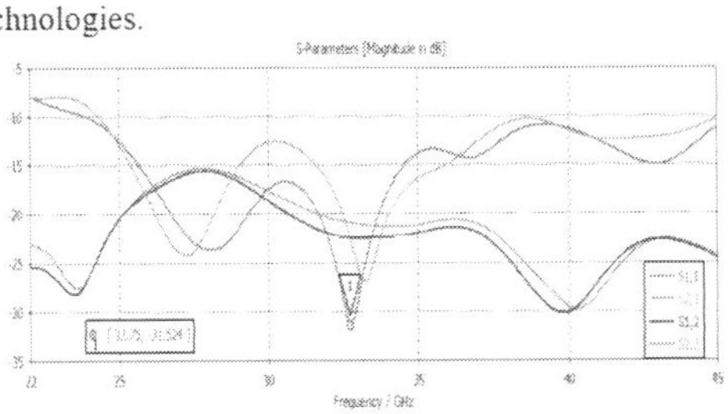

(a)

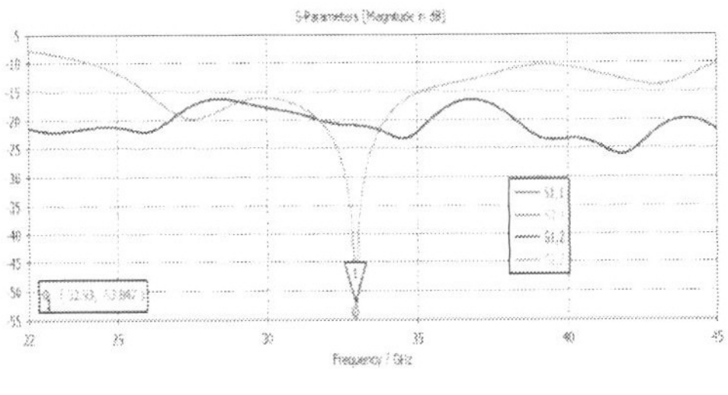

(b)

Fig.6. Simulated Sij result of the proposed antenna vs. Frequency; (a) of the first MIMO configuration and (b) of the second MIMO configuration.

Fig. 7 shows the variation gain in dBi vs frequency from 22 GHz to 45 GHz of both 1X2 MIMO configurations antenna, that the maximum gain is 5.1 dBi at 32 GHz for the first configuration and 4.6 dBi from 39.6 GHz to 33.4 GHz for the second configuration. Both proposed configurations suggest good gain characteristics.

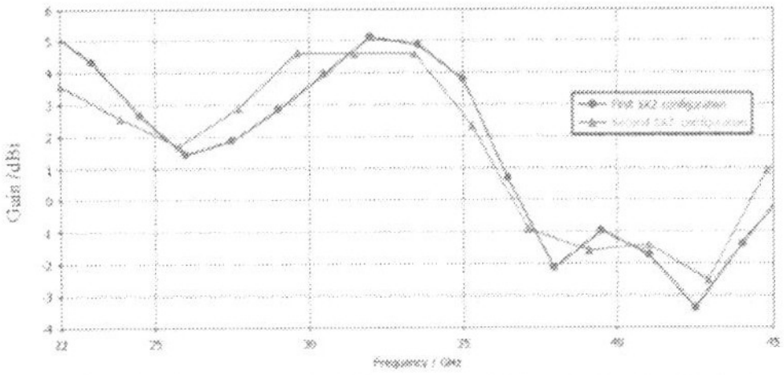

Fig.7. Simulated gain results of the proposed (a) and (b) 1X2 MIMO antenna.

3.2 second configuration of four elements MIMO antenna

In this step, the last 2X2 configurations are chowed in fig.8 (a) and (b) of 2X2 in order to achieve the UWB from 22 GHz to 60 GHz and covers K, Ka, Q, U and V bands Applications. The global size of this model is 16X16 mm, with four 50 Ω feeding cables exiting the antenna.

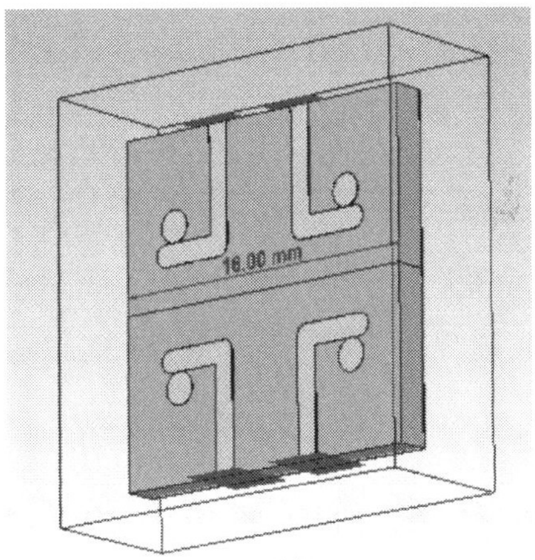

(a)

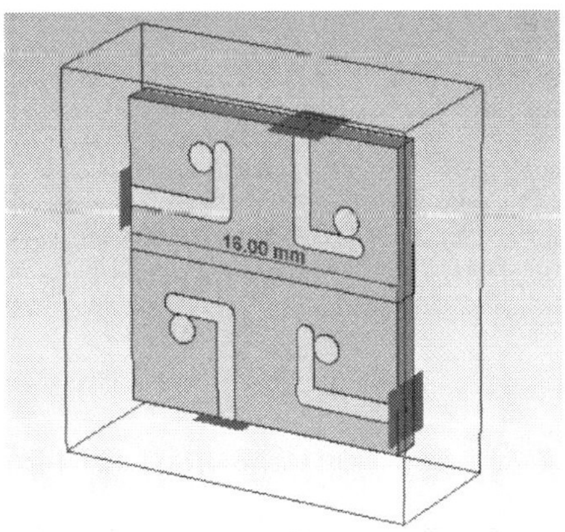

(b)

Fig.8. The proposed MIMO 2X2 antenna, (a) the first MIMO model and (b) the second MIMO model.

Fig. 9 (a) and (b) show the simulated characteristics Sij of the four radiating elements MIMO antenna. The obtained S11 reveals a good performance from 25 GHz

https://doi.org/10.1051/itmconf/20224801006

to 60 GHz for both configurations (a) and (b). Otherwise, the isolation in Fig. 9 (a) indicates better performance compared to Fig. 9 (b). Fig. 10 shows the VSWR variation of the configuration fig 8 (b) which is less than 2.

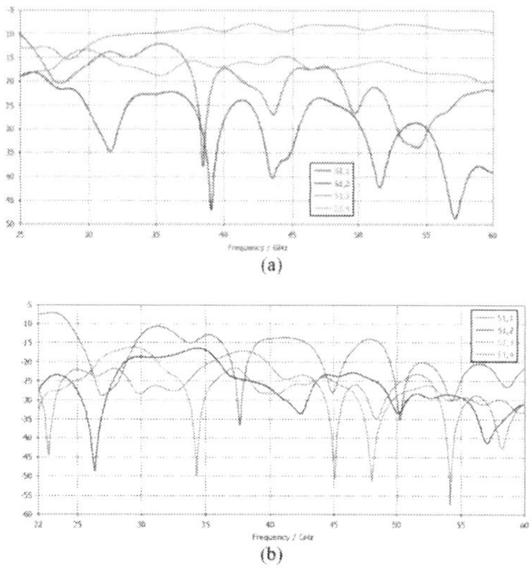

Fig9. Simulated Sij result of the proposed MIMO 2X2 antenna vs. Frequency; (a) the first MIMO antenna model and (b) of the second MIMO antenna model.

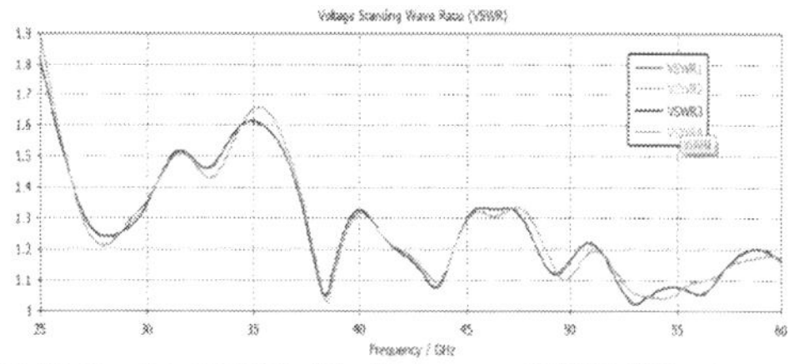

Fig10. Simulated VSWR of the first proposed MIMO 2X2 antenna .

Fig. 11 (a), Fig. 11 (b), Fig. 11 (c) and Fig. 11 (d) show the simulated far-field 3D and both polar radiation patterns E-plane and H-plane of the 2X2 MIMO configuration proposed in Fig. 8 (a) and of the second 2X2 MIMO configuration proposed in Fig 8. (b) at 38 GHz and 37.4 GHz respectively. The main beam in E-Plan of Fig. 11 (b) deviates from the z-axis by 24° for 4.26 dBi of gain with a 3-dB beamwidth of 66.6°. And the main beam in E-Plan of Fig. 11 (d) deviates from the z-axis by 33° for 6.91 dBi of gain with a 3-dB beamwidth of 66.3°. For H-Plan, the main beam of Fig. 11 (b) deviates from the z-axis by 282° for 3.51 dBi of gain with a 3-dB beamwidth of 52.5°. And, the main beam in H-Plan of Fig. 11 (d) deviates from the z-axis by 316° for 6,15 dBi of gain with a 3-dB beamwidth of 31.2°. In the present patterns, the radiation patterns of Fig. 11 (d) for the second 2X2 MIMO configuration dominate. The full-wave simulation was carried out using CST software. These results in Fig. 11 (a), and Fig. 11 (d) are simulated at 37.4 GHz and 38 GHz giving a maximum gain of 7.59 dBi and 5.25 dBi respectively.

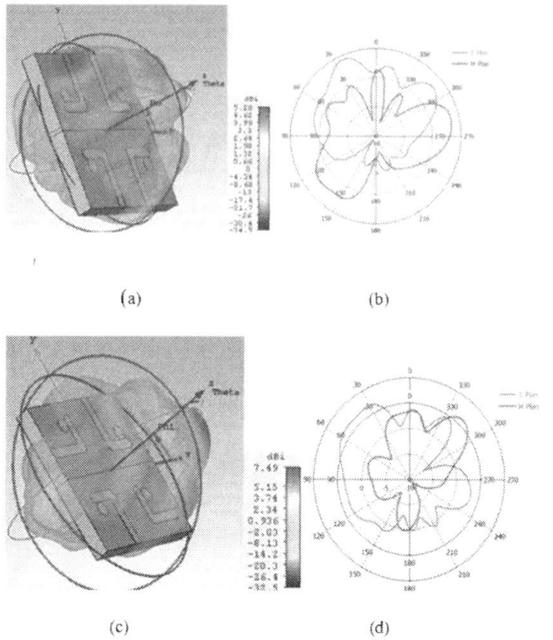

Fig11. Simulated radiation patterns at 38 GHz and 37.4 GHz of the proposed MIMO 2X2 antenna, (a) 3D, (b) polar radiation, (c) 3D, (d) polar radiation of first and second configurations.

Table 1. Performance comparisons between the proposed and some previously reported MIMO antennas.

Ref.		Bandwidth (GHz)	Applications	Antenna size (mm²)	Configuration	Average isolation	Max Gain
[11]		From 2.4 to 11	Automotive applications	24X29	1X2	15	2 dBi
[12]		From 2.1 to 20	WiMAX and military/radar applications	37X40	2X2		5.8 dB
[13]		From 2.6 to 4.2 and from 5.11 to 10	5G and X band	35X25	1X2	-	10 dBi
This Work	1X2 config (a)	from 25 to 45	K,Ka,Q bands	16X08	1X2	20	5.1 dBi
	1X2 config (b)	from 22 to 45	K,Ka,Q bands	16X08	1X2	20	7.6 dBi
	2X2 config (a)	from 22 to 60	K,Ka,Q,U,V bands	16X16	2X2	20	5.28 dBi
	2X2 config (b)	from 22 to 60	K,Ka,Q,U,V bands	16X16	2X2	20	7.49 dBi

In order to evaluate the usefulness and advantages of this work, the performance of the simulated prototype is outlined in Table 1 to compare with the previously reported antennas. These common approaches are based on MIMO technology and UWB antennas. The comparison focuses on the bandwidth, antenna size, maximum gain, configuration, and application bands. In [11], the antenna mentioned has a small size and good average isolation, but low gain. It can be seen in [12] that although the antenna has a WB, there is still a large size. Ref [13] appeared as a low dual-band. However, compared with this work, the scheme of the proposed antenna is low cost, cited, for example, single substrate layer, and small size. The comparisons show, too, that the proposed antenna model has achieved a good performance in UWB and gain value.

Conclusions

With very compact size, a dual element MIMO antenna for K, Ka ,Q bands and another four-element MIMO antenna for K, Ka, Q , U, V bands with average isolation of 20 dB are designed, analyzed, and discussed. The results demonstrate that the proposed dual-element MIMO antenna operates from 22 GHz to 45 GHz, and the other four elements MIMO antenna operates from 22 GHz to 60 GHz. Moreover, other characteristics such as good gain value, and acceptable far-field radiation patterns, make both proposed configurations of MIMO antenna suitable for several wireless communication systems like 5 G and satellite communications systems.

References

1. Ling Wu1, Xia Cao, and Bing Yang,"Design and Analysis of a Compact UWB-MIMO Antenna with Four Notched Bands",Progress In Electromagnetics Research M, Vol. 108, 127-137, 2022
2. Kommanaboyina V. Babu and all, "Compact Dual-Band Printed MIMO Antenna with Very Low Mutual Coupling for WLAN, Wi-MAX, Sub-6GHz 5G and X-Band Satellite Communication Applications",Progress In Electromagnetics Research C, Vol. 117, 99-114, 2021
3. Beijia Liu and all,"Sharply Defined UWB Rectangular Dielectric Resonator Antenna"IEEE International Symposium on Antennas and Propagation and North American Radio Science Meeting,2020
4. Peilin Zhong and all,"A UWB antenna for 5G millimeter wave frequency band", International Conference on Microwave and Millimeter Wave Technology, IEEE, 2020
5. Youcef Braham Chaouche and all,"Optimization of dual-band notched UWB TrapezoidalAntenna using Genetic Algorithm Approach", International Symposium on Antennas and Propagation and North American Radio Science Meeting, IEEE, 2020
6. Alaa M. Hediya and all,"5G MIMO Antenna System Based on Patched Folded Antenna with EBG Substrate",Progress In Electromagnetics Research M, Vol. 109, 149{161, 2022
7. Zhong Yu, and all,"A Compact Microstrip Four Port Dual Circularly Polarized MIMOAntenna for Sub-6G Application",Progress In Electromagnetics Research C, Vol. 119, 145{159, 2022

8. Hongmei Li and Naicheng Gong,"An SRR and CSRR Based UWB-MIMO Antenna",IEEE International Symposium on Antennas and Propagation and North American Radio Science Meeting,IEEE,2020
9. Kunming Zhang and all,"UWB-MIMO antenna decoupling based on a wideband parasitic unit structure",nternational Conference on Microwave and Millimeter Wave Technology (ICMMT),2020.
10. A. A. Deshmukh, S. Pawar and K. P. Ray, "Multi-band Configurations of L-shape and F-shape Microstrip Antennas,", International Conference on Computing, Communication, Control and Automation (ICCUBEA), 2017, pp. 1-5, doi: 10.1109/ICCUBEA.2017.8463875.
11. D. Potti et al., "A Novel Optically Transparent UWB Antenna for Automotive MIMO Communications," in IEEE Transactions on Antennas and Propagation, vol. 69, no. 7, pp. 3821-3828, July 2021, doi: 10.1109/TAP.2020.3044383.
12. V. S. D. Rekha, P. Pardhasaradhi, B. T. P. Madhav and Y. U. Devi, "Dual Band Notched Orthogonal 4-Element MIMO Antenna With Isolation for UWB Applications," in IEEE Access, vol. 8, pp. 145871-145880, 2020.
13. Kommana Vasu Babu, Sudipta Das, Soufian Lakrit, Shobhitkumar Kiritkumar Patel, Boddapati Taraka Madhav, and Hicham Medkour, "Compact Dual-Band Printed MIMO Antenna with Very Low Mutual Coupling for WLAN, Wi-MAX, Sub-6 GHz 5G and X-Band Satellite Communication Applications," Progress In Electromagnetics Research C, Vol. 117, 99-114, 2021.

https://doi.org/10.1051/itmconf/20224801007

A New Design of 5G Planar Antenna based on metamaterials with a high gain using array antenna

Abderrahim Bellekhiri[1] , *Noha Chahboun*[1], *Yassin Laaziz*[1], *and Ahmed El Oualkadi*[1]

[1]Laboratory of Technology of Information and Communication
National school of Applied Sciences ENSA of Tangier, Abdelmalek Essaadi University, Tetouan Morocco

Abstract. In this paper, a symmetrical monolayer metamaterial superstrate is placed on a microstrip patch antenna resonant at 3.5 GHz. MTM metamaterial based CSRR Complementary Split Ring Resonators, constituting two rectangular rings is opposite splits on both ends, which are printed on Rogers RT 5880 type substrate with relative permittivity of 2.2, tangential loss of 0.0009, and a thickness of 0.508mm. This resonator is capable of exhibiting an ε-negative permittivity. Moving an MTM superstrate loaded with a 2 × 3 grating on a single radiating element, provides a gain increase from 4.66 dBi to 5.67 dBi was found at its operating frequency, and the radiation pattern was directional and stable over the entire frequency band. For a 4×1 antenna array loaded with MTM, the gain reaches about 9.84 dBi. the proposed antenna is suitable for 5G mobile applications.

Keywords: metamaterial, CSRR, superstrate, 5G, antenna array

1 Introduction

Metamaterials (MTM) play a very important role in wireless communication systems by improving the gain of the antenna's performance, the radiation pattern. There are several enhancement techniques, using them. For example in work [1] a technique is used to improve antenna gain and bandwidth by combining triangular and rectangular patches with two types of metamaterial structure, which includes two complementary split ring resonators (CSRR), printed on the top layer, and 3×3 MTM slots on the bottom layer. A gain increased from 6.84dB to 12.8dB by adopting a MIMO antenna array configuration [2]. Another technique for improving performance of the proposed antenna, due to metasurfaces (MS), consisting of CSRR, and used as a ground plane [3]. In [4-5] a decoupling element is realized by the combination of the EBG and the SRR. A complementary split ring resonator placed between the antenna array elements exhibits better isolation [6-8]. Adding a superstrate consisting of SRR printed on both sides of a dielectric plate on a microstrip patch antenna (MPA), improves antenna gain and radiation efficiency [9]. MTM cells loaded with D-SSRRP double-layer symmetry single ring resonator pairs are inserted around the conventional patch antenna, improving the gain and directivity of the antenna [10]. In [11], a technique for improving antenna gain by 4 dB, with widening its bandwidth from 2.9% to 4.98%, by placing left-handed metamaterial (LHM) structure above the patch antenna.

In this paper a rectangular patch antenna inspired by metamaterials operating at 3.5GHz. In the first section, we will present a complementary split ring resonator, sized in this work to obtain a ε-negative permittivity at the same resonant frequency as that of the conventional antenna, with a presentation of reflection coefficient and transmission in dB, and the variations of the effective permittivity's real and imagined parts, determined by the parameter S. The second section discusses the application of metamaterial for the reference antenna, by moving a dielectric layer containing CSRRs above the patch antenna, to see its influence on gain improvement for a single element, and for an antenna array also. The simulation of these structures is carried out with the CST Microwave Studio software.

2 Geometry of metamaterial unit cell and its performance characteristics

The rectangular dielectric rings constituting the CSRR are printed on an Rogers RT 5880 type substrate with a relative permittivity ε= 2.2, a loss tangent tan(δ)=0.0009, and a height of 0.508 mm. Figure 1(a) shows the geometry of the proposed complementary split ring resonator with dimension 9mm×9mm. Figure 1(b) shows a resonance peak at 3.5 GHz and a transmission of -52 dB as a result of frequency analysis of parameters S11 and S12 (reflection and transmission coefficients). In table 1, the resonator's geometric parameters are presented.

Table 1. The optimized parameters of the CSRR.

Parameter	mm
L	9
W	7
Wt	0.75
G	0.5
S	0.65

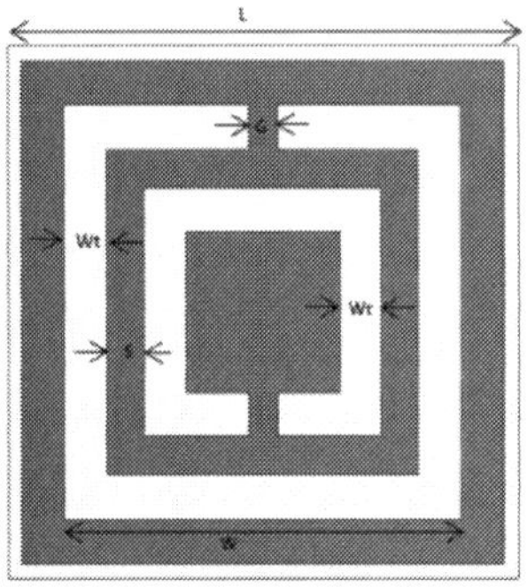

(a) CSRR

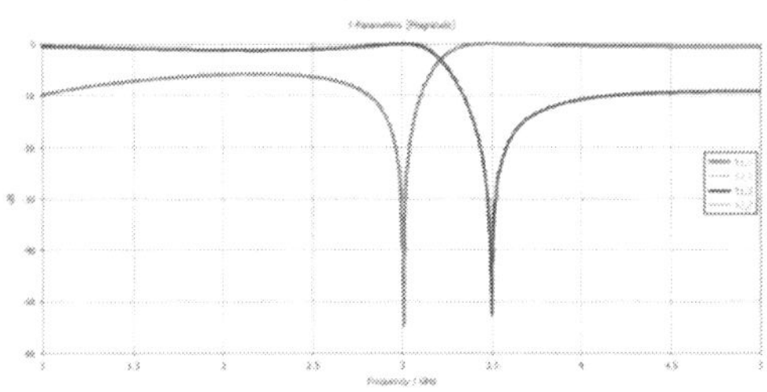

(b) Reflection coefficient

Fig. 1. (a) Representation and parameters of a square CSRR cell unit. (b) S Parameters in dB.

The suggested metamaterial unit cell's permittivity (ε) and permeability (μ) are calculated using the method (Nicolson-Ross-Weir (NRW)) [12-13]. Ziolkowski offered an approximation [14], which we illustrate below. [15]

The transmission (T) and reflection coefficient (Γ) are calculated as follows:

$$T=\frac{S_{11}+S_{22}-\Gamma}{1-(S_{11}+S_{22})\Gamma} \tag{1}$$

$$\Gamma=X\pm\sqrt{X^2-1} \tag{2}$$

Where $$X=\frac{S_{11}^2-S_{21}^2+1}{2S_{11}} \tag{3}$$

The permeability (μ) is given by

$$\mu=\frac{1+\Gamma}{\Lambda(1-\Gamma)\sqrt{\frac{1}{\lambda_0^2}-\frac{1}{\lambda_c^2}}} \tag{4}$$

Where $$\frac{1}{A^2}=(\frac{\varepsilon_\rho*\mu_\rho}{\lambda_0^2}-\frac{1}{\lambda_\chi^2})=-(\frac{1}{2\pi\Lambda}\lambda\nu(\frac{1}{T}))^2 \tag{5}$$

The permittivity (ε) is given by

$$\varepsilon=\frac{\lambda_0^2}{\mu_r}(\frac{1}{\lambda_c^2}-(\frac{1}{2\pi L}\ln(\frac{1}{T}))^2) \tag{6}$$

The variations of the real and imaginary parts of the effective permittivity (ε), and effective permeability (μ) of the resonator's are depicted in Figure 2, and were calculated using the parameters S. It was obtained that the designed resonator presents a permittivity ε-negative at 3.5GHz.

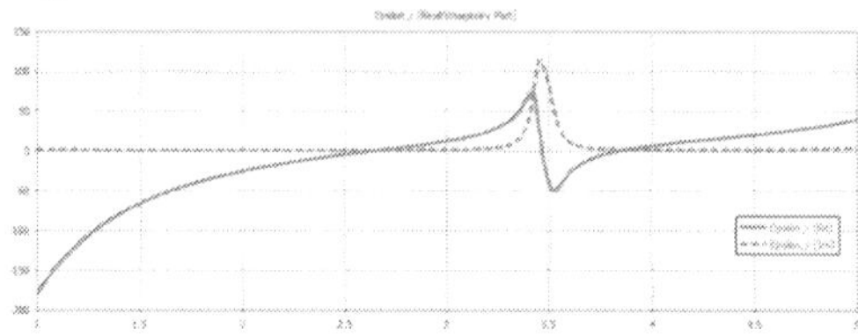

(a) Permittivity (ε)

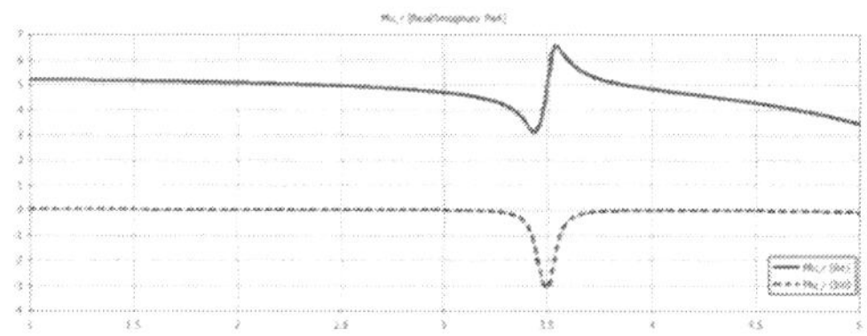

(b) Permeability (μ)

Fig. 2. (a) Effective permittivity's real and imaginary parts. (b) Effective permeability's real and imaginary parts.

3 Design of a planar antenna based on metamaterials

3.1 A single radiating element integrated with a metamaterial-inspired superstrate.

Initially, we chose a printed antenna of rectangular shape fed by microstrip line resonating at 3.5GHz. The rectangular patch of size 19.44 mm × 26.08 mm in PEC of thickness t, printed on an FR4 type substrate with a relative permittivity ε= 4.4, a loss tangent tan(δ)=0.025, a height of 1.6 mm, and a size of 36 mm x 45 mm. The final optimized parameters of the rectangular antenna are shown in Table 2.

Table 2: The optimized parameters of the conventional patch antenna

Parameter	mm	Parameter	mm
Ls	45	Wp	26.08
Ws	36	H	1.6
Lp	19.44	Wf	0.35

https://doi.org/10.1051/itmconf/20224801007

The geometry of this antenna is given in Fig. 3(a). The patch antenna has a good input impedance match at 3.5GHz with a reflection coefficient of -28dB, a frequency band of 190MHz, and a gain of 4.68dBi. The patch antenna is simulated using a finite integration-based electromagnetic solver (FIT).

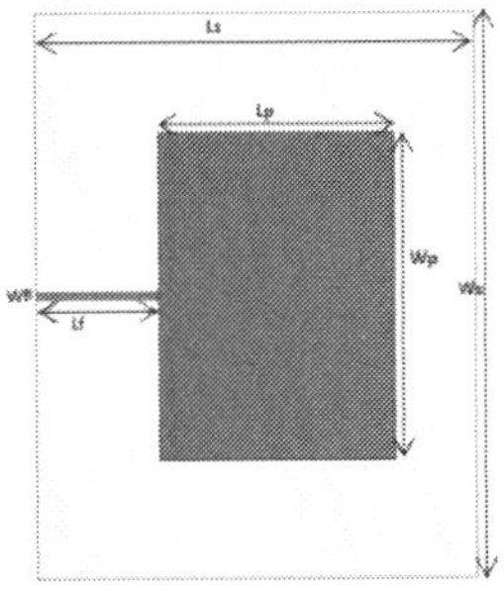

(a)

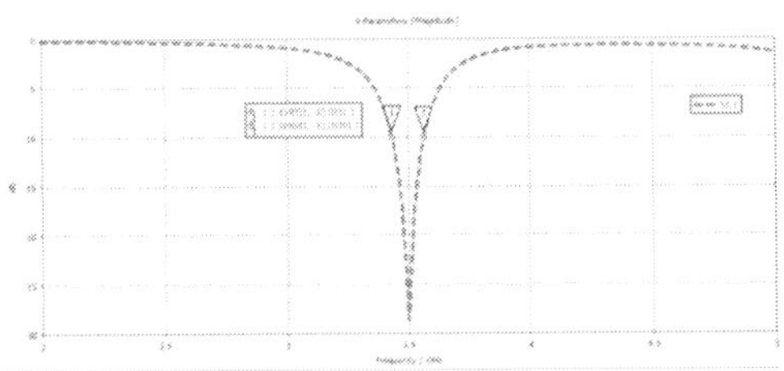

(b)

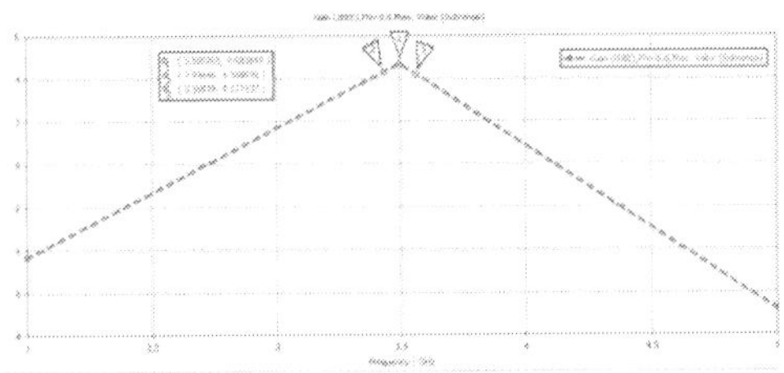

(c)

Fig. 3. (a) Basic antenna geometry, (b) reflection coefficient as a function of frequency, (c) gain as a function of frequency

To improve the gain of the antenna, the same previous antenna is used, and a two-dimensional periodic structure of the complementary split resonators is applied, printed on a Rogers's type superstrate with a size of 19.3 mm × 28.8 mm, and placed 5.092 mm above the patch. Figure 4 (a) shows the geometry of the rectangular antenna embedded with a metamaterial-inspired superstrate. A network of 2×3 metamaterial unit cells is printed on the superstrate. The distance between the resonators is x=0.1mm. The addition of proposed metamaterial superstrate above the previous patch improves the gain from 4.68dBi to 5.68dBi of the proposed antenna compared to the conventional antenna without MTMs as shown in Figure 4(b). This is because the addition of MTM superstrate suppresses surface waves due to its ε-negative characteristics and therefore results in a significant improvement in gain.

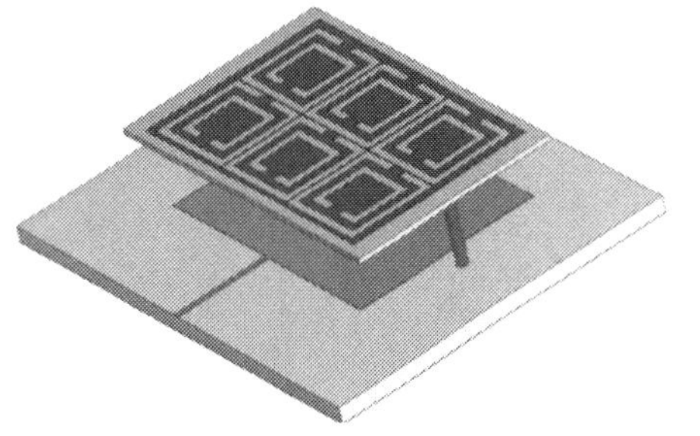

(a)

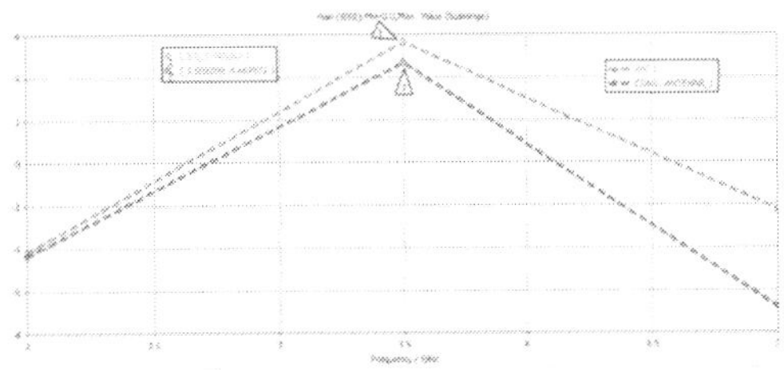

(b)

Fig. 4. (a) The geometry of the antenna with MTM, (b) the gain of the single patch microstrip antenna with and without MTM.

3.2 Design of a 1×4 antenna array integrated with a metamaterial-inspired superstrate.

To further improve the antenna gain, a 1×4 antenna array is designed as can be seen in Figure 5(a). The distance between the radiating elements is $\frac{\lambda}{2}$.

Figure 5 (b) shows that the 1*4 antenna array exhibits a good reflection coefficient match at 3.5 GHz of -17dBi, generating a second frequency band at 4.4 GHz, with a significant improvement in gain at 3.5GHz which reaches 9dBi (Figure 5(c)).

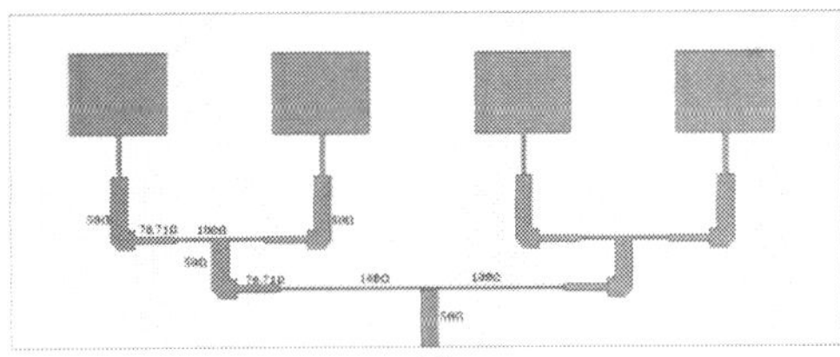

(a)

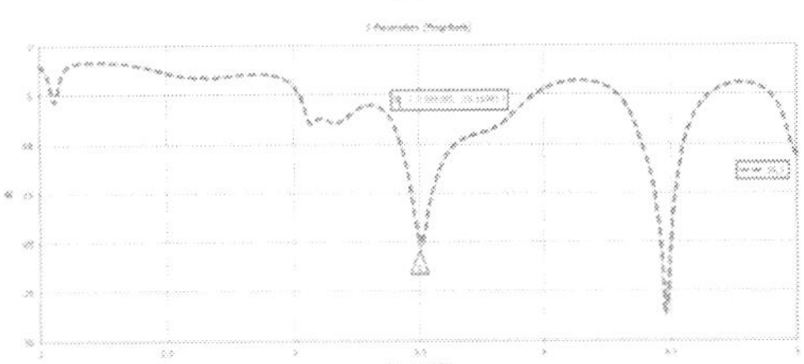

(b)

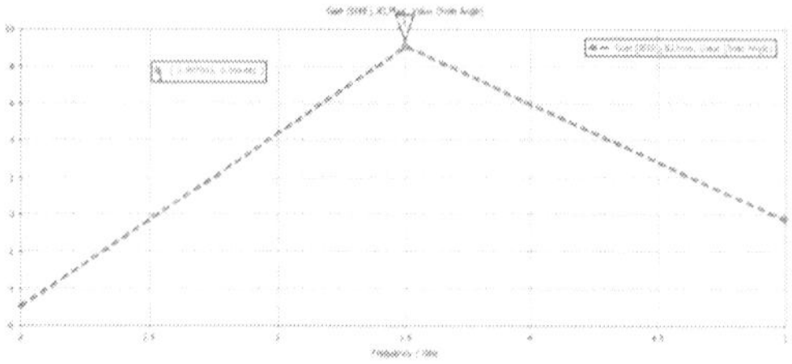

(c)

Fig. 5. (a) 1×4 antenna array, (b) reflection coefficient as a function of frequency, (c) gain as a function of frequency

Figure 6(a) shows the 4×1 antenna array, adding the preceding metamaterial superstrate above each radiating element. There is an improvement in the gain of the proposed antenna compared to the standard antenna that can be seen in Figure 6 (b).

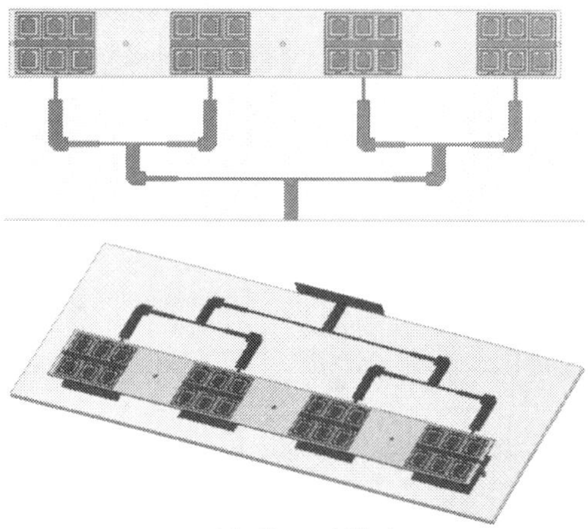

(a) Top and 3D view

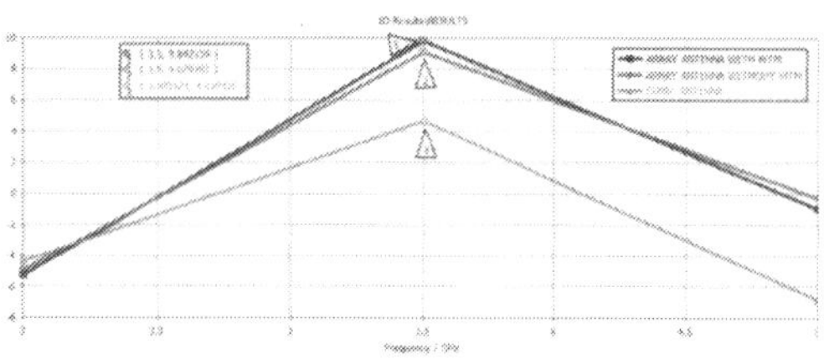

(b)

Fig. 6. (a) The 1×4 antenna array with MTM, (b) The gain of a single basic radiating element, the gain of an antenna array without MTM, and with MTM

Table 3 presents a comparison of the proposed antenna with other antennas published in the literature in terms of gain.

Table 3: Comparison between the proposed antenna and some published antennas

Ref no.	size (mm^3)	Freq. (GHz)	Band. (MHz)	Gain (dB)
11	117×127×41.6	2.4	4.98%	4.22
15	61.2×61.2×4.8	2.4	48	7.94
16	137 × 77 × 3	5.8	$1.78\ 10^3$	9.2
17	20.1× 183× 1.6	2.41 3.61	5.72% 7.3%	7.48 2.383
Pr.	79.22×222×6.6	3.5	170	9.84

4 Conclusion

In this work, we validated a novel antenna structure inspired by metamaterials suitable for 5G applications that resonates at 3.5 GHz. By moving a dielectric layer containing CSRRs above the conventional antenna, the gain is reached to the value of 5.66dBi for a single radiating element, and 9.84dBi for a 4×1 antenna array. This structure was verified by a series of optimizations utilizing the electromagnetic solver's Random technique. This proposed antenna is certified for 5G mobile applications. In the same time as perspectives, we are thinking about associating some components like varactor diodes in order to have a reconfigurable antenna that can be suitable for many standards of communication

References

1. K. Vallappil, B. A. Khawaja, M. K. A. Rahim, M. N. Iqbal, and H. T. Chattha, Metamaterial-Inspired Electrically Compact Triangular Antennas Loaded with CSRR and 3x3 Cross-Slots for 5G Indoor Distributed Antenna Systemsa, Micromachines 2022, 13, 198. https://doi.org/10.3390/mi13020198.
2. J. Khan, S. Ullah, Usman Al, F. A. Tahir, I. Peter, and L. Matekovits, Design of a Millimeter-Wave MIMO Antenna Array for 5G Communication Terminals, Sensors 2022, 22, 2768. https://doi.org/10.3390/s22072768
3. T. Yue, Z. Hao Jiang, and D. H. Werner, Miniaturized Low Profile Antenna Enabled by a Complementary SRR Loaded Metasurface, 978-1-4799-7815-1/15/$31.00 ©2015 IEEE.
4. A. KHAN, S. BASHIR, S. GHAFOOR, AND K. K. QURESHI, Mutual Coupling Reduction Using Ground Stub and EBG in a Compact Wideband MIMO-Antenna, IEEE Access, vol. 9, pp. 40972_ 40979, 2021, doi: 10.1109/ACCESS.2021.3065441.
5. J.Y. Lee, S. H. Kim, and J.-H. Jang, Reduction of Mutual Coupling in Planar Multiple Antenna by Using 1D EBG and SRR Structures, doi : 10.1109/TAP.2015.2447052, IEEE Trans. Antennas Propag.
6. NADEEM AND D.-YOU CHOI, Study on Mutual Coupling Reduction Technique for MIMO Antennas, doi :

10.1109/ACCESS.2018.2885558, IEEE Trans. Antennas Propag., vol. 7, pp. 563_586, Jan. 2019.

7. R. Selvaraju, M. H. Jamaluddin, M. R. Kamarudin, J. Nasir, and M. H. Dahri, Complementary Split Ring Resonator for Isolation Enhancement in 5G Communication Antenna Array, Progress In Electromagnetics Research C, Vol. 83, 217–228, 2018
8. P. S. Bhat, P. C. Kashyap, S. Naik, Mutual Coupling Reduction Using Double Negative Metamaterial Based Dual-Line Split-Ring Resonator (DLSRR) Arrays in Microstrip Patch Antenna Arrays For L-Band and X-Band Applications, 2018 I. C. On Advances in Communication and Computing Technology (ICACCT), India. Feb 8-9, 2018
9. H. Attia, O. Siddiqui, L. Y. and O. M. R, Metamaterial for Gain Enhancement of Printed Antennas: Theory, Measurements and Optimization, 978-1-4577-0069-9/11/$26.00 ©2011 Crown.
10. X. Gao, T. Cai, Li Zhu, Enhancement of Gain and Directivity for Microstrip Antenna Using Negative Permeability Metamaterial, DOI: http://dx.doi.org/10.1016/j.aeue.2016.03.01, International Journal of Electronics and Communications
11. H. A. Majid, M. K. A. Rahim, and T. Masri, MICROSTRIP ANTENNA'S GAIN ENHANCEMENT US-ING LEFT-HANDED METAMATERIAL STRUCTURE, Progress In Electromagnetics Research M, Vol. 8, 235{247, 2009
12. R. W. Ziolkowski, Design, fabrication, and testing of double negative metamaterials, IEEE Transactions on Antennas and Propagation, vol. 51, no. 7, pp. 1516-1529, July 2003.
13. [A. M. Nicolson and G. F. Ross, Measurement of the intrinsic properties of materials by time domain techniques, IEEE Trans. Instrumentation and Measurement, vol. IM-19, no.4, pp. 377-382, Nov. 1970.
14. W. B. Weir, Automatic measurement of complex dielectric constant and permeability at microwave frequencies, Proc. IEEE, vol. 62, pp. 33–36, Jan. 1974.
15. M. Saravanan, V. Beslin Geo, S.M. Umarani, Gain enhancement of patch antenna integrated with metamaterialinspired superstrate, S. M. et al. / Journal of Electrical Systems and Information Technology 5 (2018) 263–270
16. N. Ngoc Lan, and Vu Van Yem, Gain enhancement for MIMO antenna using metamaterial structure, International Journal of Microwave and Wireless Technologies, doi.org/10.1017/S175907871900059X
17. A. Azman, M. Z. A. Abd Aziz, M. K. Suaidi, A. Salleh, H. Nornikman, Design 2 x 2 Patch Array with L Slot Antenna for WiMAX and WLAN, 2015 IEEE 2015 Inter. Conf. on Computer, Communication, and Control Technology (I4CT 2015), April 21 – 23.

A New Compact and Low loss UHF RFID TAG

Latifa El ahmar[1,2*], *Ahmed Errkik*[1], *Ilham Bouzida*[2], *Anass El Mamouni*[3], *Ridouane Er-Rebyiy*[4]
1 Hassan First University of Settat, Faculté des sciences et techniques Settat, LMIET, Morocco
[2]Digitalization & Microelectronics Smart Devices laboratory, MAScIR, Morocco
[3] LABTIC ENSA of Tangier, University of Abdelmalek Essaadi, Tetouan Morocco
[4]EMSI Marrakech, Morocco

Abstract. This paper aims to present a new compact and low loss UHF RFID TAG using meander technique to obtain a miniaturized Design validated to operate in Moroccan UHF band. The proposed RFID TAG is designed on PI substrate with a relative dielectric constant of 3.5, tangent loss of 0.0027 and height0.0508mm. Using an EM solver, the proposed design is matched to ALIEN H3 microchip with $Z_{Microchip}$=30.39-i*211 Ohm, it's simulated and confirmed in Moroccan UHF band with a return loss around -20.49dB at 868MHz with a bandwidth of 16.8MHz, a simulated gain 1.82dBi and a good theoretical read range about 2.6m. The global dimension of the structure is 25×70 mm² which is $\lambda/14 \times \lambda/5$ mm², it's very suitable for RFID applications that required a compact and integrated RFID TAG with a middle-read range.

1 Introduction

The RFID technology is an automatic technology which uses radio waves to identify physical items. A RFID system comprised of two parts: a hardware part composed of (RFID TAG, RFID reader and antennas) and software part consists of a RFID middleware. The interaction of RFID tag and reader/antennas permits to collect an electronic product code (EPC) in real time associated to physical item, as shown in Fig.1.

The RFID TAG is divided in three types: passive TAG, active TAG and battery assist passive TAG, this classification is based on the power mode (internal or external source power). The passive TAG is more used in RFID applications for their properties: low cost and high read range. Passive tags can use three bands of frequencies: Ultra high frequency (UHF) band [860-960MHz], High frequency (HF) at 13.56MHz and Low frequency band (LF) [125-134KHz].

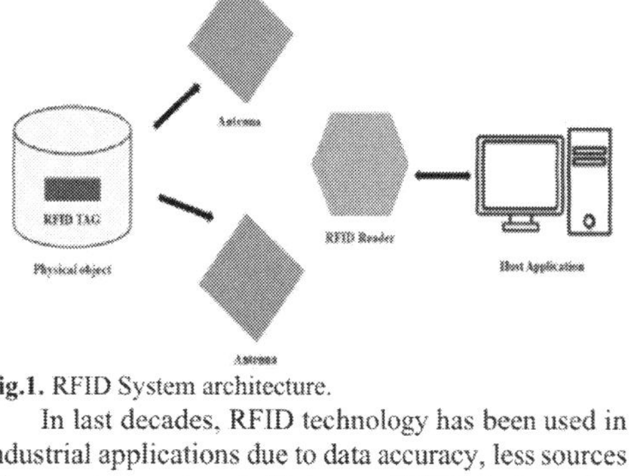

Fig.1. RFID System architecture.

In last decades, RFID technology has been used in industrial applications due to data accuracy, less sources of errors, low cost, and efficiency of RFID devices, also, RFID TAG could have more functionality and give more information associate to other sensors. [1-2]

This present work aims to study a new structure of RFID passive tag validated to operate in Moroccan UHF band using meander technique. The meander technique shows considerable interests due to capability to reduce the length of a dipole antenna specially in small frequency bands such UHF bands which theoretically gives antennas with large dimensions. This paper is organized as a following outlines: The 2[end] section is devoted for presentation of meander line technique, the proposed RFID TAG Design and its characteristics presented in 3rd section, the 4th section is dedicated to present and analyse the EM simulation and in last section a conclusion and perspectives in future.

2 Meander line technique

The meander line technique is one of the techniques used to reduce the antenna sizes from length λ_0 to a length less than 0.1 $\times\lambda_0$.

A meander line antenna (MLA) is a microstrip antenna, it consists of horizontal and vertical segments, this combination forms turns. The MLA behaves as a resonant LC circuit, the horizontal segment acts as capacitor and vertical segments reacts as an inductor. The number of turns N is related directly to the resonance frequency, enhancing the number of turns decrease the resonance frequency. As illustrated in Fig.2. [3-7]

The total inductance L and capacitance C are computed from equation (1):

* Corresponding author: l.elahmar@uhp.ac.ma

- Total inductance L_T:

$$L_T=\frac{L\times l}{2} \quad (1)$$

With L is inductance per unit length and l is the length of meander line.

- Total capacitance C_T:

$$C_T=C\times l \quad (2)$$

With C is capacitance per unit length and l is the length of meander line.

The theoretical length of MLA id defined in equation (3):

$$N\times S=\frac{\lambda}{10} \quad (3)$$

With: N number of turns, S spacing between two parallel vertical segments.

The characteristics impedance of MLA is identified in equation (4):

$$Z_0=27log(\frac{2\times S}{d}) \quad (4)$$

With d dipole diameter.

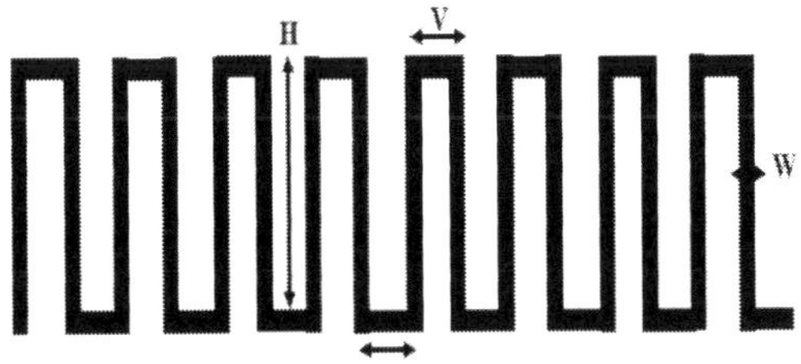

Fig.2. Meander line antenna structure.

3 RFID TAG Design

The art and science of UHF passive TAG Design has been known as a very motivated field of research. An understanding of basics RF physics is obligate to select the best RFID TAG for customer's requirements for their applications.

The RFID TAG has several characteristics that must be validated and considered during the design flow and the applications requirements. Those parameters are listed: Radiation pattern, Gain, Bandwidth, input impedance, Maximum reading distance and impedance coupling. A passive RFID TAG consists of an RFID TAG antenna coupled to a RFID microchip. To ensure an excellent power transmission between RFID TAG antenna and microchip, a highly matching impedance is required as verified in equation (5):

$$Z_{Antenna}=Z^*_{Microchip} \quad (5)$$

The purpose of this paper is focusing on a miniaturization of the RFID TAG with low loss. The proposed RFID TAG is designed at 868MHz using meander technique, it's composed of two inverted meander line arms with 8 turns. This design aims to have a TAG RFID highly suitable for difficult application conditions due to the excellent properties of PI, such as: Flexibility, low density, strength, Corrosion resistance and durability. Fig.3 presents the geometry of the proposed RFID TAG and Table 1 lists the electrical specifications of used material.

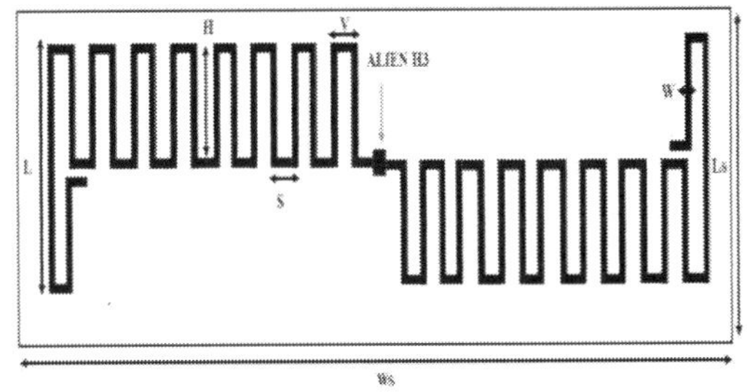

Fig.3. The proposed RFID TAG design.

Table .1. Material Specification

	Dielectric constant	Loss	Thickness [mm]
PI	3.5	0.0027	0.0508
Silver	PEC		0.01

A parametric study and optimisation are carried out to reach the proposed TAG RFID design, which is matched to ALIEN H3 microchip with a complex impedance $Z_{Microchip}$=30.39-i*211 Ohm at 868MHz. The proposed RFID TAG is consisted of two inverted meander arms with 8 turns in each arm. To attempt optimal parameters at 868MHz an EM simulation is done using CST MWS. Table 2 summarizes the optimum parameters at 868MHz.

Table 2. Optimal parameters of the proposed RFID TAG at 868MHz.

	Parameter	Value [mm]
Meander line antenna	W	0.5
	V	2.27
	H	11.99
	S	4
	L	22.51
Substrate	Ls	25
	Ws	70

4 Simulation results and Discussion

 https://doi.org/10.1051/itmconf/20224801008

An EM simulation is done using CST MWS to evaluate the parameters of the RFID TAG. In Fig.4, presents a simulated return loss versus Frequency /GHz, the proposed RFID TAG demonstrates at desired frequency 868MHz a S11=-20.49dB with a bandwidth of 16.8MHz.

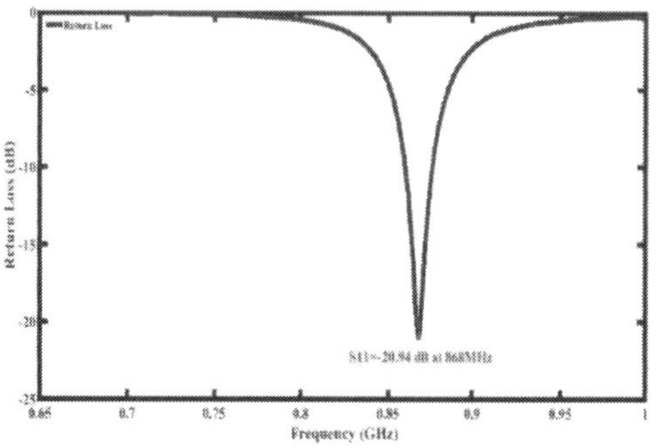

Fig.4. Return loss Vs Frequency /GHz of proposed RFID antenna.

Fig.5 presents a simulated VSWR of the proposed RFID TAG versus Frequency/GHz. The simulated VSWR is around 1.19 at 868MHz, which describes that 99.68% of the power is transmitted to the microchip and 0.32% is reflected. The designed RFID TAG shows a peak directivity of 1.8dBi and an angular width 89.9°, as presented in Fig.6.

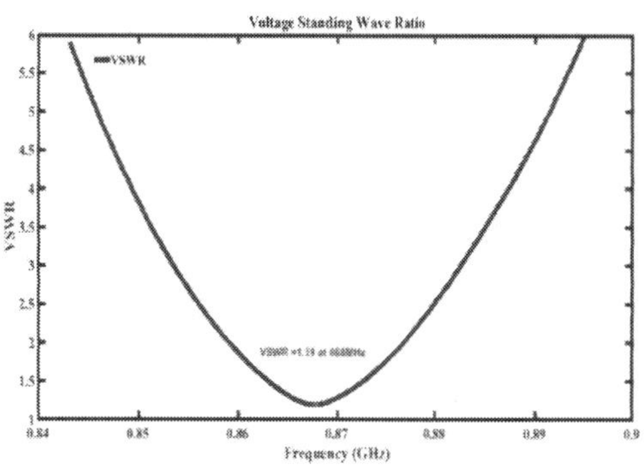

Fig.5. VSWR Vs Frequency /GHz of proposed RFID TAG.

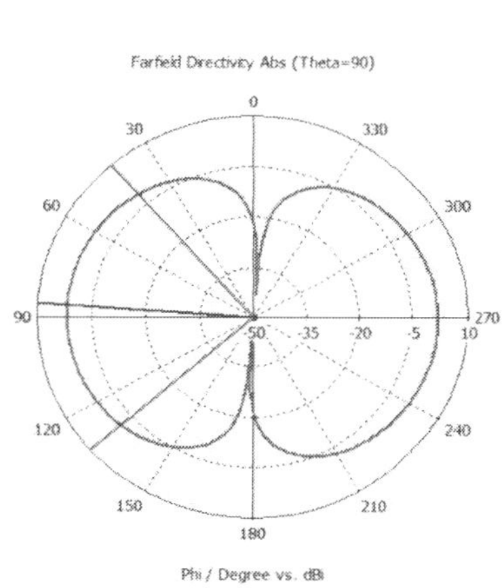

Fig.6. Radiation pattern Vs Frequency /GHz of proposed RFID TAG.

The proposed RFID TAG is designed to ALIEN H3 microchip with impedance $Z_{Microchip}=30.39\text{-}i*211$ Ohm at 868MHz. As presented in Fig.7, the simulated input impedance of the proposed RFID TAG antenna demonstrated is about $Z_{RFID\ TAG\ Antenna}=26.69+i*216.2$ Ohm, it is closely around the conjugate of the microchip impedance. The proposed RFID TAG has been shown an excellent peak gain around 1.82dBi at 868MHz, as shown in Fig.8.

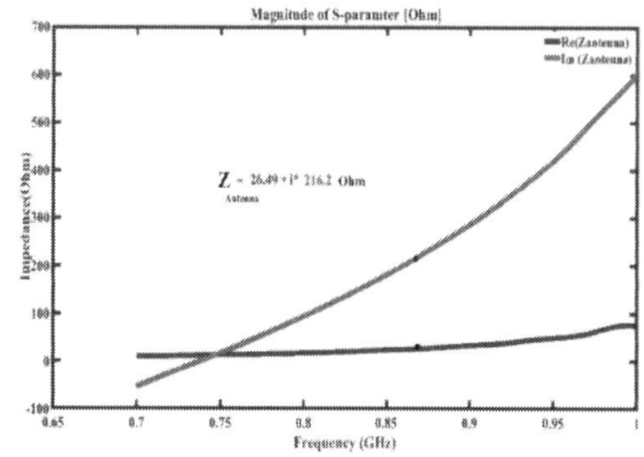

Fig.7. Simulated input impedance Vs Frequency /GHz of proposed RFID antenna.

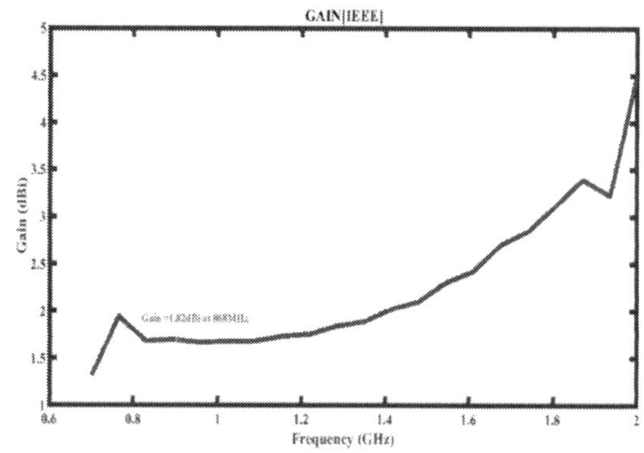

Fig.8. Simulated gain Vs Frequency /GHz of proposed RFID TAG.

Passive UHF RFID TAGs can achieve a long-read range, this criterion is required in RFID application, that demands a large distance of reading. theoretically is defined from FRIIS uplink formula, which is defined in equation (6).

$$\text{Distance}=\frac{\lambda}{4\pi}\sqrt{\frac{P_t G_t G_r}{P_{th}}} \qquad (6)$$

Using equation (6), the reading distance is plotted as a function of the frequency/GHz. As illustrated in Fig.9, the proposed RFID TAG offered a read range about 2.6 m at 868MHz, which is a meddle read-range more suitable for applications such control access for employees or tracking metallic products that needs flexible and adhesive RFID TAG.

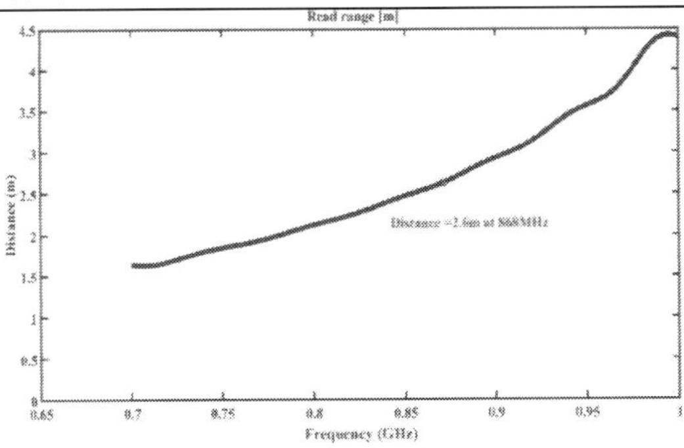

Fig.9. Simulated read range Vs Frequency /GHz of proposed RFID TAG.

5 Conclusion

In this paper, a new RFID TAG is presented using meander technique. The proposed RFID TAG is validated to operate at UHF band especially the Moroccan band[867,5-868MHz], with a return loss -22.49dB at 868MHz and bandwidth of 16.8MHz. It is offered an excellent gain around 1.82dBi and a good read range 2.6m suitable for applications, that's required a middle-read range. The proposed RFID TAG proved a good matching impedance with an input impedance closely around the conjugate of the microchip impedance $Z_{RFID\ TAG\ antenna}$=26.49+i*216.2 Ohm and with conjugate impedance of the microchip $Z^{*}_{Microchip}$=30.29+i*211Ohm. The miniaturized size of the structure is 25×70mm^2 which is $\lambda/14 \times \lambda/5$mm^2, the proposed RFID TAG is suitable for RFID application which required an integrated RFID TAG. The future perspective is to fabricate this structure using ink jet method and test the validation of the simulation results by a reading test using an RFID base station.

References

1. J. Kwangho and S. Lee. *A systematic review of RFID applications and diffusion: key areas and public policy issues.* JOItmC. **1** (2015)
2. Nambiar, Arun N. *RFID Technology: A Review of its Applications.* WCECS (2009).
3. Gonzalez, R. Meneses and L. Montes Peralta, *Antenna Design based on Meander Line Technique to be applied to RFID tag*, WSN, **44** (2016).
4. L.Ying et al, *Design of a Meander Line Dipole RFID Tag Antenna for Temperature Sensor*, ICE2ME (2019).
5. S.Bhaskar,& A.K. Singh, *A Compact Meander Line UHF RFID Antenna for Passive Tag Applications*.PIER, 99, **57-67** (2021).
6. P.Wongsiritorn, C.Phongcharoenpanich,, D.Torrungrueng,& M.Krairiksh ,*Meander-Line UHF RFID Tag Antenna with Semi-Circular Structure*, ISAP (2009).
7. X.L.Travassos, A.C .Lisboa, & D.A. Vieira, *Design of Meander-Line Antennas for Radio Frequency Identification Based on Multiobjective Optimization* ,Int J Antennas Propag, **1-5**(2012).
8. B.D.Braaten, M.T.Reich & J. Glower, A *Compact Meander-Line UHF RFID Tag Antenna Loaded With Elements Found in Right/Left-Handed Coplanar Waveguide Structures*, EEE Antennas Wirel. Propag. Lett. *8*, 1158-1161(2009).
9. K.Rao, P.V.Nikitin, & S.Lam, *Antenna design for UHF RFID tags: a review and a practical application*, IEEE Trans. Antennas Propag, *53*, **3870-3876** (2006).
10. M.Rokunuzzaman, M.T.Islam, W.S.Rowe, S.Kibria, M.Jit Singh, &N.Misran, Design *of a Miniaturized Meandered Line Antenna for UHF RFID* Tags, *PLoS* ONE,**11**(2016)

https://doi.org/10.1051/itmconf/20224801009

Front-to-Back Ratio Enhancement of an Aperture-stacked patch antenna by an Improved circular slots and New Reflector element.

Salaheddine Aourik[1*], *Ahmed Errkik*[2]
[1] *Hassan First 1 University of Settat, Faculté Sciences et Technique, Laboratoire Mechanics, IT, Electronics and Telecommunications, Morocco*

Abstract. In this paper, we investigate an aperture-stacked patch microstrip antenna with three slots in a circular shape that feeds the antenna in order to have an improvement in its performance. The structure optimized and analyzed in this study has an upper substrate of foam dielectric material with a thickness of 0.254 mm and a dielectric constant of 1, a lower substrate of rogers RT3003 with a thickness of 0.5 mm and a loss tangent of 0.001. A cost-effective technique of improving the front-to-back ratio used in this work is to install a reflector which take shape like an aperture-stacked, printed on a thin dielectric layer (0.254 mm) with a dielectric constant of 4.4 and a loss tangent of 0.025. As a result of using the proposed reflector to increase the directivity, the front-to-back ratio of the proposed antenna improved by about 10 dB. With 7.79 dB and matching S11 -40 dB, the structure provides an intriguing gain value.

1. Introduction

An antenna is a tool for transmission and reception electromagnetic waves. This process can be done using the same antenna or using one for transmission and the other for reception. One of the most widely used antennas is the microstrip antenna technology, which became established in the late 1970s. By the beginning of 1980s, basic microstrip antenna elements and arrays had a good foundation in terms of design and modeling. and researchers were focusing on increasing antenna performance attributes (e.g., radiation, matching, bandwidth ...) and expanding the technology's applicability.

The inherently narrow bandwidths and low directivity limited Microstrip patch antennas although they were a lot of appealing features like low cost, and low profile, facility of solid-state device integration. Previous research attempted to correct these flaws which have yielded significant results. In order to achieve larger bandwidths, shaped slots and electromagnetically coupled patches have been used, while so as to achieve higher directivities parasitic patches or high permittivity substrates have been used [1][2]. The majority of previous attempts, on the other hand, were aimed at improving either bandwidth or directivity. To increase the patch antenna's versatility, it would be advantageous to improve their gain and directivity through of front to back ratio at the same time [3].

A multi-layer microstrip patch antenna configuration can be employed to obtain higher directivity performance [4]. The top side of a microstrip antenna is a metal conducting patch that can be printed on a thin grounded dielectric surface known as the substrate. A single radiating element can be fed through coaxial line, microstrip line, or electro-magnetic coupling in the simplest situations [5]. The slotted power shown in Figure 1 is a much more interesting and practical method of powering the multi-layer antenna.

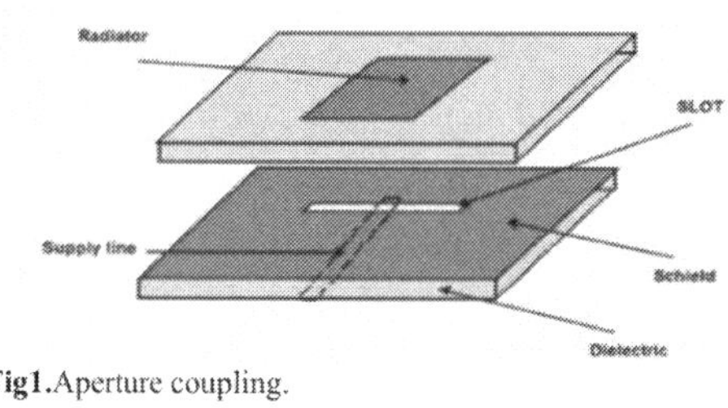

Fig1.Aperture coupling.

Patch antennas are made up of a dielectric substrate on an optically planar ground plane and a radiating element on the other side of the substrate, which is made up of conducting materials [1]. In this paper the goal is on matching enhancement technique of a multilayer patch antenna for Satellite applications. In order to enhance the directivity, antenna losses are contained by controlling those slots feeding which can have a significant impact on the matching, bandwidth for a given permittivity and thickness of the substrate. In this work, to improve the radiation characteristics of the slot excited patch antenna, a new shielding plane was placed behind the antenna, this method eliminates any unnecessary radiation in the back and improve the front-to-back ratio.

2. Aperture Coupled Patch Antenna Design

The aperture coupled feed technique is used in this paper in order to take advantage of the benefits of this type of antenna and improve the defects. The ground plane separates the radiating patch and the microstrip feed line in this technique. A slot or an aperture in the ground plane is used to connect the patch to the feed line. Microstrip feed line is placed on the bottom side of the lower substrate and is connected to the radiating patch via a slot in the ground plane [2]. This feeds technique possesses several advantages over other feeding methods, we cite that the most interesting in terms of realization is the separation of the feed line from the radiating element to avoid parasitic radiation generated by the feed line [7]. These antennas are more advantageous in arrays since the feed and phase shifting circuits are electrically isolated from the patch antennas.

2.1 Antenna Parameters

The bottom substrate is usually made of a high dielectric constant material, while the top substrate is made of a thick low dielectric constant material [3]. Aperture coupling [6] is also useful for fabricating antennas with two or more dielectric layers; however, the resonant frequency must be determined which is dependent on the layers used in the antenna [4]. The substrate's Dielectric constant (ε_r), its operating frequency (f_r), and the height of the substrate (h) from the ground plane are the three design parameters of a microstrip antenna, and they affect the patch's dimensions, radiation pattern, return loss, and other parameters [3]. The problem of effective dielectric constant of multilayer structures for antenna applications has been studied by a number of researchers [4,5,11], to obtain expressions for the effective filing fractions and effective dielectric constant. In this work we have based on Wheeler's [8] transformation from the physical z plane to the complex flux-potential g plane ($g = u + jv$), these areas are depicted in Figure 2. The ratio of the area occupied by each individual dielectric in the g plane defines the degree of filling or filling fraction q_i of each of the dielectric materials S_i (i=1,2,3) and the entire area of the cross section S_c. (1) – (3) are Svacina's filling fraction q equations, with q_1 and q_2 equal to the effective filling fractions of the substrate dielectric (relative permittivity ε_{r1}), and dielectric material above the superstrate (relative permittivity ε_{r2}), respectively.

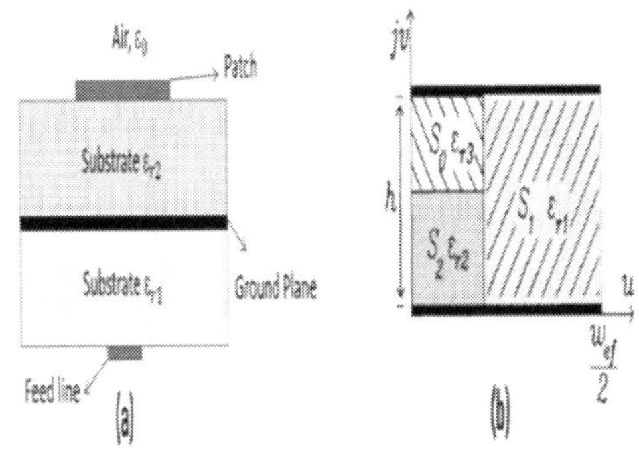

Fig 2. (a) Section of the under-review microstrip antenna structure, (b) Approximately how dielectric materials are distributed between parallel plates, as shown in [9] and [10].

$$q_1 = \frac{S_1}{S_C} = 1 - \frac{h}{2w_{ef}} \ln\left(\frac{\pi}{h} w_{ef} - 1\right) \quad (1)$$

$$q_2 = \frac{S_2}{S_C} = 1 - q_1 - \frac{1}{2}\frac{h - v_c}{w_{ef}} \times \ln\left[\pi \frac{w_{ef}}{h} \frac{\cos\left(\frac{v_c \pi}{2 h}\right)}{\pi\left(\frac{h_2}{h} - \frac{1}{2}\right) + \frac{v_c \pi}{2 h}} + \sin\left(\frac{v_c}{2}\frac{\pi}{h}\right)\right] \quad (2)$$

the effective line width taken from [18] is

$$w_{ef} = w + \frac{2h}{\pi} \ln\left[17.08\left(\frac{w}{2h} + 0.92\right)\right] \quad (3)$$

and the quantity v_c is

$$v_c = \frac{2\pi}{\pi} \tan^{-1}\left[\frac{\pi}{\frac{\pi}{2}\frac{w_{ef}}{h} - 2}\left(\frac{h_2}{h} - 1\right)\right] \quad (4)$$

The effective permittivity of the wide microstrip line is calculated by [9] and [10] using the approximate capacitance of an equivalent parallel-plate structure with the filling fractions arranged as shown in Figure 2.

$$\varepsilon_{eff} = \varepsilon_{r1} q_1 + \varepsilon_{r2} \frac{(1 - q_1)^2}{\varepsilon_{r2}(1 - q_1 - q_2) + q_2} \quad (5)$$

The resonant frequency of a rectangular microstrip antenna is given by [1]:

$$f_r = \frac{c}{2(L + 2\Delta L)\sqrt{\varepsilon_{eff}}} \quad (6)$$

The starting of the calculation has been made to have a resonance frequency at 28 GHZ, and for the bottom substrate is rogers RT3003 with 0.5 mm thickness with dielectric constant of 3 and loss tangent equal to 0.001, for this design, the top substrate is made of dielectric foam with a thickness of 0.254 mm and a dielectric constant of (εr=1). For the calculation of the basic antenna elements (width, length, feed line) we have used the formulas that are already published on the work [11]. Aperture coupling is a non-contact feeding mechanism that has been studied to overcome problems related to matching, directivity and stray radiation from the feed

 https://doi.org/10.1051/itmconf/20224801009

lines [12]. A general schematic diagram of this ASP (aperture-stacked patch) microstrip antenna is shown in Figure 3.

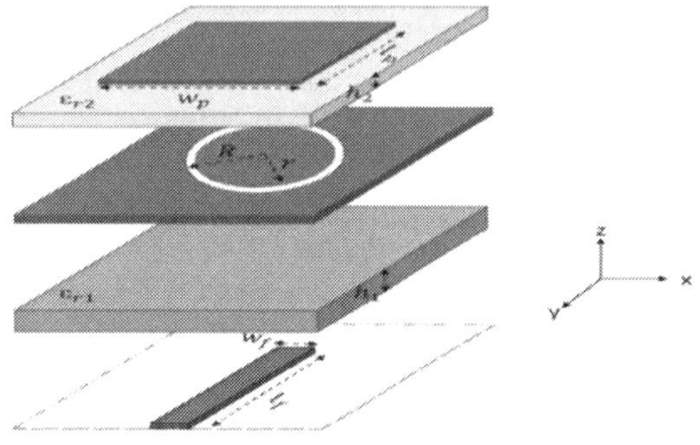

Fig 3. ASP (aperture-stacked patch) microstrip antenna designed with circle slot, The dimensions are $l_p = 5.43\ mm$, $w_p = 4.83\ mm$, $h_1 = 0.5\ mm$, $h_2 = 0.254\ mm$, $w_f = 0.58\ mm$, $l_f = 7.147mm$, $\varepsilon_{r1} = 3$, $\varepsilon_{r2} = 1$.

2.2 The Aperture Slot Proposed

Design of the proposed antenna consists of circular slot in the ground plane; it was designed by an outer radius R and an inner radius r, their centre is midpoint of the ground plan. The circle slot was studied by the effect of changing the deference between outer and inner radius (ΔR=R-r) in order to have a good current flow on the surface. The optimization procedure used here begins with a random set of antenna parameters. It is possible to start with a known solution, but we prefer to start with purely random parameters to avoid biasing the solution. CST Microwave Studio is used to simulate the proposed antenna, and the simulated results are obtained, the Table 1 shows the results of different radius value.

Table 1. Effect of changing the deference between outer and inner radius (ΔR=R-r)

ΔR (MM)	$f_r(GHz)$	ε_{eff}	$w_{ef}\ (mm)$	*S11(dB)*
0.08	NA	-	-	NA
0.1	26.8	4.1431	6.72	-14
0.12	27.2	3.9534	6.45	-12
0.14	26.4	4.3176	6.86	-21
0.16	27.4	3.7856	6.34	-23
0.18	28.9	3.0648	5.92	-13
0.2	28	3.2012	6.02	-19
0.22	28.6	2.8721	5.34	-17
0.24	NA	-	-	NA
0.26	29.2	2.3952	5.02	-16
0.28	NA	-	-	NA

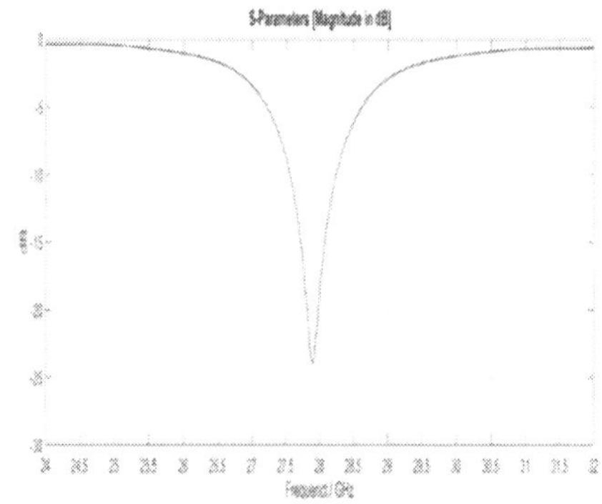

Fig 4. Return loss response of the antenna with ΔR= 0.2 mm.

2.3 Optimization Approach

In order to optimize the obtained result, we choose to improve the coupling slot by increasing the number of circles, with the same radius difference (ΔR). The figure 5 shows the summary of the parametric study that we have done to have a result well matching to the frequency we want (28 GHz). The parametric study carried out clearly shows the number of circles effective to matching the antenna to have a good surface current flow, the use of only one circle of opening is not sufficient to have a good antenna adaptation. As well as the increase in the number of circles is not always effective as an example two and four circles. On the other hand, the use of three circles presents an acceptable result at the level of adaptation and at the level of the surface current. The number of circles cannot be increased more than four due to the global antenna size and the parameters for calculating this size [13].

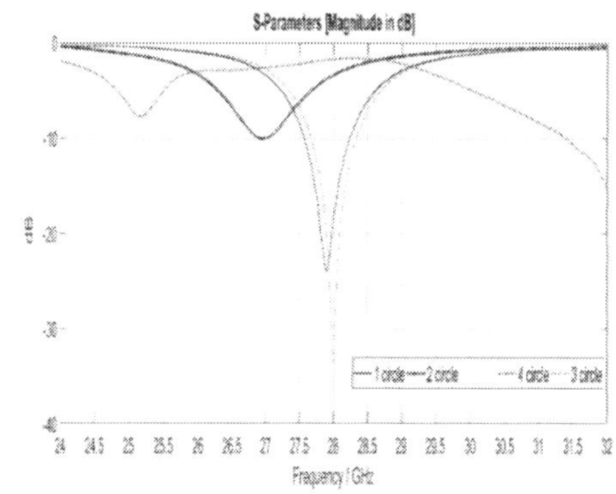

Fig 5. Return loss variations of antenna in different aperture circles.

The figure 6 (a) shows a general schematic of the aperture-coupled patch antenna with three slot couplings in the form of circles with equal slot widths ΔR. The current surface shown in Figure 6 (b), it is important to note that the current flow has been well distributed over the constructed slots which guarantees to get increased antenna performance and minimized losses caused by slots.

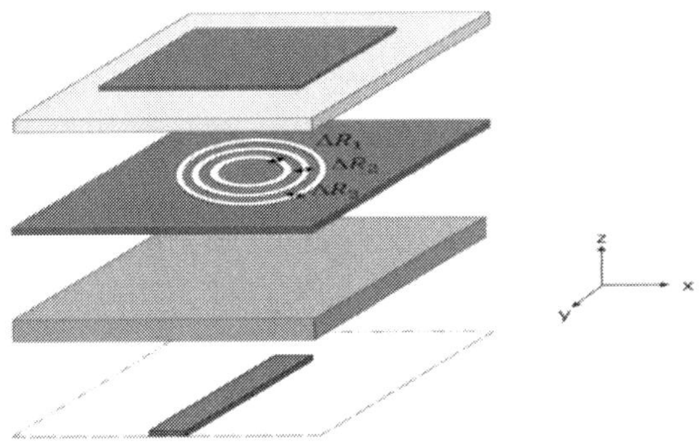

(a)

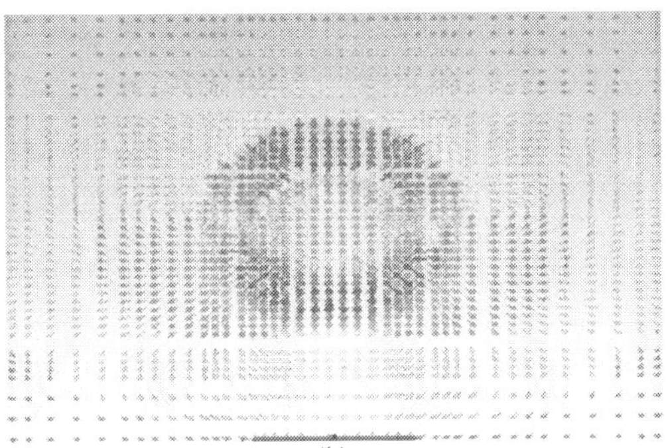

(b)

Fig 6. (a) General schematic of the antenna with three slot couplings in the form of circles, (b) Current distribution through the slots

3. Backward directed radiation reduction techniques

In this section we study technique to reduce the backward directed radiation for proposed microstrip antenna. We proposed and experimentally verified the simple technique of using a microstrip patch antenna element directly behind the microstrip line that fed aperture which acts as a reflector to substantially decrease radiation from an aperture coupled antenna. It is worth nothing that both the suggested reflector patch and the traditional radiating/directive patch utilise a similar ground plane.

We need also to mention the reflector element principal of operation is not as similar as the mechanism used in [14.15] for reflect array's led printed antennas. Reflected arrays like traditional reflectors depend on far-field results to attain the desired properties. A near field interaction between the excitation slot and the reflector patch in the printed reflector element achieved the desired far-field cancellation of the radiated fields.

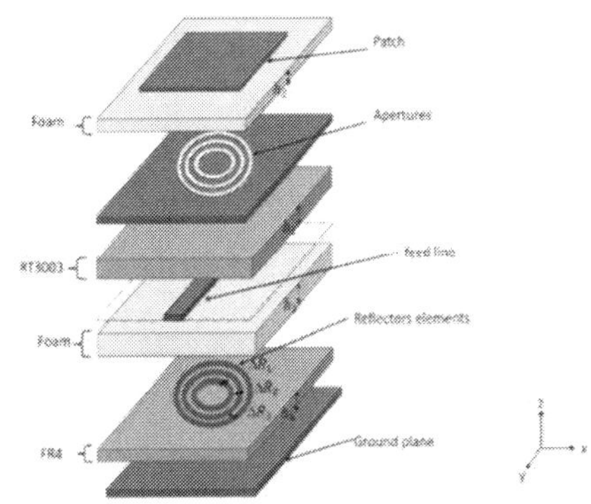

Fig 7. Aperture stacked patch antenna with reflectors patch elements.

We show a schematic diagram of a proposed ASP antenna with a reflector element in Figure 7 which consists of a reflector plate in the form of circles reflecting the shape of the aperture in the ground plan. So as to ensure that the surface wave losses related with this structure have been decreased. A low dielectric constant laminate of thickness h3 separated the reflector microstrip patch. As shown in Figure 7, a thin dielectric layer (h_4=0.254 mm) is utilised to etch the conductor of the reflector patch with dielectric constant of 4.4 and loss tangent equal to 0.025. The dielectric material that divides the feed-line and the reflector can be different thicknesses, however, it is usually electrically thick (greater than $0.1\lambda_0$, where λ_0 is the wavelength corresponding to the center frequency within the antenna's operation band). Therefore, direct coupling between this patch and the feed line is minimal.

Several parameters can be altered to reduce backward directed radiation. The relative phase between the slot and the reflectors patch controls the radius of the patch, whereas the radius of the patch and the distance between the slot and the reflectors control the relative magnitudes. This latter variation makes intuitive sense since the distance between the slot and the reflector patch increases the power coupled to this patch is expected to decrease resulting in the change in relative magnitude. Variating the radius of the reflector patch will provide the finer phase control required to provide effective field cancellation because there are only a few foam thicknesses compatible with the Ka-band frequency range. We have chosen to fix the radius of the reflectors in this work. Figure 8 shows the effect of varying the parameter (h_3) of substrate thicknesses on the Front to Back ratio performance for proposed antenna incorporating reflector elements.

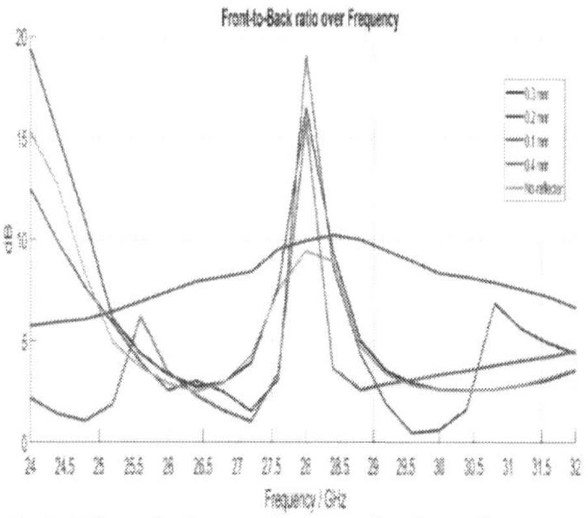

Fig 8. Effect of substrate thickness (h_3) for Reflector Patch on Front-to-back Ratio.

As can be seen from Figure 8, the more spacer thickness increases, the more F/B ratio ameliorates until it reaches certain point where its performance starts to degrade. In this experiment increasing the thickness (h_3) improves F/B ratio until a value of 0.4 mm where at the ratio degrade for the proposed antenna, we selected h_3 = 0.2 mm with 19.92 dB.

4. Antenna performances

In this section, we study the performance of the proposed antenna with reduced back radiation, which we show the antenna efficiency at the front to back ratio, gain, directivity and matching at 28 GHz.
A matching analysis for the designed patch antenna by using the reflector elements is presented in Figure 9 It can be observed that the center frequency is kept when we use technique to reduce the backward directed radiation. The addition of reflector elements affects mostly on the bandwidth but also the desired frequency is kept and well matched.

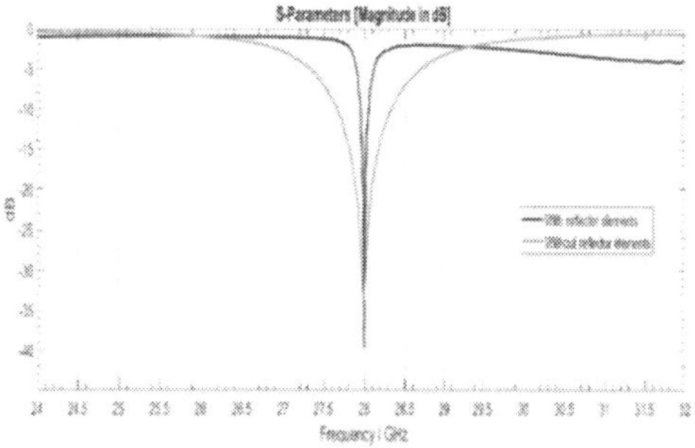

Fig 9. Reflection coefficient simulated with and without reflector elements.

Figures 10 (a), (b), and (c) demonstrate how adding a reflector element to the proposed ASP microstrip antenna reduces field radiation in the antenna's back half plane. The reduction in total power radiated into the antenna's rear hemisphere is visible across the entire impedance bandwidth, emphasizing the technique's matching ability. The most significant field cancellation occurs near backfire (180°), which is especially noticeable in the E-plane. The total radiated power has increased in the forward hemisphere of the H-plane, and the indentations in the pattern that have been smoothed over. The measurement of the radiated fields had a few minor alignment issues.

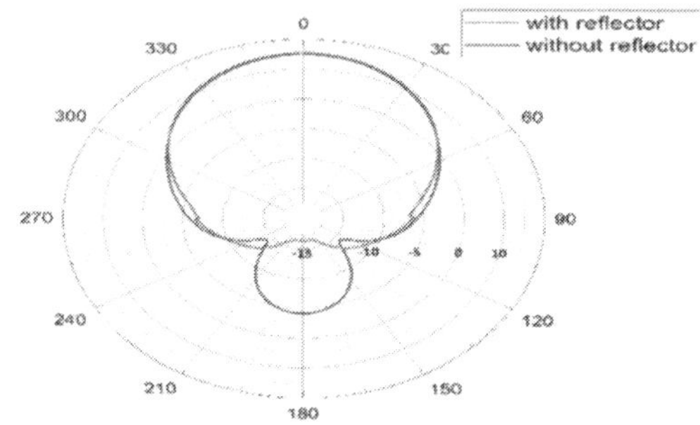

(a)

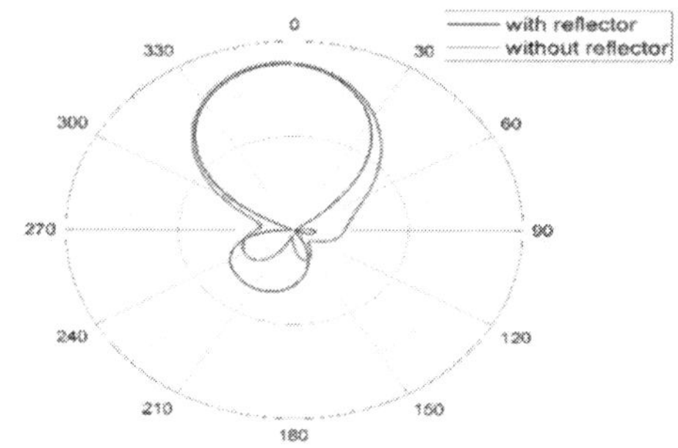

(b)

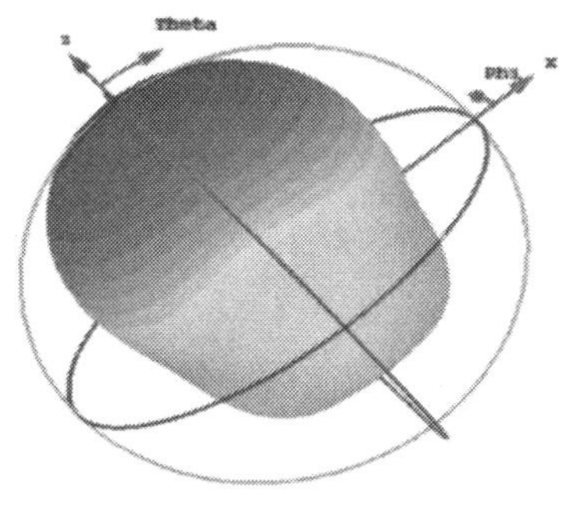

(c)

Fig 10. Radiation pattern of the proposed antenna at 28 GHz (expressed in dB). (a) E-Plane. (b) H-Plane. (c) 3D radiation pattern.

The gain with or without reflector, both are presenting the figure 11 which show us that there is no big change among them except some increase when we used the reflector. The antenna gives an interesting gain value with 7.79 dB. For the directivity the figure 12 shows an interesting increase with the reflector, which is normal due to the reduction of the total radiated power in the rear hemisphere of the antenna, which underlines the ability to increase the directivity by this technique.

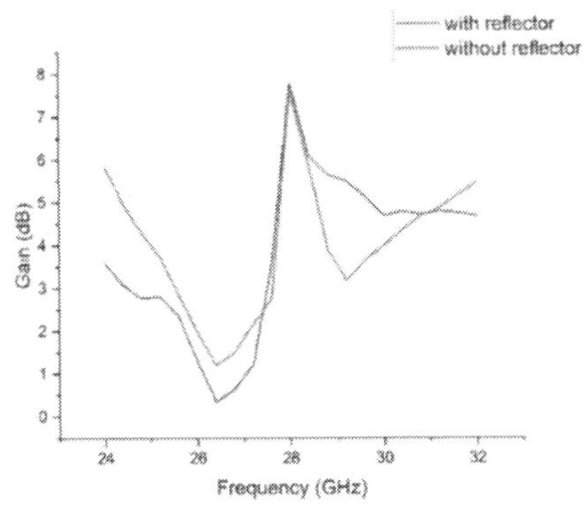

Fig 11. Comparison of gain antenna with and without reflector

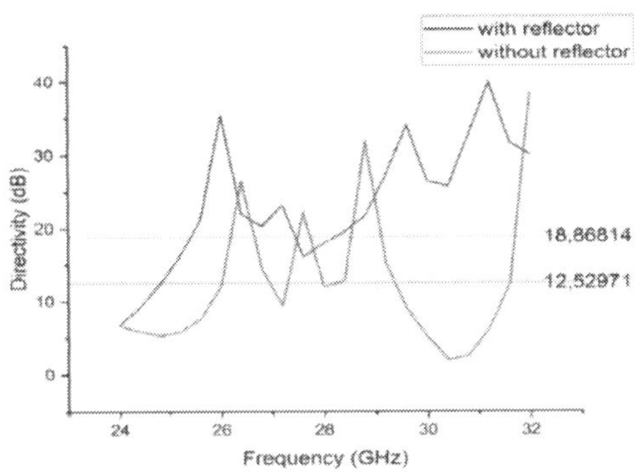

Fig 12. Comparison of directivity antenna with and without reflector

In order to validate the results obtained, a comparative study was carried out which compared the work carried out with previous research. The table illustrates the specification and performances of this comparison that shows the interest of the results obtained at the different levels matching, directivity, gain and F-to-B ratio.

Table 2. Comparisons between the proposed antenna and other works.

Ref	Resonance Frequency (GHz)	Matching at resonance Frequency (dB)	Gain (dB)	F-to-B ratio (dB)
[16]	6	< -25	3.5	8.2
[17]	1.7-2.7	<-35	7.5	18.6-25
[18]	10.5	<-25	7.6	15
[19]	30	<-30	5	20
[20]	3.4	<-30	8	9.4
This work	28	<-40	7.9	19.92

5. Conclusion

Design and characterization of an aperture-stacked patch microstrip antenna with reflector elements is presented in this paper. By an optimization approach, a new slot aperture has been used to have a good surface current flow at resonance frequency (28 GHz). In the same way, a study was carried out to improve the antenna performance in order to concentrate the radiation power and eliminate the power radiated in the rear hemisphere of the antenna, which is shown in the results presented in the different parameters. As a result, the improvement of the front-to-back ratio of a patch antenna was studied and simulated by an improvement of more than 10 dB. The antenna has a good frequency matching at 28 GHz <-40 dB and has an average gain of 7.79 dB. without forgetting to mention the increase in directivity caused by the improvement of the front-to-back ratio up to 18.8 dB and proved by the radiation plane E and H. The proposed antenna is a real contender for uses in low-profile, high directivity, and high gain satellite systems and high-resolution microwave radar.

REFERENCES

1. Hemant Kumar Gupta, P.K.Singhal,Pavan Kumar Sharma,Veerendra Singh Jadon "Slotted Circular Microstrip Patch Antenna Designs for multiband Application in Wireless Communication, " International Journal of Engineering and Technology, Vol 1Mar 2012, pp 158-167.
2. Sapna Verma, J.A. Ansari.," Analysis of U-slot loaded truncated corner rectangular microstrippatch antenna for broadband operation," International Journal of Electronics and Communications, Vol. 69, 2015, pp 1483–1488.

3. Debatosh Guha.," Microstrip and Printed Antennas: Recent Trends and Developments," IEEE Transactions on Microwave Review, Dec 2003, pp 10-15.
4. W. H. Syed and A. Neto, "Front-to-Back Ratio Enhancement of Planar Printed Antennas by Means of Artificial Dielectric Layers," in IEEE Transactions on Antennas and Propagation, vol. 61, no. 11, pp. 5408-5416, Nov. 2013, doi: 10.1109/TAP.2013.2275915.
5. Ajay Mohan, Jitendra Kumar, Vinay Kumar Verma, "Multi-Layer Aperture Coupled Micro Strip Antenna with Variation in Dielectric Constants" International Journal of Science and Research, ISSN: 2319-7064, Volume 2 Issue 1, January 2013, pp 208-2011.
6. Kazimierz Piwowarczyk, Marian Wnuk, Leszek Nowosielski, Rafal Przesmycki, and Marek Bugaj, "The Algorithm of Design Multi-layer Microstrip Antenna", Progress in Electromagnetics Research Symposium Proceedings, Moscow, Russia, August pp 19-23, 2012.
7. Sudarshan Kumar Jain, "Design and Analysis of Multilayer Substrate Structure Microstrip Patch Antenna", SSRG International Journal of Electronics and Communication Engineering (SSRG-IJECE) – Volume 2 Issue 11–November 2015.
8. H. A. Wheeler, "Transmission-line properties of parallel strips separated by a dielectric sheet," IEEE Trans. Microwave Theory Tech., vol. 13, pp. 172–185, Mar. 1965.
9. J. Svaˇcina, "Analysis of multilayer microstrip lines by a conformal mapping method," IEEE Trans. Microwave Theory Tech., vol. 40, pp. 769–772, Apr. 1992.
10. S.-S. Zhong, G. Liu, and G. Qasim, "Closed form expressions for resonant frequency of rectangular patch antennas with multidielectric layers," IEEE Trans. Antennas Propagat., vol. 42, pp. 1360–1363, Sept. 1994.
11. Salaheddine Aourik, Ahmed Errkik, Jamal Zbitou, Ahmed Lakhssassi, Abdelali Tajmouati, Larbi Elabdellaoui "Microstrip Antenna Array with Dumbbell Defected Ground Structure for Ka-Band Radar Application" CPI 2019, October 14-16, 2019, Fez, Morocco
12. D. M. Pozar, "A Microstrip Antenna Aperture Coupled to a Microstrip Line," Electronic Letters,Vol. 21, pp. 49 - 50, January 1985.
13. W. Menzel and W. Grabherr, "A microstrip patch antenna with coplanar feed line", IEEEMicrowave and Guided Wave Letters, Nov. 1991, vol. I, pp. 340-342.
14. D. M. Pozar, S. D. Targonski and H. D. Syrigos, "Design of millimeter-wave microstrip reflectarrays,' IEEE Trans. Antennas Propagat. , vol. 45, pp. 287 - 296, February 1997.
15. W. Menzel, D. Pilz and R. Leberer, "A 77 GHz FM/CW radar front-end with a lowprofile low-loss printed antenna," IEEE Trans. Microwave Theory Techn. , vol. 47, pp. 2337 - 2241, December 1999.
16. Y. Cheng, X. Ding, B. Wang and W. Shao, "An Azimuth-Pattern-Reconfigurable Antenna with Enhanced Gain and Front-to-Back Ratio," in IEEE Antennas and Wireless Propagation Letters, vol. 16, pp. 2303-2306, 2017, doi: 10.1109/LAWP.2017.2715373.
17. I. Govindanarayanan and N. Rangaswamy, "Asymmetric Folded Dipole Antenna With High Front-to-Back Ratio for LTE Base Stations," in IEEE Antennas and Wireless Propagation Letters, vol. 15, pp. 869-872, 2016, doi: 10.1109/LAWP.2015.2478197.
18. W. H. Syed and A. Neto, "Front-to-Back Ratio Enhancement of Planar Printed Antennas by Means of Artificial Dielectric Layers," in IEEE Transactions on Antennas and Propagation, vol. 61, no. 11, pp. 5408-5416, Nov. 2013, doi: 10.1109/TAP.2013.2275915.
19. J. Row and L. Kuo, "Pattern-Reconfigurable Array Based on a Circularly Polarized Antenna with Broadband Operation and High Front-to-Back Ratio," in IEEE Transactions on Antennas and Propagation, vol. 68, no. 5, pp. 4109-4113, May 2020, doi: 10.1109/TAP.2019.2948679.
20. L. Gong, K. Y. Chan and R. Ramer, "Substrate Integrated Waveguide H-Plane Horn Antenna with Improved Front-to-Back Ratio and Reduced Sidelobe Level," in IEEE Antennas and Wireless Propagation Letters, vol. 15, pp. 1835-1838, 2016, doi: 10.1109/LAWP.2016.2538823.

A Two-Stage Support Vector Machine and SqueezeNet System for Range-Angle and Range-Speed Estimation in a Cluttered Environment of Automotive MIMO Radar Systems

Zakaria Benyahia[1], *Mostafa* Hefnawi[2], *Mohamed* Aboulfatah[1], *Hassan* Abdelmounim[1], and *Taoufiq* Gadi[1]

[1] Hassan I University Settat, Morocco
[2] Royal Military College of Canada, Kingston, Ontario, Canada

Abstract. This paper proposes a two-stage deep-learning approach for frequency modulated continuous waveform multiple - input multiple - output (FMCW MIMO) radar embedded in cluttered and jammed environments. The first stage uses the support vector machine (SVM) as a feature extractor that discriminates targets from clutters and jammers. In the second stage, the angle, range, and Doppler estimations of the extracted targets are treated by the SqueezeNet deep convolutional neural network (DCNN) as a multilabel classification problem. The performance of the proposed hybrid SVM-SqueezeNet method is very close to the one achieved by the SqueezeNet only but with the advantage of identifying the type of targets and reducing the training time required by the SqueezeNet.

1 Introduction

Multi-input and multi-output (MIMO) radar systems have recently received considerable research interest in the automotive industry [1]-[7]. Unlike traditional phased-array radars in which a steered beam is used at the transmitter to scan a sector, MIMO radar systems transmit orthogonal waveforms from each omnidirectional antenna element simultaneously, allowing for a sector scan rate that is several times faster than steered beam radars [8]-[15]. The orthogonality of the waveforms can be achieved using multiplexing techniques such as time-division multiplexing (TDM), frequency division multiplexing (FDM), code division multiplexing (CDM), or space division multiplexing (SDM) [16]-[19]. Moreover, MIMO radar with collocated antennas can achieve a better spatial resolution and clutter interference rejection capability by combining all the transmitting paths in a virtual array with an extended aperture. MIMO radar signal processing techniques can be based on traditional techniques such as Fast Fourier Transform (FFT), Multiple Signal Classifier (MUSIC) [20], the Estimation of Signal Parameters via Rotational Invariance Techniques (ESPRIT) [21], and Minimum Variance Distortionless Response (MVDR) [22]. On the other hand, Deep neural networks (DNNs) have been proposed for MIMO radar systems as an alternative to traditional techniques that are computationally expensive and very sensitive to clutters and interferences [23]-[28]. In [23], a transfer learning-based deep convolutional neural network (DCNN) was proposed for distributed Frequency Modulated Continuous Wave (FMCW) MIMO radar systems. The DCNN was trained using 2D range-azimuth angle maps of stationary targets. Simulation results showed that the proposed DCNN gives high-resolution radar images compared to a single MIMO radar system. Their experiment was restricted to non-moving targets and did not consider any clutters or interferences. In [24], the focus was on targets classification only, which was achieved using the first stage of unsupervised pretraining followed by supervised fine-tuning of the convolutional layers to extract spatially localized features. In [25], a deep learning model was proposed to simultaneously estimate the range and direction of arrival of stationary targets. The system was built on existing architectures for DoA estimation and took into consideration the low-resolution ADCs constraints.

The You Only Look Once YOLO networks have also been actively applied to automotive radar in recent years [26]-[28]. In [26], the YOLO network was applied to the radar data in the range-angle domain to detect stationary targets and classify them with over 90% accuracy. The high angle resolution MUSIC was used to generate the range-angle images in a free clutter and interference environment.

In [27], the YOLO network detection and classification was based on the range-Doppler domain. In [28], the training of the YOLO network was based on the high-resolution radar sensor data to determine the type of detected object and estimate its moving direction as well.

Most of these methods were based on either the range-angle or the range-Doppler images that were generated using high-resolution techniques and did not consider clutters and interferences in their environment.

In this paper, we propose a deep learning network (DLN) to classify moving targets and estimate their range-angle-doppler in the presence of clutters and jammers. The proposed DL network is performed in two stages. In the first stage, radar cross-section (RCS)-based target classification is used with the support vector machine (SVM) to identify the type of targets and, at the same time, acts as a filter by extracting clutters and jammers. In the second stage, the range, angle, and Doppler estimations of the classified targets are treated by SqueezeNet as a multilabel classification problem. Since the clutters and jammers were rejected at the first stage, the SqueezeNet was trained in a free clutter and jammer environment, speeding up its training time. The training of the SqueezeNet was based on a low-resolution 2D FFT applied to the MIMO radar data cube.

2 SYSTEM MODEL

An FMCW MIMO radar with two transmitting (TX) antennas and four receiving (RX) antennas, as shown in Fig. 1, is considered. TDM waveforms are assumed where only one transmit antenna is activated at once. This system can be mounted on a vehicle to track the range, angle, and speed of the vehicles in front of it as shown in Fig. 2.

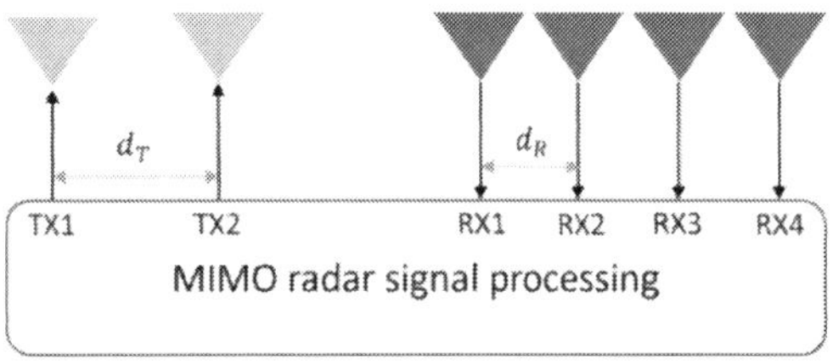

Fig. 1. Automotive radar(vehicle ego) and two target vehicles [29]

Fig. 2. MIMO radar with 2 transmitters and 4 receivers.

2.1 Virtual Array

When a MIMO radar transmits M independent orthogonal signals and receives echo returns with N receivers, a total of $M \times N$channels is processed, creating a virtual array of up to $M \times N$ receive elements. A received signal $\mathbf{x}(t)$ by the $MN-element$ receive virtual array can be expressed as [12]-[13]

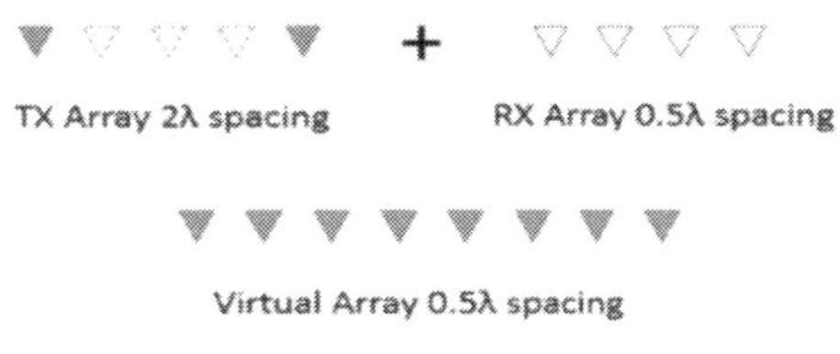

Fig. 3. Virtual array with 2 transmitters and 4 receivers [28]

$$\mathbf{x}(t) = \sum_{k=1}^{K} \boldsymbol{a}_k^v s_k(t) + \boldsymbol{n}(t) \qquad (1)$$

where $\boldsymbol{a}_k^v = \boldsymbol{a}_k^T \otimes \boldsymbol{a}_k^R \in \mathbb{C}^{MN\times 1}$ is the steering vector of the virtual array due the k-th target, the Kronecker product is denoted by $\otimes$, the elements of $\boldsymbol{n}(t)$ are assumed to be independent and identically distributed complex Gaussian random vectors , $\boldsymbol{a}_k^T$ and $\boldsymbol{a}_k^R$ are the TX and RX steering vectors, respectively, and can be expressed as:

$$\boldsymbol{a}_k^T = \left(1, e^{j\left(\frac{2\pi}{\lambda}\right)d_T \sin\theta_k'}, \dots, e^{j(M-1)\left(\frac{2\pi}{\lambda}\right)d_T \sin\theta_k'}\right)^T \qquad (2)$$

$$\boldsymbol{a}_k^R = \left(1, e^{j\left(\frac{2\pi}{\lambda}\right)d_R \sin\theta_k}, \dots, e^{j(N-1)\left(\frac{2\pi}{\lambda}\right)d_R \sin\theta_k}\right)^T \qquad (3)$$

where d_T and d_R are the inter-element spacing of the TX and RX, respectively. θ_k' and θ_k are the direction of the k-th target with respect to transmit and receive arrays, respectively.

Using a virtual array allows the system to increase the resolution and accuracy beyond the physical number of elements present in the transmit and receive arrays. According to (1), the location of an element of the virtual can be expressed as follows

$$Virtaul\ Array(x) = \sum_{m=0}^{M-1}\sum_{n=0}^{N-1} \delta\big(x-(x_m+x_n)\big) \qquad (4)$$

where x is the position across the array, x_m are the location of the transmit elements and x_n are the locations of the receive elements.

Fig. 3 shows the concept of the virtual array for the case of two TX elements and four RX elements. The aperture of the virtual array is dependent on the inter-element spacing of the TX and RX arrays. Therefore, to achieve the maximum angular resolution (i.e. 8-element virtual array) the interelement spacing of the TX array and the RX array are fixed at two wavelengths ($d_T = 2\lambda$) and half-wavelength ($d_R = \lambda/2$), respectively.

2.2 Radar Data Cube

The radar data cube is a convenient way to represent space-time processing conceptually. Fig. 4 shows the organization of the radar data cube. To extract the range and the velocity, the processor must perform an FFT along the fast-time axis and along the slow-time axis (for each pulse and each RF channel), respectively. To extract

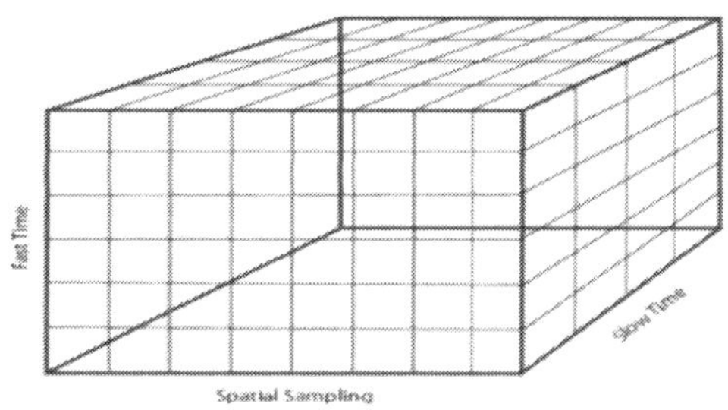

Fig. 4. Organization of Data Cube

https://doi.org/10.1051/itmconf/20224801010

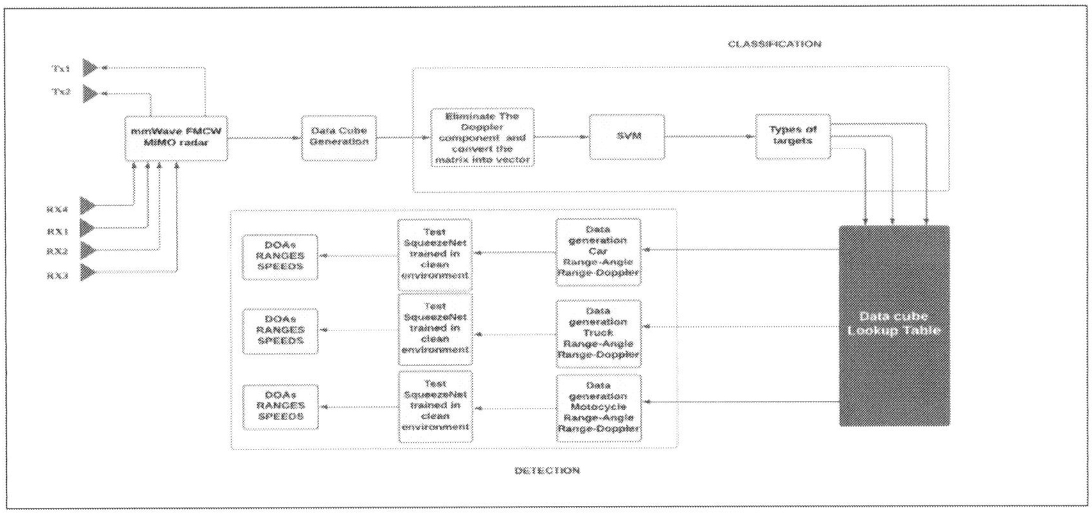

Fig. 5. Two-stage SVM-SqueezeNet Network for classification, Range-Angle Range-Doppler estimation, , and clutter/jammer rejection

the DOA, the processor must perform an FFT across the multiple array elements. For higher angular resolution, the FFT can be replaced with MUSIC, ESPRIT, or MVDR algorithms.

2.3 Proposed two-stage SVM-SqueezeNet Network

Fig. 5 shows the proposed two-stage SVM-SqueezeNet Network. The first stage uses the support vector machine (SVM) as a classifier. The SVM is a supervised learning algorithm used for classification and pattern learning problems. The objective of the SVM algorithm is to seek iteratively the hyperplanes that divide the datasets into classes and choose the optimal one that maximizes the margin between data points that belong to different classes and are closest to the hyperplane (Fig. 6). The data points that are closest to the hyperplane are called support vectors. SVM uses a set of mathematical functions called kernels that transforms input datasets into the desired form that makes them separable. Kernel functions can be of different types, summarized in Table I [29]. In this paper, the polynomial kernel function is used.

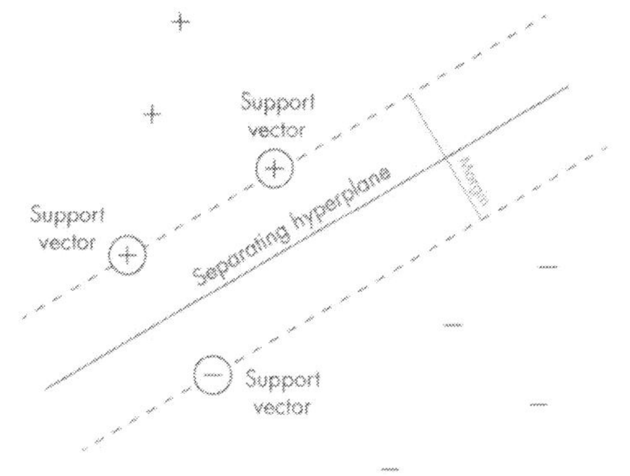

Fig. 6. Defining the "margin" between classes – the criterion that SVMs seek to optimize [29]

TABLE I. DIFFERENT TYPES OF KERNEL FUNCTIONS

Type of SVM	Mercer Kernel	Description
Gaussian or Radial Basis Function (RBF)	$K(x_1,x_2) = e^{(-\frac{\|x_1-x_2\|^2}{2\sigma^2})}$	One class learning. σ is the width of the kernel
Linear	$K(x_1,x_2) = x_1^T x_2$	Two class learning.
Polynomial	$K(x_1,x_2) = (x_1^T x_2 + 1)^p$	ρ is the order of the polynomial
Sigmoid	$K(x_1,x_2) = \tanh(\beta_0 x_1^T x_2 + \beta_1)$	It is a mercer kernel for certain β_0 and β_1 values only

The classification of the detected targets as trucks, cars, motorcycles or pedestrians was based on their RCSs, modelled according to [38] and summarized in Table II. Therefore, the Doppler dimension of the data cube was squeezed to provide a 2D matrix that is then entirely vectorized by concatenating the real and imaginary parts of each column. The size of the data cube used in this paper is $550 \times 8 \times 32$, and the specific converting process can be summarized as follows

$$Data\ cube^{(550\times8\times32)} \rightarrow A^{(550\times8)} = \begin{pmatrix} a_{11} & \cdots & a_{1\,8} \\ a_{21} & \cdots & a_{2\,8} \\ \vdots & \ddots & \vdots \\ a_{550\,1} & \cdots & a_{550\,8} \end{pmatrix}$$

$$\rightarrow B^{(4400\times1)} = \begin{pmatrix} a_{11} \\ \vdots \\ a_{550\,8} \end{pmatrix}$$

$$\rightarrow C^{(8800\times1)} = \begin{pmatrix} \Re e(B) \\ \Im m(B) \end{pmatrix}$$

TABLE II. RCS OF TARGETS

Category	RCS
Car	$\min(10\log10(R) + 5\ \mathrm{dBsm}, 20\ \mathrm{dBsm})$
Truck	$\min(10\log10(R) + 5\ \mathrm{dBsm}, 45\ \mathrm{dBsm})$
Motorcycle	7 dBsm
Pedestrian	-10 dBsm

The data set contains realizations of the following nine scenes:

- Two cars driving in a free clutter/jammer environment
- Two cars driving in a cluttered/jammed environment
- Two trucks driving in a free clutter/jammer environment
- Two trucks driving in a cluttered/jammed environment
- Two motorcyclists driving in a free clutter/jammer environment
- Two motorcyclists driving in a cluttered/jammed environment
- Two pedestrians walking in a free clutter/jammer environment
- Two pedestrians driving in a cluttered/jammed environment
- Clutters and jammers only

In each category, about 80% is reserved for the training data set while 20% is reserved for the test data set.

In the second stage, the SqueezeNet network predicts the range, angle, and Doppler of the already classified targets, namely "Car", "Motocycle," and "Truck". The SqueezeNet uses fire modules, as shown in Fig. 7 [30-31] to achieve high accuracy with smaller convolution neural network architecture.

Since the clutters and jammers are filtered out by the first stage, the SqueezeNet is trained in a clean environment, using pairs of DOAs, pairs of ranges and pairs of speeds for each type of target as shown in Table III, IV , and V.

The SqueezeNet requires data images for training and testing. Therefore, the range-angle and range-doppler images are generated by applying the 2D FFT to the MIMO radar data cube. Fig. 8 and Fig. 9 show the range-angle and range-speed plots, respectively, of two cars moving in a free clutter and interference environment with a pair of $azimuths = (2^o, 5^o)$, a pair of $ranges = (100\ \mathrm{m}, 130\ \mathrm{m})$ and a pair of $speeds = (36.111\ \mathrm{m/s}, 41.66667\mathrm{m/s})$. For the range-angle labelling, it is corresponding to the mth pair of ranges and the nth pair of angles, while it is referring to the mth pair of ranges and the nth pair of speeds for the range-speed labelling.

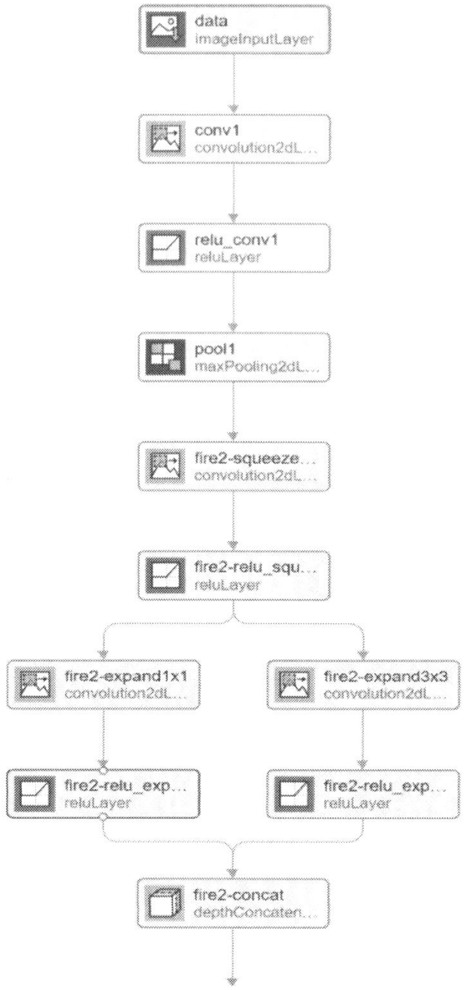

Fig. 7. SqueezeNet Fire Module [30]

The range-speed response of Fig. 9 shows that the targets are moving at the relative speeds of (–16.66 m/s, -22.22 m/s) with respect to the speed of the radar. It should be noted that the negative speeds indicate that the targets are moving away from the ego car.

Referring to Fig. 5, the data cube used for the training are indexed and placed in a lookup table. This indexation is used by the SVM predicted labels to allow tracing back the corresponding data cube. Table VI explains the indexation operation for the car predicted label case. The label "car" without clutter-jammer is found in indexes 1, 3 and 5. The first index 1 refers to the first pair of DOAs, the third index to the pair of ranges, and the sixth index to the pair of speeds. These indexes allow us to identify the data cube corresponding to these pairs, applied to the 2D FFT to generate the range-angle and range speed images. These images are then applied to the SqueezeNet to predict the ranges, DOAs, and speeds. The same process is applied to each type of target.

TABLE III. PAIRS OF RANGES, ANGLES, AND SPEEDS CAR

Pair	DOA (degrees)		RANGE (m)		SPEED (m/s)	
1	2	5	100	130	36.11	41.66
2	-1	5	70	130	30.55	41.66
3	-4	5	40	130	25	41.66
4	-7	5	70	100	19.44	41.66
5	-1	2	40	100	30.55	36.11
6	-4	2	40	70	25	36.11
7	-7	2			19.44	36.11
8	-4	-1			25	30.55
9	-7	-1			19.44	30.55
10	-7	-4			19.44	25

TABLE IV. PAIRS OF RANGES, ANGLES, AND SPEEDS MOTOCYCLE

Pair	DOA (degrees)		RANGE (m)		SPEED (m/s)	
1	6	9	80	100	22.22	27.77
2	3	9	60	100	16.66	27.77
3	0	9	40	100	11.11	27.77
4	3	6	20	100	5.55	27.77
5	0	6	60	80	16.66	22.22
6	0	3	40	80	11.11	22.22
7			20	80	5.55	22.22
8			40	60	11.11	16.66
9			20	60	5.55	16.66
10			20	40	5.55	11.11

TABLE V. PAIRS OF RANGES, ANGLES, AND SPEEDS TRUCK

Pair	DOA (degrees)		RANGE (m)		SPEED (m/s)	
1	3	6	110	140	33.33	38.88
2	0	6	80	140	27.77	38.88
3	-3	6	50	140	22.22	38.88
4	-6	6	80	110	16.66	38.88
5	0	3	50	110	27.77	33.33
6	-3	3	50	80	22.22	33.33
7	-6	3			16.66	33.33
8	-3	0			22.22	27.77
9	-6	0			16.66	27.77
10	-6	-3			16.66	22.22

TABLE VI. INDEXATION METHOD FOR THE LABEL CAR

Index	Predicted label	Angle	Range	Speed
1	Car	1	3	6
2	Truck	2	3	5
3	Car	1	3	4
4	Motorcycle	3	5	2
5	Car	7	2	1
6	Car with jammer	2	6	9

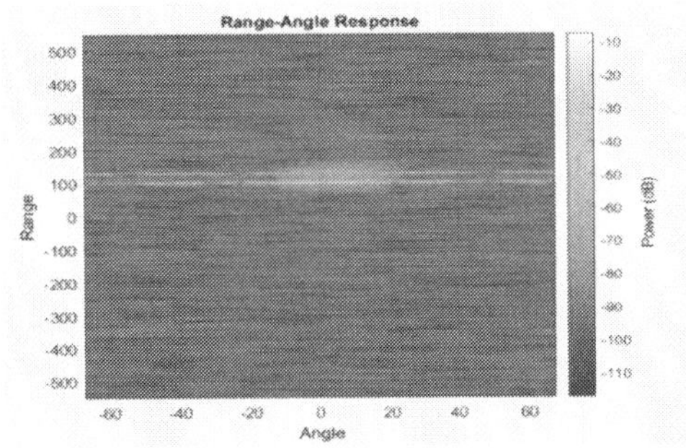

Fig. 8. Range-angle response of two cars moving in a free clutter/jammer environment.

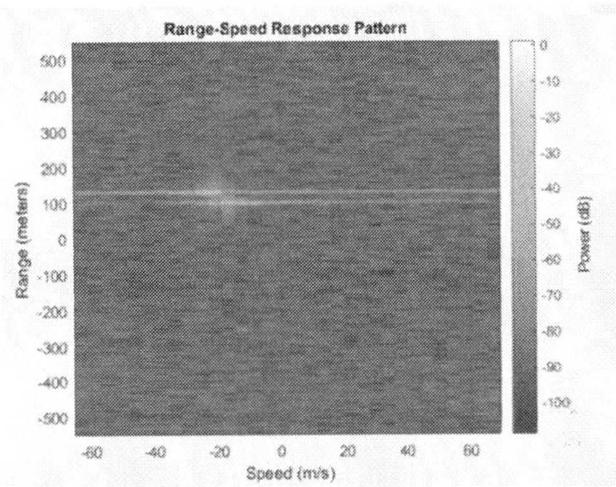

Fig. 9. Range-speed response of two cars moving in a free clutter/jammer environment.

3 EXPERIMENTS AND RESULTS

Table VII summarizes the data set and the labelling for the classification by the SVM. Eight labels are used for the training, and for each category, 80% is reserved for the training data set, while 20% is reserved for the test data set. The confusion matrix of Fig. 10 and Table VIII show the performance of the SVM. As it can be concluded, almost all of the eight types of targets are correctly identified by the network.

TABLE VII. SVM DATA SET AND THE LABELLING

Labels	Data Set	Training	Test
Car in a clean environment	600	480	120
Truck in a clean environment	600	480	120
Motocycle in clean environment	600	480	120
Car with Jammer and Clutter	3000	2400	600
Truck with Jammer and Clutter	3000	2400	600
Motocycle with Jammer and Clutter	3000	2400	600
Pedestrian	360	240	120
Pedestrian with Jammer and Clutter	1800	1200	600
Jammer and Clutter	21	16	5

 https://doi.org/10.1051/itmconf/20224801010

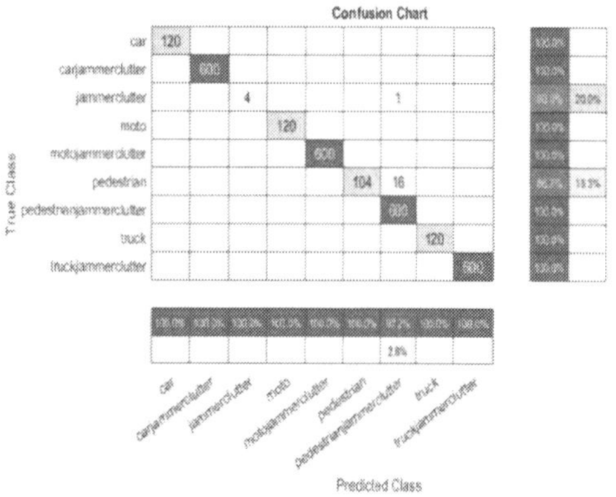

Fig. 10. Confusion chart for classification

TABLE VIII. PERFORMANCE OF SVM FOR CLASSIFICATION AND NOISE REJECTION

Architecture	SVM
No. of Classes	9
No. of Training Samples	10096
No. of Test Samples	2885
Validation Accuracy	99.6343%
Accuracy	99.4107%

On the other hand, Table IX shows the performance of the SqueezeNet, trained in a clean environment, for the range-angle and range-speed estimation of each type of target.

TABLE IX. PERFORMANCE OF SQUEEZENET IN A CLEAN ENVIRONMENT

Category		Car	Truck	Motocycle
Range-angle	Validation accuracy	100%	100%	100%
	Accuracy	98.33%	99%	99.1667%
Range-doppler	Validation accuracy	100%	100%	100%
	Accuracy	99%	99.23%	98.33%

The performance of the proposed hybrid SVM-SqueezeNet network is compared to the SqueezeNet only, which is trained in an environment with clutters and jammers. In this comparison, we focus on the detection aspect only, i.e. the range-angle and range-speed estimation of targets in the presence of clutters and jammers. According to Table III, the training is achieved with 60 labels. In the case of range-angle estimation, this corresponds to 6 pairs of ranges and 10 pairs of angles. Similarly, in the case of range-speed estimation, this corresponds to 6 pairs of speeds and 10 pairs of angles. For each type of target, there are 48 images (8 for clean environments, 32 for targets with jammer and 8 for targets with clutter).

Fig. 11 and Fig. 12 show the range-angle and range-doppler plots of two cars moving, in the presence of clutter and a jammer environment, with a pair of $azimuths = (2^o, 5^o)$, a pair of $ranges = (100\text{ m}, 130\text{ m})$ and a pair of $speeds = (36.11\text{ m/s}, 41.66\text{m/s})$. The clutter signal was generated using the constant gamma model [39], with the gamma value set to -20 dB, typical for flatland clutter. For each range, the clutter return is a combination of the returns from many small clutter patches on that range ring, and each patch has an azimuth width of 10 degrees. On the other hand, the jammer signal was generated using the barrage jammer model, which is a strong, continuous white noise directed toward the radar receiver. It is assumed that the jammer is at a specific location, and the jamming signal is associated with a specific direction.

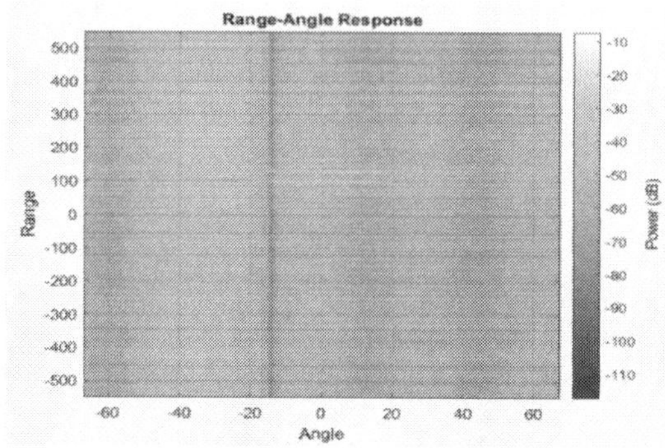

Fig. 11. Range-Angle response with clutter and jammer

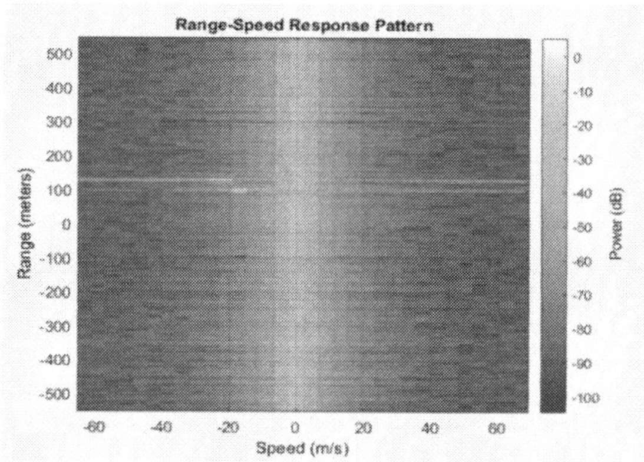

Fig. 12. Range-Speed response with clutter and jammer

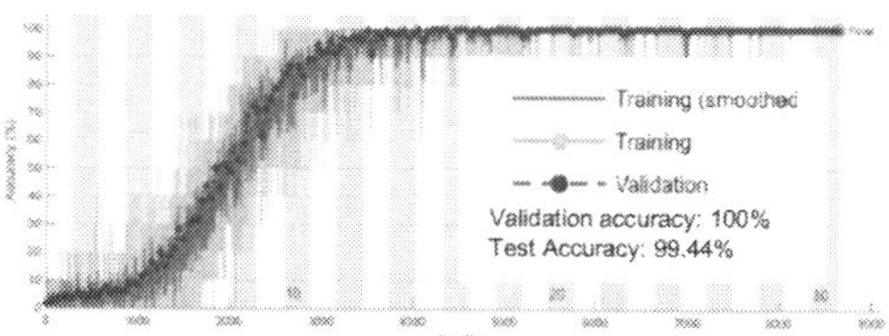

Fig. 13. Validation Accuracy of SqueezeNet Range-Speed

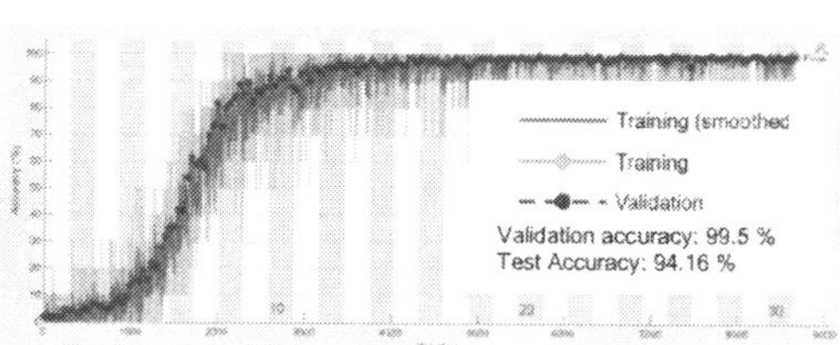

Fig. 14. Validation Accuracy of SqueezeNet Range-Angle

Fig. 13 and Fig. 14 show the validation accuracy of the SqueezeNet in the case of range-angle and range-speed estimation, respectively. The SqueezeNet model was trained with 2880 samples and tested with 720 samples for all 60 classes for 30 epochs. The training converges with a test accuracy of 94.16% for range-angle estimation and 99.44% for range-speed estimation. It is noted that in both cases, range-angle and range-speed estimation, the performance of the SqueezeNet is very close to the performance of the SqueezeNet trained in a clean environment but combined with the SVM.

4 Conclusion

A two-stage SVM-SqueezeNet approach was proposed for FMCW MIMO radar to classify and detect moving targets in a cluttered and jammed environment. The proposed approach reached an average accuracy of 98.33% in the range-angle estimation and 99 % in the case of range speed estimation. The performances are very close to the one achieved by the SqueezeNet only (94.44% for range-angle and 99.44% for range-speed) but with the advantage of identifying the type of targets and reducing the training time required by the SqueezeNet.

References

[1] O.Bialer , A.Jonas, and T.Tirer "Code Optimization for Fast Chirp FMCW Automotive MIMO Radar", IEEE Transactions on Vehicular Technology 70 (8), 7582-7593, July, 2021.

[2] M.Gottinger , M.Hoffmann, M. Christmann, M.Schutz, F. Kirsch, P.Gulden, and M.Vossiek Coherent Automotive Radar Networks: The Next Generation of Radar-Based Imaging and Mapping. IEEE Journal of Microwaves, vol. 1, no. 1, pp. 149 - 163, Januar, 2021.

[3] B.Li , S.Wang, Z.Feng , J.Zhang , X.Cao , and C.Zhao. Fast Pseudospectrum Estimation for Automotive Massive MIMO Radar. arXiv: 1911.07434v3, Dec 2021.

[4] M.Gottinger , M.Hoffmann, M. Christmann, M.Schutz, F. Kirsch, P.Gulden, and M.Vossiek Coherent Automotive Radar Networks: The Next Generation of Radar-Based Imaging and Mapping. IEEE Journal of Microwaves, vol. 1, no. 1, pp. 149 - 163, Januar, 2021.

[5] C.Pfeffer, R.Feger, C.Wagner and A.Stelzer. FMCW MIMO Radar System for Frequency-Division Multiple TX-Beamforming. IEEE transactions on microwavetheory and techniques vol. 61, no. 12, decembre 2013

[6] A.Bose , B.Tang , M.Soltanalian and J.Li. Mutual Interference Mitigation for Multiple Connected Automotive Radar Systems. IEEE Transactions on Vehicular Technology

[7] N.Amani , F.Jansen, A.Filippi , M V.Ivashina, and Rob Maaskant. Sparse Automotive MIMO Radar for Super-Resolution Single Snapshot DOA Estimation With Mutual Coupling IEEE Access, vol. 9 ,2021

[8] P. Sévigny, "MIMO Radar: Literature survey of papers between 2003 and September 2008," DRDC, Ottawa, ON, TM 2008-333, Mar. 2009.

[9] J. Li and P. Stoica, MIMO Radar Signal Processing. Hoboken, NJ: Wiley-IEEE Press, 2008.

[10] P. W. Moo and Z. Ding, "Tracking Performance of MIMO Radar for Accelerating Targets," IEEE Trans. Signal Process., vol. 61, no. 21, pp. 5205–5216, Nov. 2013.

[11] I. Bekkerman and J. Tabrikian, "Target Detection and Localization Using MIMO Radars and Sonars," IEEE Trans. Signal Process., vol. 54, no. 10, pp. 3873–3883, Oct. 2006.

[12] E. Fishler, A. Haimovich, R. Blum, D. Chizhik, L. Cimini, and R. Valenzuela, ``MIMO radar: An idea whose time has come," in Proc. IEEE Radar Conf., Apr. 2004, pp. 71-78.

[13] F. C. Robey, S. Coutts, D. Weikle, J. C. McHarg, and K. Cuomo, ``MIMO radar theory and experimental results," in Proc. 28th Asilomar Conf. Signals, Syst. Comput., Nov. 2004, pp. 300-304.

[14] M. Hefnawi, J. Bray, J. Bathurst, and Y. Antar, "MIMO Radar Using a Vector Network Analyzer," Electronics, vol. 8, no. 12, Art. no. 12, Dec. 2019, doi: 10.3390/electronics8121447.

[15] J. Bergin and J. R. Guerci, MIMO Radar Theory and Application. Boston, MA, USA: Artech House, 2008.

[16] H. Sun, F. Brigui, and M. Lesturgie, "Analysis and comparison of MIMO radar waveforms," in 2014 International Radar Conference, 2014, pp. 1–6.

[17] Olivier Rabaste, Laurent Savy,Mathieu Cattenoz, and Jean-Paul Guyvarch. Signal waveforms and range/angle coupling in coherent colocated MIMO radar. In Radar (Radar), 2013 International Conference on, pages 157–162. IEEE, 2013.

[18] Ming Xue, Jian Li, and Peter Stoica. MIMO radar waveform design. In Fulvio Gini, Antonio DeMaio, and Lee Patton, editors,Waveformdesign and diversity for advanced radar systems, pages 89–120. IET Press, 2012.

[19] Hongbo Sun, Frédéric Brigui, and Marc Lesturgie. Analysis and comparison of MIMO radar waveforms. In 2014 International Radar Conference. IEEE, 2014.

[20] Ralpph O. Acmidt, "Multiple Emitter Location and Signal Parameter Estimation, "IEEE Trans. Antennas & Prop., vol. AP-34, no. 3, Mar. 1986, pp. 276-280

[21] R. Roy and T. Kailath, "ESPRIT-estimation of signal parameters via rotational invariance techniques," IEEE Trans. Acoust. Speech Signal Process., vol. 37, no. 7, pp. 984–995, Jul. 1989.

[22] J. Capon, "High-resolution frequency-wavenumber spectrum analysis," Proc. IEEE, vol. 57, no. 8, pp. 1408-1418, Aug. 1969.

[23] Jiho Seo , Yunji Yang , Yong - gi Hong , Jaehyun Park, "Transfer learning - based radar imaging with deep convolutional neural networks for distributed frequency modulated continuous waveform multiple - input multiple - output radars," IET Radar, Sonar & Navigation, 2021;15:1209–1220; doi: 10.1049/rsn2.12105.

[24] M. S. Seyfioglu, A. M. Ozbayoglu, and S. Z. Gurbuz, "Deep convolutional autoencoder for radar-based classification of similar aided and unaided human activities," IEEE Transaction on Aerospace and Electronic Systems. Vol. 54, no. 4 Aug. 2018.

[25] Ramakrishna Sai Annaluru , Khurram Usman Mazher, Robert W. Heath, "Deep Learning Based Range and Doa Estimation using low Resolution FMCW Radars," 2021 IEEE Statistical Signal Processing Workshop (SSP), DOI: 10.1109/SSP49050.2021.9513759.

[26] Woosuk Kim 1, Hyunwoong Cho , Jongseok Kim, Byungkwan Kim, and Seongwook Lee , "YOLO-Based Simultaneous Target Detection and Classification in Automotive FMCW Radar Systems," Sensors 2020, 20(10), 2897.

[27] Pérez, R.; Schubert, F.; Rasshofer, R.; Biebl, E. Deep learning radar object detection and classification for urban automotive scenarios. In Proceedings of the 2019 Kleinheubach Conference, Miltenberg, Germany, 23–25 September 2019.

[28] Jin-Cheol Kim, Hwi-Gu Jeong, and Seongwook Lee," Simultaneous Target Classification and Moving Direction Estimation in Millimeter-Wave Radar System," Sensors 2021, 21(15), 5228.

[29] Support Vector Machine (SVM) MATLAB, [Online]. Available:https://www.mathworks.com/discovery/support-vector-machine.html?s_tid=srchtitle_Support%20Vector%20Machine_1

[30] F. N. Iandola, S. Han, M. W. Moskewicz, K. Ashraf, W. J. Dally, and K. Keutzer. SqueezeNet: AlexNet-level accuracy with 50x fewer parameters and< 0.5 MB model size. arXiv preprint arXiv:1602.07360, 2016.

[31] SqueezeNet,[Online]. Available:https://www.mathworks.com/help/deeplearning/ref/squeezenet.html

[32] Radar Target Classification Using Machine Learning and Deep Learning-MATLAB,[Online]. Available:https://www.mathworks.com/help/radar/ug/radar-target-classification-using-machine-learning-and-deep-learning.html

[33] T.Cheng , B.Wang , Z.Wang , R.Dong , and B.Cai. Lightweight CNNs-Based Interleaved Sparse Array Design of Phased-MIMO Radar, IEEE sensors journal, vol. 21, no. 12, june 15, 2021

[34] T.Wan , X.Fu , K.Jiang , Y.Zhao and B.Tang. Radar Antenna Scan Pattern Intelligent Recognition Using Visibility Graph, IEEE access, volume 7, 2019

[35] F.Meng , K.Tian, and C.Wu Deep Reinforcement Learning-Based Radar Network Target Assignment, Electronics, vol. 11, no. 3, January 2022, doi: 10.3390/electronics11030311

[36] Increasing Angular Resolution with Virtual Arrays-MATLAB, [Online]. Available:https://www.mathworks.com/help/phased/ug/increasing-angular-resolution-with-mimo-radars.html.

[37] Automotive Adaptive Cruise Control Using FMCW Technology [Online].Available:https://www.mathworks.com/help/radar/ug/automotive-adaptive-cruise-control-using-fmcw-technology.html.

[38] Karnfelt, Camilla, et al. "77 GHz ACC Radar Simulation Platform." 9th International Conference on Intelligent Transport Systems Telecommunications, (ITST), IEEE, 2009 , pp. 209–14.

[39] Barton, David. "Land Clutter Models for Radar Design and Analysis," Proceedings of the IEEE. Vol. 73, Number 2, February, 1985, pp. 198–204

ITM Web of Conferences **48**, 01011 (2022)
ICCWCS 2022

https://doi.org/10.1051/itmconf/20224801011

Design of flower-shaped high-efficiency compact UWB antenna

Aicha Bembarka[1*], *Larbi* Setti[1], *Abdelwahed* Tribak[2], and *Hafid* Tizyi[2]

[1]LSTA, FPL, Abdelmalek Essaadi University, Tetouan, Morocco
[2]Department of Electronics Microwaves and Optics, National Institute of Posts and Telecommunications (INPT), Rabat, Morocco

Abstract. The present paper describes a novel planar antenna based on the fractal structure designed for Ultra-Wide Band (UWB) applications. The antenna exceeds the UWB band by covering the frequency range between 7.36 and 19.57 GHz with S11 < -10 dB (the fractional bandwidth is 90.68 %) which involves four applications, the C-band (4-8 GHz), the X-band (8-12 GHz), the Ku-band (12-18 GHz) and the K-band (18-26.5 GHz). The low-cost Rogers RO4003C substrate is employed to print the proposed antenna in a small dimension of 12 x 16.9 x 0.81 mm^3. This antenna is invented from the shape of a flower with three elliptical leaves. The antenna simulated radiation efficiency is greater than 97 % with an omnidirectional radiation pattern in the E-plane and a bidirectional radiation pattern in the H-plane, across the entire bandwidth. The maximum realized gain of the proposed antenna is about 5.25 dBi. Therefore, the obtained results demonstrated that the flower-shaped antenna is a good candidate for practical UWB 5G applications.

1 Introduction

The new wireless technologies are needing to transmit and receive more data in a faster way. More than that, they need sufficient exploitation of the spectrum, which demands emerging new concepts. The UWB and cognitive radio (CR) paradigms are the best options for overcoming these issues. The CR contributes to better usage of frequency band channels. Furthermore, they aid in the avoidance of signal interference while the frequency channels are in use [1]. Several antennas have been developed for the CR [2–4].

In addition to the CR paradigm, UWB antennas are thought to be viable prospects for future wireless communication systems. They are notable for their resilience to multipath propagation, low power consumption, and ability to transmit a large quantity of data in a short period at local distances ranging from 1 to 10 m. For decades, traditional antennas such as conical, volcano smoke antennas, and coaxial horns were well-known in the design of ultra-wideband antennas, but the advancement of wireless communications necessitated antennas that are compact, inexpensive, and have less complexity [5]. As a result, printed antennas are becoming increasingly popular. As in [6] a UWB Vivaldi antenna is presented that has an overall dimension of 49 x 48 x 0.8 mm^3 and covers the frequency from 2.9 to 9 GHz. In order to improve the antenna bandwidth, a combination between Hemi cylinder slots and directors is used. Moreover, planar monopole antennas based on Metamaterial, Octagonal-shaped, and Circular-shaped radiated elements have been reported in [7–10] for UWB applications. The antenna in [8] has a wideband covering the range 3.1-10.6 GHz, has a good gain that varies from -2 to 6 dBi, and is characterized by an efficiency greater than 70 %. In [9] the gain of the antenna is of 5.8 dBi when it resonates between 3.1 and 12 GHz. Authors in [10] proposed an antenna that has a 183.91 % of the fractional bandwidth and has a high efficiency that goes between 83 and 95 %, the antenna gain is between 2.58 and 5.05 dBi. Although these advantages they occupy a large space of 40 x 40 x 1.6 mm^3, 34.5 x 22 x 1.6 mm^3 and 47 x 47 x 0.675 mm^3 respectively. Another method for creating UWBs is to employ fractal geometries.

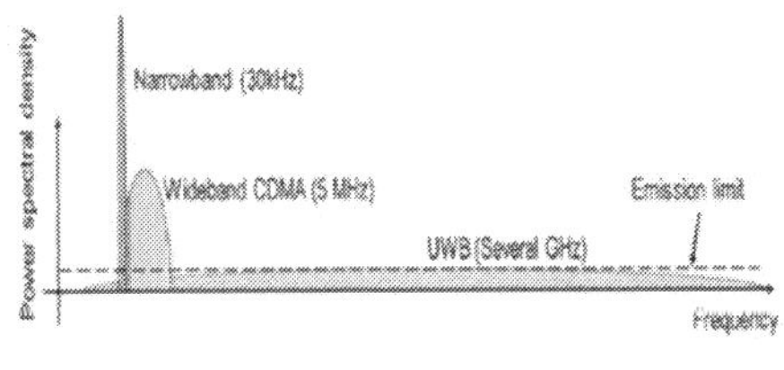

Fig. 1. Power spectral density versus frequency of the Narrowband, the Wideband, and the UWB [1].

The fractal structures have gained a lot of interest in the realization of UWB sensing antennas in miniaturized space. This type of antenna is known for its ability to create multiple resonant frequencies with their repetitive topology and ease installation in RF circuits. Various ultra-wideband antennas based on fractal geometry were introduced in the literature. The authors in [11] proposed a planar UWB antenna based on the fractal Sierpinski carpet to achieve a bandwidth of 9.6 GHz in a space of 24 x 30 x 0.8 mm^3. The hexagonal-shaped fractal antenna [12] and the Spanner Shaped fractal for

* Corresponding author: aichabembarka@gmail.com

 https://doi.org/10.1051/itmconf/20224801011

cognitive radio antenna [13] have been also designed in the UWB range.

In this work, a flower-shaped fractal UWB antenna for Cognitive radio is presented and analyzed with two simulation tools CST Microwave Studio and verified using HFSS. The proposed geometry is planar and simply printed on the substrate's top layer, where the ground plane is partially below the substrate. From one ellipse considered fractal petal and two other ellipses scaled and rotated symmetrically to the petal, the flower shape of the antenna has been formed. The ellipses will be named also in this paper by "leaves". The antenna's most important characteristics are simulated to evaluate the performance of the proposed antenna. The antenna radiation efficiency is greater than 97 % and the peak realized gain attains 5.25 dBi. This paper is organized in this way; in section 1, we gave the geometry of the proposed fractal flower antenna. After section 1, we discuss the difference made in the reflection coefficient of the antenna radiated element with one leaf and with three leaves. The radiation efficiency, the realized gain, and also the radiation pattern is presented in this section. Finally, we conclude the work done.

2 Design of proposed antenna

The proposed antenna follows the principle of fractal geometry. The term "fractal" was introduced for the first time by the mathematician Benoit Mandelbrot in 1975, but it was discovered earlier by the mathematicians Pierre Fatou and Gaston Julia. The name "fractal" is derived from the Greek word "fractus," which means "broken" or "fractured." It is a structure made up of similar fragments of different sizes and orientations. The geometry and configuration of the proposed antenna are illustrated in Figure 2. It is constructed on Rogers RO4003C substrate ($\varepsilon_r = 3.38$, $h = 0.81$, and δ=0.0027). The proposed antenna has a planar structure having a dimension of 12 x 16.9 mm^2. This antenna takes its shape from a flower, it was designed in accordance with the fractal concept in the interest of getting broad bandwidth. Herein, the flower patch is composed of three elliptical leaves. A central leaf (petal) surrounded by two similar leaves. The two leaves are scaled and rotated by $\boldsymbol{\theta}$=15° and $\boldsymbol{\theta}$=-15° in symmetry to the petal. The effective (electrical) semi-major axis of the central ellipse a_{eff} takes into account stored energy in the fringing field of elliptical edge which can be calculated using equation (1):

$$a_{eff} = \frac{15}{\pi e (f_r)_{11}^{e,0}} \sqrt{\frac{q_{11}^{e,0}}{\varepsilon_r}} \tag{1}$$

e is eccentricity of elliptical patch.
ε_r is permittivity of dielectric substrate.
$(f_r)_{11}^{e,0}$ is dual resonant frequency.
$q_{11}^{e,0}$ is called the Mathieu function, of the dominant $TM_{11}^{e,0}$ mode, which value can be obtained approximately from [14]:

$$q_{11}^{e} = -0.0049e + 3.7888e^2 - 0.7278e^3 + 2.314e^4 \tag{2}$$
$$q_{11}^{0} = -0.0049e + 3.8316e^2 - 1.1351e^3 + 5.2229e^4 \tag{3}$$

After the calculation of the effective semi-major axis, the physical semi-major axis a is derived from equation (4):

$$a_{eff} = a\left\{1 + \frac{2h}{\pi \varepsilon_r a}\left[\ln\left(\frac{a}{2h}\right) + (1.41\varepsilon_r + 1.77) + \frac{h}{a(0.286\varepsilon_r + 1.65)}\right]\right\}^{\frac{1}{2}} \tag{4}$$

Where h is the substrate height
Using the equation $b = a\sqrt{1 - e^2}$ that connects the two semi-axes of an ellipse, we may determine the physical semi-minor axis b, then:

$$b_{eff} = b\left\{1 + \frac{2h}{\pi \varepsilon_r a}\left[ln\left(\frac{b}{2h}\right) + (1.41\varepsilon_r + 1.77) + \frac{h}{b}(0.286\varepsilon_r + 1.65)\right]\right\}^{\frac{1}{2}} \tag{5}$$

The major axis of the central leaf is **Lc**, as illustrated in Figure 2 and which is equal to 4.25 mm, while its minor axis length is **Wc**=1 mm. The scaled rotated leaves have a major axis length and a minor axis length **Le** and **We** respectively of 3.6 and 1.38 mm. The similar leaves helped to create more resonant frequencies at higher frequencies above 11 GHz and that is what fractal is known by. They are known by repeating shapes that give more resonant frequencies and these resonant frequencies can be controlled by changing the size of the repeated structures.

The ground plane is a critical part of antennas because it connects the distributed current to the power source ground. The proposed antenna ground plane is located at the other side of the substrate and to ensure that there is no electromagnetic field is kept between the radiated element and the ground plane as in [15]. A partial ground plane has been employed in this design. The proposed antenna ground is rectangular and its upper limit is far from the feed origin by 8 mm. Moreover, an inverted T-shaped slot truncated from its middle top edge helped to enhance the gain of the flower-shaped antenna in the band 11-16 GHz to higher values as can be shown in Figure 6. A microstrip feed line with 8.75 mm length provides a 50 Ω input impedance. Table 1 summarizes various dimensions of the proposed antenna geometry.

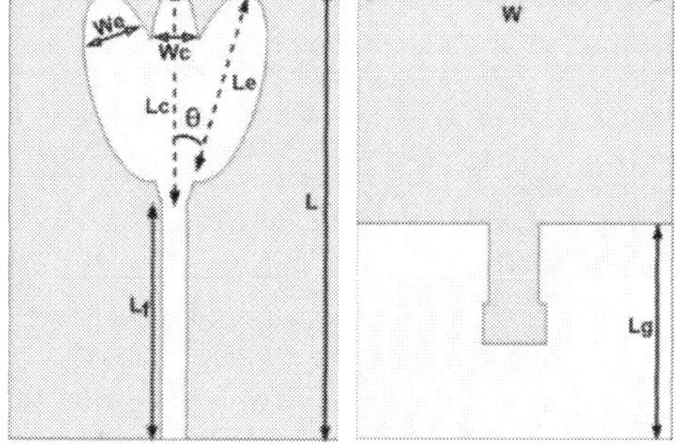

Fig. 2. Geometry of the proposed antenna (a) Top View (b) Botton View.

Table 1. Parameters of the fractal flower-shaped antenna.

Dimension	Value(mm)	Dimension	Value(mm)
L	16.9	W	12
Lg	8	Lf	8.75
Lc	4.25	Wc	1
Le	3.6	We	1.38
θ	15		

3 Results discussion

The results of the simulated reflection coefficient versus frequency are presented in Figure 3. A study has been done on the effect of the leaves on the reflection coefficient. Hence, we studied two cases; radiated element with only one ellipse and radiated element with three ellipses. The straight graph presents the reflection coefficient when a single elliptical patch is used, and the dotted curve provides the response of the radiated element with three ellipses. It can be clearly seen that the reflection coefficient increased when three ellipses were employed in comparison to one ellipse taking the -10 dB criteria. At first, the antenna was dual-band resonating at 9.62 GHz and 19.65 GHz. It can be clearly noticed that the reflection coefficient at the high band is a little high of -17 dB, which is not sufficient. On the other hand, after adding the two other leaves to the first leaf the antenna extended its resonant frequency from 7.36 to 19.57 GHz corresponding to nearly 49.64 % of the band added. Consequently, repeating the number of leaves proved that this technique helps to generate more resonant frequencies, then enhancing the antenna working band and improving its matching.

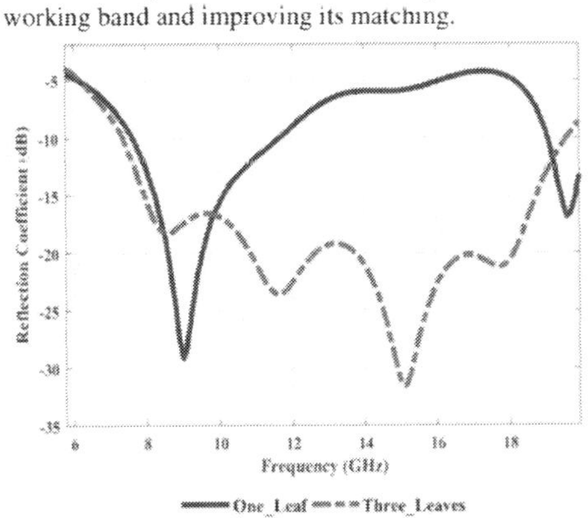

Fig. 3. Comparison of the simulated reflection coefficient between antenna with one ellipse and with three ellipses.

For further verification, we simulated the proposed structure with another software that uses a different calculating method which is HFSS. The design made in the 3D simulator CST was exported as a "Sat" file to the other 3D software (HFSS). The reflection coefficient in CST and HFSS of the proposed antenna is depicted in Figure 4. As can be seen that the response obtained from HFSS is almost the same reached previously in CST. The antenna has a minimum reflection coefficient of -31.59 dB at a frequency of 15.1 GHz. Moreover, the antenna is well matched all over the resonant band. Both responses proved that the proposed flower antenna exceeds the UWB spectrum. It has a relative bandwidth of 90.68 % (89.22 %) covering frequencies from 7.36 to 19.57 GHz (from 7.45 to 19.45 GHz) in CST and HFSS respectively.

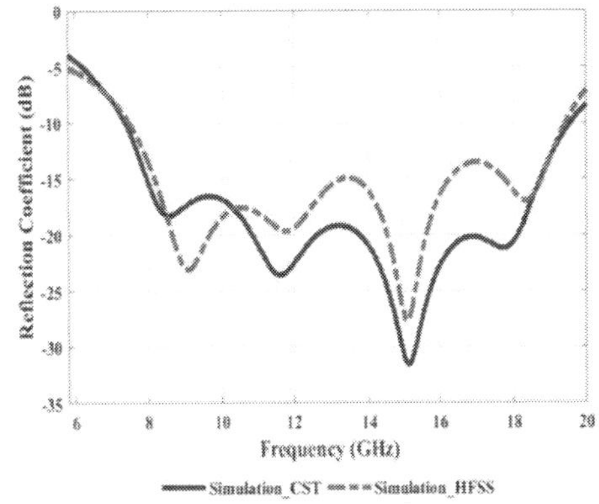

Fig. 4. Simulated reflection coefficient of the proposed antenna.

Two of the fundamental antenna parameters have been addressed in Figure 5, Figure 6, and Figure 7. The first one is the efficiency, also recognized as radiation efficiency, and the second one is realized gain. The last one is the radiation pattern.

The antenna efficiency ε_R is proportional to the power radiated from the antenna P_R and the power delivered to it P_{in}, as shown in the below equation:

$$\varepsilon_R = \frac{P_R}{P_{in}} \quad (6)$$

Figure 5 illustrates the simulated radiation efficiency of the proposed fractal antenna in comparison to the conventional partial ground plane. It can be observed that the partial ground plane with the inverted T-shaped slot improves the efficiency of the antenna. Herein, the lowest efficiency of the antenna is achieved at the frequency 19.57 GHz with 97.3 % when the ground has the T-shape truncation. On the other hand, a high amount of power is transmitted via the proposed antenna that is equal to 99.05 % as shown in simulation, which means there is only 0.95 % of the dissipated power. Our flower antenna has a high efficiency all over the working band from 7.36 to 19.57 GHz.

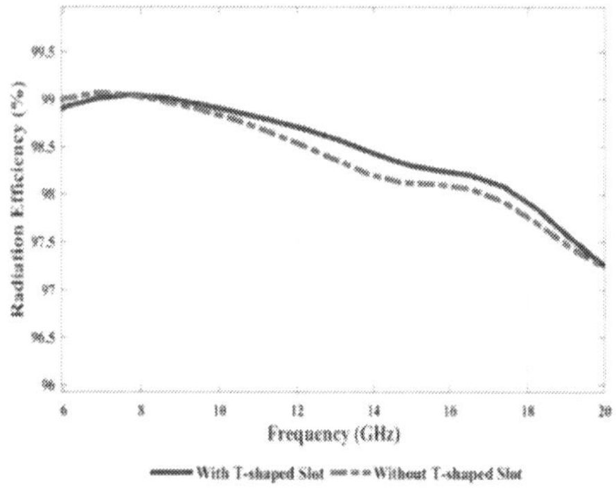

Fig. 5. Simulated radiation efficiency of the proposed flower-shaped antenna.

The simulated peak realized gain of the proposed antenna versus frequency is depicted in Figure 6. In the working band, the gain varies from 2.6 to 3.95 dBi for a maximum gain found with 4.147 dBi at the frequency of 12.2 GHz in CST. Whereas in HFSS it goes till 5.25 dBi at the frequency 19.45 GHz, and which varies from 2.67 to 5.25 dBi. Besides, the minimum value of gain is obtained at 15.6 GHz with 2.45 dBi. Making an inverted cut of a T-shape on the ground plane proved its potential to improve antenna gain for frequencies between 10 and 15 GHz. Hence, since the proposed antenna is compact and planar, the gain obtained is at an acceptable level.

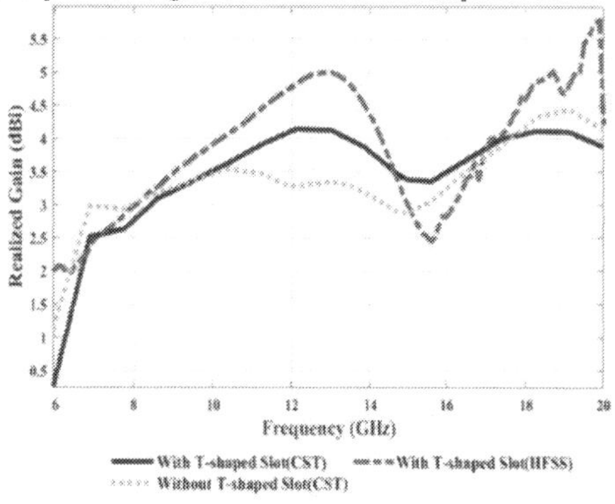

Fig. 6. Simulated realized gain of the proposed flower-shaped antenna.

For additional explanation of how the proposed antenna radiates energy out into the space and how it receives energy, the 3D and 2D view of the radiation pattern are presented in Figure 7. The calculations are performed at two frequencies 11.8 and 15 GHz to determine the antenna's stability behaviour in the operational band. According to simulations, the maximum obtained directivity of the antenna at the frequency 11.8 GHz is about 4.16 dBi, as demonstrated by the region covered in red color in Figure 7(a). At the other frequency (15 GHz), the flower antenna achieved a directivity of 3.46 dBi, as shown in Figure 7(b). The 2D pattern or also known as the polar plot of radiation pattern have been added in Figures (c)(d) in order to easily visualize the antenna radiation. It can be clearly observed that the antenna has an omnidirectional radiation in the E-plane (Co. Polarization) and in the H-plane (Cross. Polarization) it has bidirectional radiation. Figure 7(c) depicts the flower antenna polar plot at 11.8 GHz, the main lobe magnitude at this frequency is about 4.1 dB at the angle 342°, with angular 3-dB width of 57.7° for a cut done at phi component equal to 90°. Another polar plot of the antenna radiation pattern has been added in Figure 7(d) at 15 GHz. As shown in this Figure, a main lobe of magnitude 3.38 dB has been obtained at the direction 353°, while the angular 3-dB width is 54.1°.

Consequently, from the results obtained above the proposed antenna is compact and less dissipative, also it has a high gain and an omnidirectional radiation pattern stable among the working band, which makes the flower antenna very suitable for integration in UWB cognitive radio sensing applications.

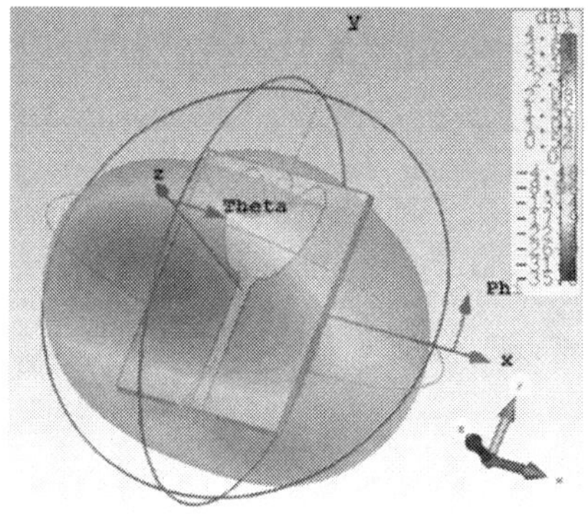

(a)

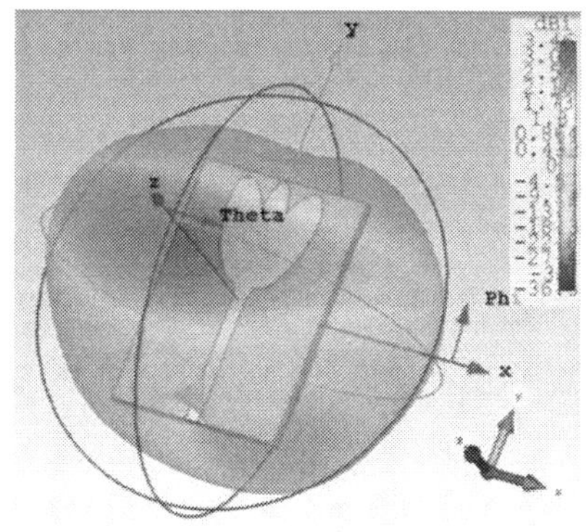

(b)

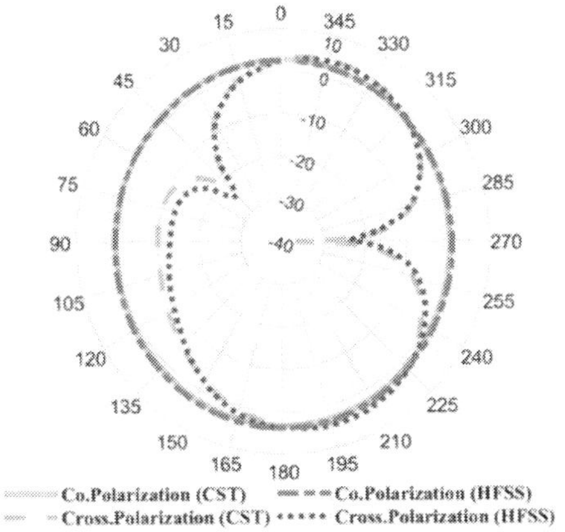

(c)

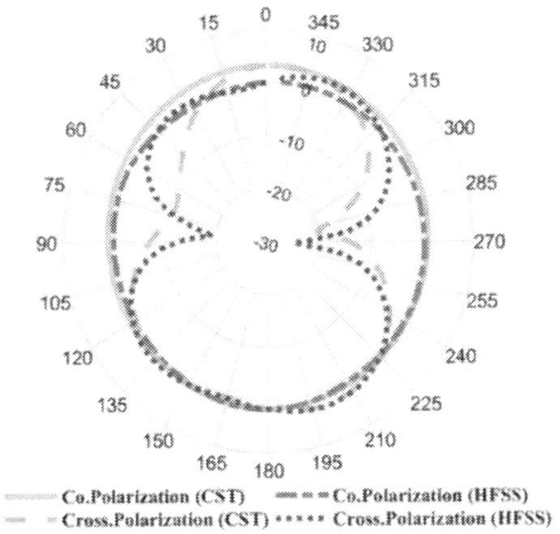

(d)

Fig. 7. Simulated radiation characteristics of proposed flower-shaped antenna (a) 3D at 11.8 GHz, (b) 3D at 15 GHz, (c) 2D at 11.8, (d) 2D at 15 GHz.

Table 2 summarizes a comprehensive comparison between the proposed antenna and previously published UWB antennas to determine the validity of the proposed design. As shown in the table, although the antenna in [18] has a high gain of 9.5 dBi, its dimension is vast, and its operating bandwidth is smaller than that of our structure. Even though the antenna in the literature [19] performs well, its size is quite large. In comparison to our antenna, we developed a miniaturized antenna in area of 12 x 16 x 0.81 mm³ while maintaining a high efficiency ranging between 97.4 and 99.05 % and reasonable gain of 5.25 dBi and more than that it covers a wide bandwidth range of 90.68 %.

Table 2. Comparison between previously published UWB antennas and the proposed antenna.

Parameter / Antenna	size [mm]	FR[a] [GHz](FB[b] [%])	Max. Gain[c] [dBi]	R. Eff.[d] [%]
Proposed antenna	**12x16.9 x0.81**	**7.36-19.57 (90.68)**	**5.25**	**> 97**
[16]	**25x38 x1.6**	**2.4-6 (86.71)**	**2.85**	**> 90**
[17]	**23x23**	**6.4-11.736 (58.7)**	**4.91**	**N/A**
[18]	**147x243 x 1.6**	**1.68-2.97 (55.5)**	**9.5**	**N/A**
[19]	**100x104 x0.5**	**2-9 (127.27)**	**6.5**	**> 98**

[a]FR= Frequency Range. [b]FB=Fractional Bandwidth. [c]Max. Gain= Maximum Gain. [d]R. Eff= Radiation Efficiency.

4 Conclusion

A UWB antenna operating in a multi-band/multi-standard; the C X, Ku and K bands, of a fractal geometry were investigated in this paper. It has achieved a -10 dB relative bandwidth of about 90.68 % at the center frequency of 13.465 GHz. The fractal antenna was inspired by the shape of a flower that has three leaves which helped to obtain an ultra-wide bandwidth without increasing the substrate size. The overall size occupied by the antenna is 12 x 16.9 x 0.81 mm³. The proposed antenna has good performances, it has high efficiency of 97 %, a good gain of 5.25 dBi at 19.45 GHz, and also it is well matched all over the operating band. Moreover, this flower antenna has an omnidirectional radiation pattern thus making it capable of spectrum sensing in cognitive radios. As future work, the research might be expanded to include the fabrication and testing of the design.

References

1. N. Anveshkumar, A. S. Gandhi, and V. Dhasarathan, Cognitive radio paradigm and recent trends of antenna systems in the UWB 3.1–10.6 GHz, Wirel. Networks **26**, 3257–3274 (2020).
2. P. Palanisamy and M. Subramani, An integrated antenna for cognitive radio wireless sensor networks and HD video transmission applications, Int. J. RF Microw. Comput. Aided. Eng **31**, 11 (2021).
3. M. Y. Sandhu, S. Afridi, M. R. Anjum, B. S. Chowdhry, and Z. H. Khand, Ultra Wide Band (UWB) Horseshoe Antenna for Future Cognitive Radio Applications, Wirel. Pers. Commun **122**, 1087–1099 (2022).
4. M. Aggarwal and A. P. S. Pharwaha, Modified Koch curve broadband fractal antenna for spectrum sensing in cognitive radio, Int. J. Microw. Wirel. Technol **13**, 947–953 (2021).

5. M. Anouar and S. Larbi, *Array Antenna for Wireless Communication 5G*, In the Proceedings of the Third International Conference on Smart City Applications, SCA, 10-11 October 2018, Tetouan, Morocco (2018).
6. S. Guruswamy, R. Chinniah, and K. Thangavelu, A printed compact UWB Vivaldi antenna with hemi cylindrical slots and directors for microwave imaging applications, AEU - Int. J. Electron. Commun, **110** (2019).
7. Z. Zerrouk and S. Larbi, *Frequency Reconfigurable Antenna Using Infinity Complementary Split Ring Resonator for Wireless Communication*, in 2021 International Conference on Electrical, Computer and Energy Technologies, ICECET, 9-10 December 2021, Cape Town-South Africa (2021).
8. S. Heydari, K. Pedram, Z. Ahmed, and F. B. Zarrabi, Dual band monopole antenna based on metamaterial structure with narrowband and UWB resonances with reconfigurable quality, AEU - Int. J. Electron. Commun **81**, 92–98 (2017).
9. R. Sanyal, P. P. Sarkar, and S. Sarkar, Octagonal nut shaped monopole UWB antenna with sextuple band notched characteristics, AEU - Int. J. Electron. Commun, **110** (2019).
10. K. Sharma, A. Karmakarb, M. Sharma, A. Chauhan, S. Bansal, M. Hooda, S. Kumar, N. Gupta, A. K. Singh, Reconfigurable dual notch band antenna on Si-substrate integrated with RF MEMS SP4T switch for GPS, 3G, 4G, bluetooth, UWB and close range radar applications, AEU - Int. J. Electron. Commun **110**, 7–9 (2019).
11. R. Gurjar, D. K. Upadhyay, B. K. Kanaujia, and A. Kumar, A compact modified sierpinski carpet fractal UWB MIMO antenna with square-shaped funnel-like ground stub, AEU - Int. J. Electron. Commun, **117** (2020).
12. S. Tripathi, A. Mohan, and S. Yadav, Hexagonal fractal ultra-wideband antenna using Koch geometry with bandwidth enhancement, IET Microwaves Antennas Propag **8**, 1445–1450 (2014).
13. C. Waghmare and A. Kothari, *Spanner shaped ultra wideband patch antenna*, in 2014 Students Conference on Engineering and Systems, SCES, 28-30 May 2014, Allahabad, India (2014).
14. M. Gupta, K. K. Mutai, V. Mathur, and D. Bhatnagar, A Novel Elliptical Ring Microstrip Patch Antenna for Ultra-Wideband Applications, Wireless Pers Commun **114**, 3017-3029 (2020).
15. C. C. Lin, H. R. Chuang, and Y. C. Kan, A 3-12 GHz UWB planar triangular monopole antenna with ridged ground-plane, Prog. Electromagn. Res **83**, 307–321 (2008).
16. A. K. Gautam, A. Bisht, and B. K. Kanaujia, A wideband antenna with defected ground plane for WLAN/WiMAX applications, AEU - Int. J. Electron. Commun **70**, 354–358 (2016).
17. A. H. Majeed, A. S. Abdullah, K. H. Sayidmarie, R. A. Abd-Alhameed, F. Elmegri, and J. M. Noras, Balanced dual-segment cylindrical dielectric resonator antennas for ultra-wideband applications, IET Microw. Antennas Propag **9**, 1478–1486 (2015).
18. X. W. Dai, X. Y. Zhou, and G. Q. Luo, Wideband directional antenna system with different polarizations for wireless communication system, AEU - Int. J. Electron. Commun **75**, 119–123 (2017).
19. Z. Yang, H. Jingjian, W. weiwei, and Y. Naichang, An antipodal Vivaldi antenna with band-notched characteristics for ultra-wideband applications, AEU - Int. J. Electron. Commun **76**, 152–157 (2017).

High-Directivity and Circularly Polarized Compact Patch Antenna for UHF RFID Reader Devices

Mohamed Latrach[1*] and Islam MD Saiful[2]

[1]RF-EMC research group, ESEO, IETR UMR CNRS 6164, 10 Boulevard Jean Jeanneteau, 49100 Angers, France
[2]RF-EMC research group, ESEO, 10 Boulevard Jean Jeanneteau, 49100 Angers, France

Abstract. This paper presents the design, construction, and testing of a high directivity and circularly polarized small patch antenna for the RFID Base Stations (Readers) operating in the European UHF RFID band. The designed and optimized patch antenna structure has a 15cm × 15 cm × 0.6 cm compact size where the -3dB beam angle is reduced by using slotted ring embedded in the patch. The gain of the antenna at 862 MHz is 3.2 dBic with the E and H plain beam angle are 61.6° and 65.2° respectively. Finally, the designed antenna is compared to a single-layer patch antenna of the same cross-sectional dimensions as the proposed antenna structure.

1 Introduction

The aim of this study is to develop a new antenna structure for UHF RFID base stations. The sought criteria are compactness and a high directivity in order to allow an optimal compromise in the implementation of RFID solutions in the UHF band applications.
Different techniques used to increase the directivity of planar antennas are creating antenna array, increase the ground plane size, increasing antenna size, miniaturizing antennas and using superstrate [1, 2]. In [2] the electric field distribution on top surface above a superstrate has been used to determine its shape and size that minimize its edge diffraction.

Usually high gain and highly efficient antenna means higher directivity. Further enhancement of directivity is hardly possible without changing the antenna size because the half-power beam width (HPBW) is a function of antenna size. For this situation placing one or more superstrates can narrow the HPBW thereby increases the directivity of the patch. HPBW can be reduced by making antennas in an array. In UHF frequency band the antenna size is already too big. Hence, making array will lead the antenna size uncomfortable. This is cannot be a solution of compact size high directive antenna.

This study deals with the process of narrowing the half-power beam width (HPBW) of patch antenna in UHF frequency band. By high "directivity" we could mean either higher gain or narrower beam width. Both of these can only be achieved by increasing the aperture of the antenna, that is to say the size. In theory we could increase the efficiency of the antenna without increasing size thus gaining in both, but most well designed antennas are already over the 50% of efficient, so even if we could somehow increase efficiency we would not gain much. Reducing the HPBW may not increase the gain but the directivity may increase.

In our approach, we consider the problem of enhancing the directivity of a patch antenna based on a new configuration antenna with three dielectric layers and modified radiating element. We begin by studying a conventional single layer patch antenna fabricated on FR4 substrate and then defined and characterized the new form of the proposed antenna.

2 Design and simulation

The construction of the patch antennas under consideration is shown in Fig. 1. In fact we will consider here two antennas of Fig. 1(a) and (b). Fig.1 (a) shows a conventional single layer patch antenna on FR4 dielectric substrate. The second version is the proposed antenna and shown in Fig. 1(b) where two layers of FR4 are used; one for ground plane support and the other is for patch (radiading element) support. As shown in Fig. 1(b) there is an air gap separating the two previous layers of FR4. In addition of those two non equal rings are etched in the radiating element; one at the centre of the patch and the other around the feed position. For both the case the used thickness of FR4 is 1.6 mm, permittivity, with dielectric substrate, of 4.3 and the loss tangent is 0.021. For the antenna of Fig. 1(a) the ground plane size is 15 cm ×15 cm and the patch size is 8.15 cm × 8.15 cm. The antenna is diagonally feed for circular polarization. For the antenna in Fig. 1(b) the ground

* Corresponding author: mohamed.latrach@eseo.fr

plane size remains 15 cm × 15 cm. A little slot ring is etched around the feed point. The second larger slot ring is etched at the center of the radiating element (patch).

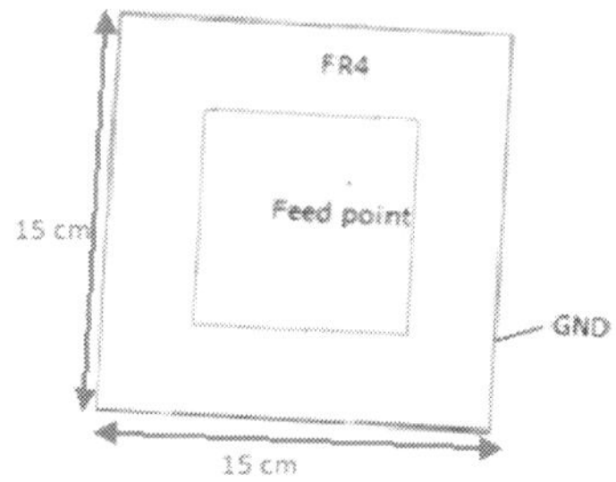

(a)

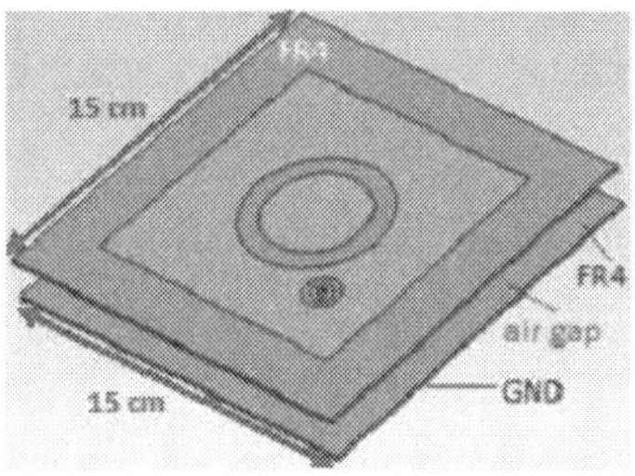

(b)

Fig. 1. Patch antennas: (a) conventional with single layer FR4 dielectric substrate and (b) proposed structure.

The optimisation of the different dimensions, such as the dimensions of the radiating elements, the height of the air gap and those of the slots, were carried out by using both CST microwave studio simulator and Ansoft HFSS simulator.
The surface current distribution, on the radiated elements, is shown in Fig. 2 at the resonant frequency of the antennas. A better distribution of the current density can be observed in the case of the proposed structure.

3 Experimental results

In order to compare and validate the properties of the proposed antenna structure, the double-sided FR4 substrate was used for the conventional patch antenna etching and the single-sided FR4 substrate for the construction of the new structure.

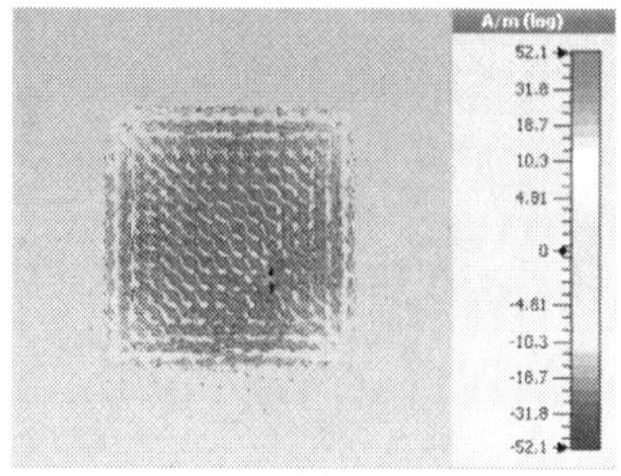

(a)

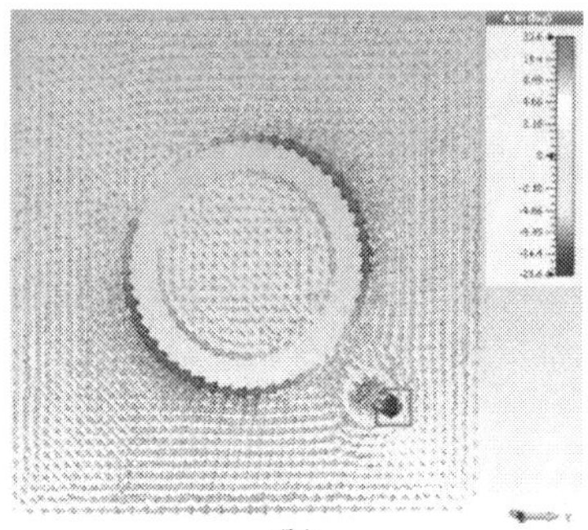

(b)

Fig. 2. Surface current distribution on the antennas radiating element : (a) conventional patch antenna and (b) proposed antenna.

The photos of the fabricated prototypes of the patch on single layer Fr4 and the proposed patch antenna are shown in Fig. 3 (a) and (b) respectively.
As shown in Figure 3, the two feed points are located along the diagonal line of the radiating elements. Their positions have been optimized to provide a circularly polarized radiation pattern and to allow an optimal matching of their input impedance.

The measurements of the both antennas are carried out in the ESEO anechoic chamber antenna testing setup.
The single layer antenna (conventional patch antenna) resonates at 866 MHz with the S_{11} value of -28.35 dB. The second antenna resonates at 862 MHz with the measured S_{11} value of -25 dB (Figure 4). The observed discrepancy between the measurement and the simulation can be partly explained by the precision of the proposed structure assembly.

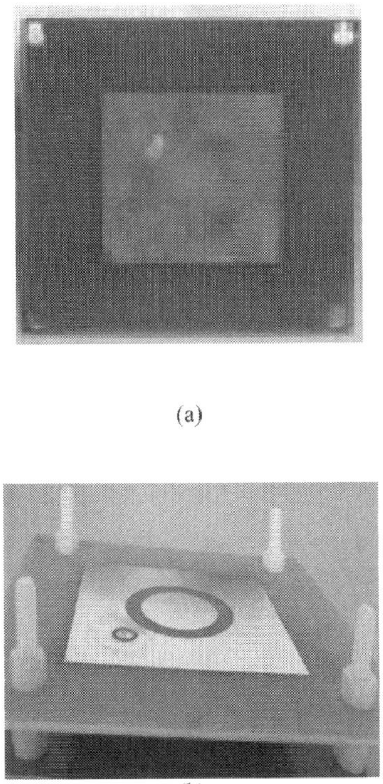

(a)

(b)

Fig. 3. The photo of the fabricated prototypes :
(a) Conventional antenna and (b) proposed structure.

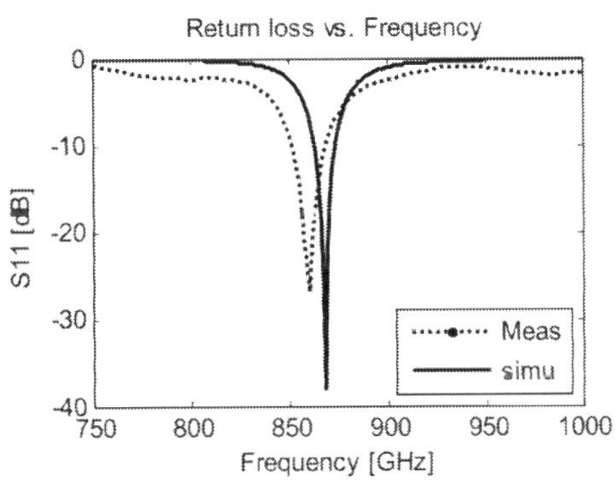

Fig. 4. Simulation and measurement comparison of the proposed antenna return loss vs. frequency.

The Fig. 5 and 6 show the measured E-plane and H-plane radiation patterns for the two fabricated antennas (Fig. 3). The E-plane – 3db beam of the conventional and the proposed antennas are 68.6 degree and 61.6 degree respectively. The H-plane – 3db beam of the conventional and the proposed antennas are 73.1 degree and 65.2 degree respectively.

As shown in Fig. 5 and 6, the proposed antenna radiation pattern is improved by 7.17 dB in E-plane and 6.53 dB in H-plane compared to conventional patch antenna radiation. Hence the directivity of the proposed antenna embedded with two slots is increased by 6 dB.

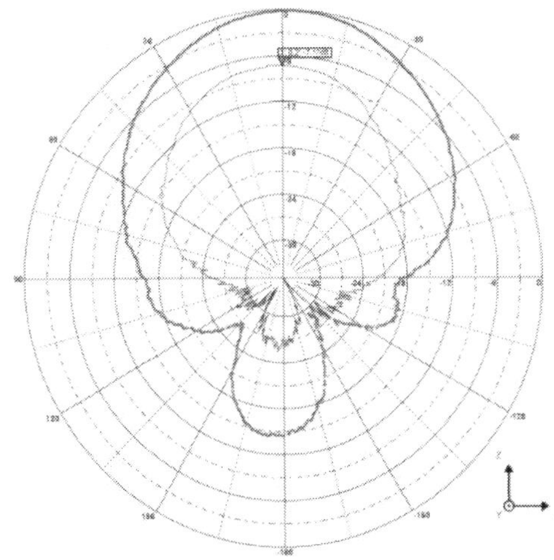

Fig. 5. Measured E-plane radiation patterns of the two antennas : (——) conventional patch antenna and (-----) proposed antenna structure.

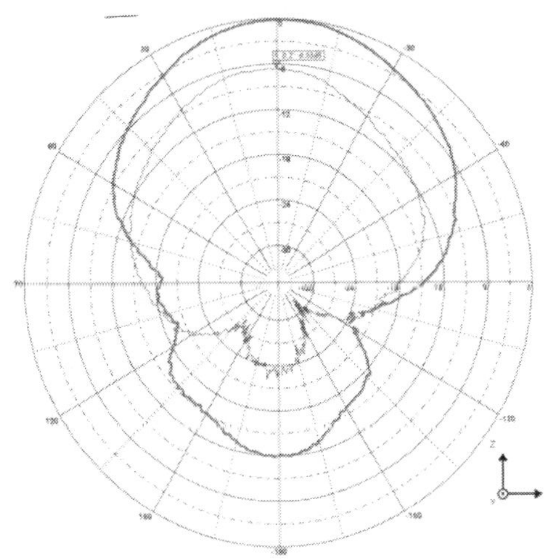

Fig. 6. Measured H-plane radiation patterns of the two antennas : (——) conventional patch antenna and (——) proposed antenna structure.

The simulation and measurement comparison of the proposed antenna gain versus frequency curves, in E and H planes, are shown in Fig. 7 and 8 respectively. Measurement results agree well with the simulation ones.

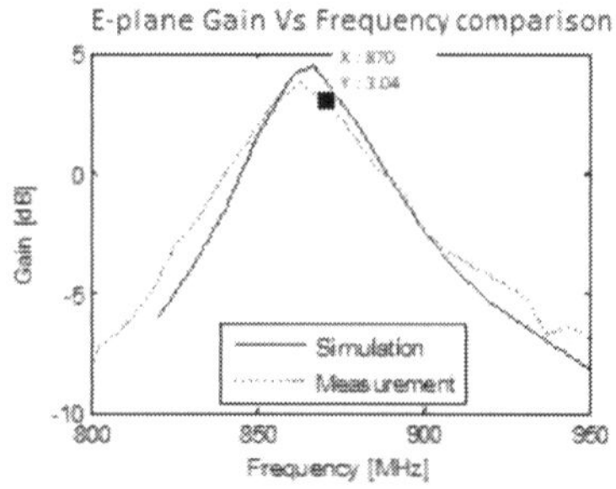

Fig. 7. Simulation and measurement comparison of the proposed antenna gain versus frequency in E plane.

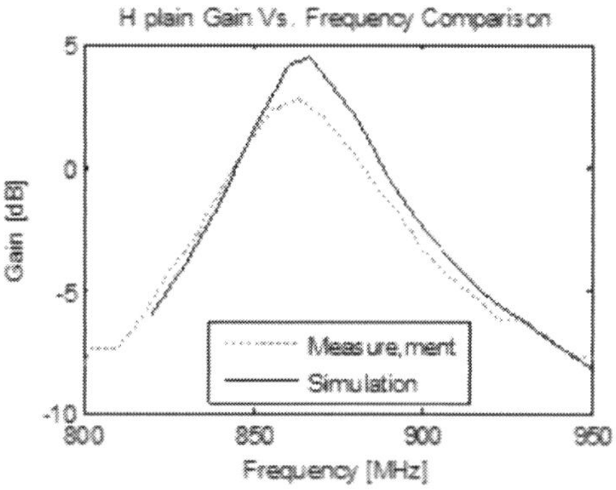

Fig. 8. Simulation and measurement comparison of the proposed antenna gain versus frequency in H plane.

Table 1 summarizes the obtained experimental results for the designed antenna structure. The various obtained values confirm the interesting characteristics of the antenna structure developed in this study.
Such a structure will make it possible to overcome certain constraints in the deployment of RFID in indoor.

Table 1. Experimental results of the proposed antenna

	Gain	HPBW	Back radiation
E plane	3.22 dBi	61.6°	-15.5 dB
H plane	3.01	65.2°	-15.8 dB

4 Conclusion

We have presented a technique to enhance the gain and directivity of patch antenna for EU UHF RFID reader applications. The proposed antenna size remains same of the conventional patch structure, taken as reference, but but increasing the height to an optimal value of 0.6 cm generates the gain enhancement of 7.17 dB in the E plain and 6.53 dB in the H plain. Hence this technique is a good candidate to enhance antenna gain and directivity without increasing the antenna volume too much as for example in the case of gain enhancement of patch antenna with superstrate layers. The measurement results show that the half power angles of the proposed patch antenna structure reduced by 7 degree in the E plane and 7.9 degree in the H plane compare to the original patch on single layer FR4. Hence the method is suitable to increase the directivity and gain of patch antenna without increasing the antenna size.
The developed structure can allow to overcome certain constraints in the deployment of RFID in indoor.

References

1. Lee, Y. J., J. Yeo, R. Mittra, and W. S. Park, Application of electromagnetic bandgap (EBG) superstrates with controllable defects for a class of patch antennas as spatial angular filter, IEEE Trans. Antennas Propagation., Vol. 53, No. 1, 224-235, Jan. 2005.
2. Shen, X.-H. and G. A. E. Vandenbosch, *Aperture field analysis of gain enhancement method for microstrip antennas,* Proc. 10th IEE Int. Conf. Antennas Propagat., Vol. 1, 186-189, Apr. 14-17, 1997.

MIMO Radar Hardware Acceleration with Enhanced Resolution

Eric Pitre[1], *Vincent* Roberge[1], *Joey* Bray[1], and *Mostafa* Hefnawi[1]

[1]Royal Military College of Canada, Department of Electrical and Computer Engineering, Canada

Abstract. This paper proposes a method for accelerating an enhanced resolution 3D (range, velocity, and azimuth) Multiple Input Multiple Output (MIMO) radar on a Graphics Processing Unit (GPU). Implementation of MIMO radars and the investigation of their performance are of interest to the research community. However, the MIMO mode of operation increases the computational requirements of the radar system and seldom permits real-time operation without performance compromises. Current methods for achieving reasonable frame rates include reducing the scope of the radar (i.e., limiting the number of dimensions, the field of view, or the ranges of interest), choosing efficient but coarse algorithms (i.e., the FFT for range, velocity, and bearing estimation), or offloading the computation on task specific hardware, DSP, or FPGA. The proposed framework enables real-time operation of the MIMO radar by performing the signal processing on a GPU without compromising the radar coverage, while replacing the widely used 3D FFT with an enhanced resolution alternative. The proposed framework is tested on a Frequency Modulated Continuous Wave (FMCW) MIMO radar using 8 transmitters, 8 receivers, and having a Coherent Processing Interval (CPI) of 256 chirps. The parallel implementation of the enhanced resolution signal processing yields an acceleration of 453.6x when compared to sequential execution on CPU.

1 Introduction

Multiple Input Multiple Output (MIMO), in the context of radar, can be interpreted as a different way of scanning a Field of View (FOV). Instead of relying on mechanically steering an antenna or adjusting the phase of an array to transmit and receive in a particular direction, the MIMO radar illuminates the entire FOV at once and performs the beamforming on receive using multiplexing techniques [1]-[2]. Although the computational demands are greater, the MIMO radar can simultaneously evaluate echoes across the FOV without having to go through the lengthy scanning process required by classical surveillance radars [1], [3].

1.1 Motivation

Using Software Defined Radios (SDRs), a MIMO radar was built and tested at the Royal Military College of Canada (RMC) in 2021 which could operate in different multiplexing modes and yielded a relatively high refresh rate when compared to similar works [3]. However, the signal processing time was still significant, taking approximately 1.25 seconds using a 3D Fast Fourier Transform (FFT). The radar took 308 milliseconds to collect the 256-chirp Coherent Processing Interval (CPI), meaning that the signal processing takes 4.1 times longer that the acquisition time, slowing the radar considerably [3]. Additionally, range and angular measurement were taken in [3] to quantify the performance of the of radar. In general, the resolutions were larger than predicted in theory.

This work aims to realize a high refresh rate MIMO radar which leverages the computational power of parallel processing, improving the radar in [3]. Additionally, the FFT-based range and azimuth estimations are replaced by other algorithms to increase the quality of the radar images. The goal is to implement an algorithm which generates high resolution images in less than 1 second per frame. Table 1 shows the resolution measurements of the radar in [3].

Table 1. Measurement Results, summarized from [3]

Type	Desired Resolution	Measured Resolution
Range	0.75 m	1.75 m
Velocity	0.1 m/s	0.2 m/s
Azimuth	2°	10.5°

2 Problem Formulation

In MIMO radar, multiplexing techniques are used so that N_{tx} transmitters can simultaneously emit their waveforms. The echoes are sampled and sorted by N_{rx} receivers which generate a virtual array containing $N_{tx} \times N_{rx}$ elements [1]-[2]. Figure 1 shows a 2×4 MIMO Uniform Linear Array (ULA) configuration where 4 receivers demultiplex the echoes from 2 transmitters and generate 8 virtual elements. Note that in the figure, $\varphi = \beta d \sin(\theta)$ where θ is measured from broadside.

MIMO radar prototypes have been implemented [3]–[7] which all take advantage of the virtual array, made possible by multiplexing the transmitter waveforms. Prototypes have varying levels of complexity either in their dimensionality (range, azimuth, elevation, and velocity) and the number of receiving and transmitting antennas. A common issue for MIMO radar prototypes, with the exception of [4] and [7], are the slow refresh rates due to the extra required computations. The signal processing execution times of [3]–[6] vary from 1.25 seconds to 11 seconds which may not be considered real-time depending on the application. The radar prototypes in [4] and [7] are only 2D radars (range and azimuth) where the required signal processing is more manageable. Despite this, [4] offloaded the computation tasks on DSP and FPGA to minimize the signal processing delay.

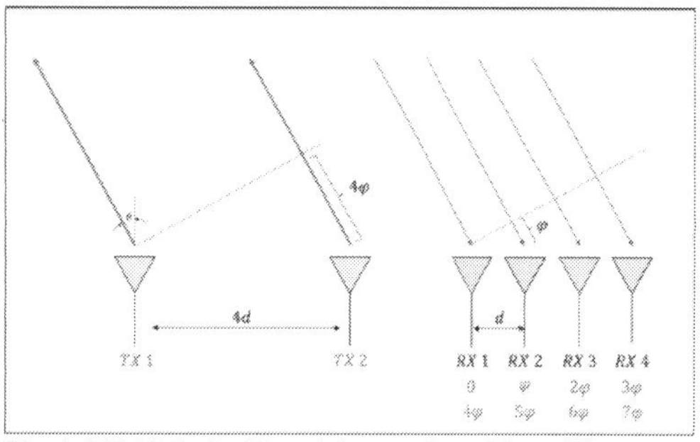

Fig. 1. MIMO Radar Virtual Array, adapted from [2]

2.1 Existing Signal Processing

In [3], the Frequency Modulated Continuous Wave (FMCW) radar uses a chirp time of 4.096 μs sampled at 250 MHz for a total of 1024 samples per chirp. With a CPI of 256 chirps and 64 virtual array elements, the collected data cube contains $1024 \times 256 \times 64$ or $\sim$16.7M complex samples. The 3D FFT (or any signal processing alternative) is performed on the data cube in order to extract the target range, velocity, and azimuth information. Following this, the 3D data cube is compressed into two 2D images to display the range-bearing and range-doppler graphs.

The 3D FFT is simply a sequence of 1D FFTs across the three dimensions of the cube. The range FFT is performed for each virtual channel and each chirp (total of 16,284 FFTs of size 1024). Since only the *positive* ranges are of interest, only half of the range FFT outputs are kept (data cube now has a size of $512 \times 256 \times 64$). Following this, the doppler FFT is performed for all range cells and all virtual channels (total of 32,768 FFTs of size 256). Finally, the FFT beamformer is performed across all range-doppler cells (total of 131,072 FFTs of size 64). Once the 3D FFT is completed, the range-doppler and range-bearing images are generated by compressing the data cube. The output of a range-doppler pixel, for example, is the power sum across all bearings of the range-doppler bin index, as expressed in (1). To generate the entire range-doppler image, (1) needs to be performed for all combinations of range and doppler bin indices.

$$Out[r,d] = \sum_{b=0}^{N_b-1} |x[r,b,d]|^2 \qquad (1)$$

where:

Out = output pixels of range-doppler image
r = range bin index
d = doppler bin index
b = bearing bin index
N_b = number of bearing bins
x = 3D data cube samples

Figure 2 shows the radar frame computation steps, as done on a sequential CPU. Figure 3 illustrates the computation of a range-doppler pixel of the *Cube Compression* function.

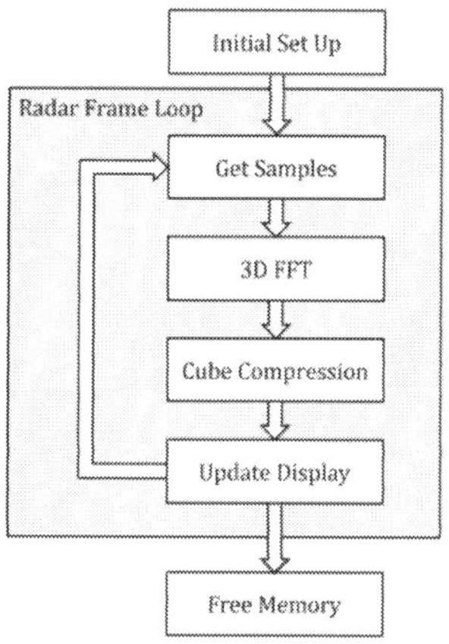

Fig. 2. Radar Frame Computations Steps using 3D FFT method

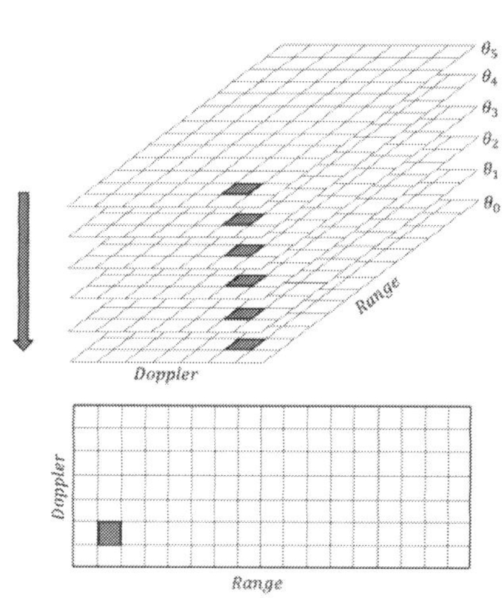

Fig. 3. Visualisation of Cube Compression for Range-Doppler Image

 https://doi.org/10.1051/itmconf/20224801013

3 GPU Architecture

Graphics Processing Units (GPU) were optimized for gaming applications where developers needed to increase the throughput of operations to generate high quality graphics [8]. Since the computations required to generate image frames are highly independent, a large amount of them could be performed concurrently in parallel [8]. The GPU is able to run several hundred to several thousand of threads simultaneously, and is composed of several building blocks containing Streaming Multiprocessors (SM) [5]-[6]. These SMs have the necessary components to perform highly threaded computations such as Streaming Processors (SP), Special Function Units (SFU), and Shared Memory. The SM is interconnected to a Global Memory which receives and transfers data to and from the host computer [8]. Figure 4 shows the architecture of a typical GPU.

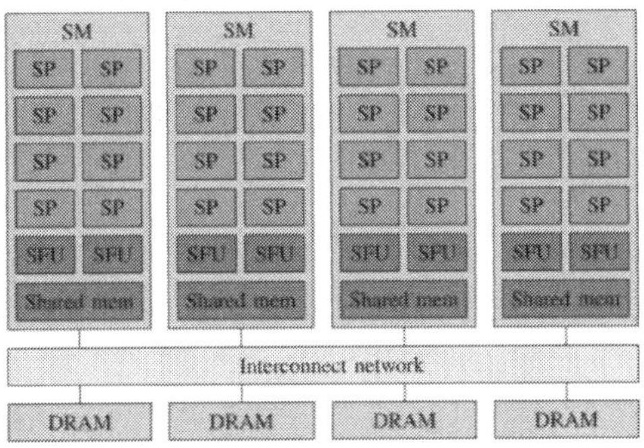

Fig. 4. GPU Architecture, reproduced from [10]

With Compute Unified Device Architecture (CUDA), a programmer can code sequential instructions which are to be executed on the CPU, and *kernels* which are functions that run in parallel on the GPU [9]. When a kernel is called from the CPU, threads are launched on the GPU which will execute the functions. Threads are organized in blocks and are mapped to SMs, which schedule and perform the parallel computations of up to 1024 threads per block [9]. All threads have a unique ID which locates them within the grid (all launched blocks form the grid).

4 Proposed Algorithm

In order to have more control over the algorithm and its parallel implementations, the existing GNU Radio 3D FFT-based signal processing chain implemented in [3] was entirely replaced by a C++ program.

4.1 Range Estimation

To increase the resolution of the range estimates, the FFT is replaced by the Chirp Z Transform (CZT) [11]. The CZT is a frequency analysis tool which is typically used as a *zoom* of the FFT. By selecting the frequency band of interest $f_0 \rightarrow f_1$, the CZT uses all N samples to generate an N-point DFT from f_0 to f_1. The CZT can be expressed as a convolution in time, which is then performed as a multiplication in the frequency domain [8]. Due to this, the FFT can be used for the transformations and its efficiency can be leveraged to reduce the execution time of the CZT. Figure 5 illustrates the CZT algorithm executed as a fast convolution, where $W = \exp(-j2\pi\Delta f)$

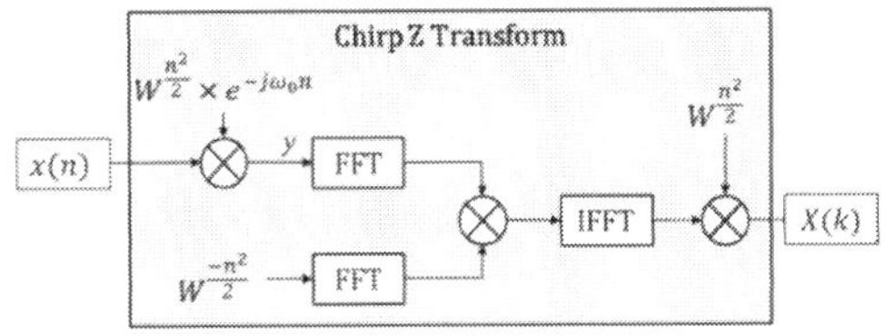

Fig. 5. Fast CZT Algorithm: Block Diagram

4.2 Azimuth Estimation

The Bartlett DOA, or traditional beamformer, is used as a replacement for the angular FFT. Although it is not a *high-resolution* DOA estimator, the calculations in (2) need to be performed for all angles of interest. Where an N-point spatial FFT provides N equally spaced frequency outputs, the Bartlett DOA performs the computation for M angles. Generally, M equally spaced angles are selected to generate the steering vectors [12], which contain coefficients $a[\theta, n] = e^{-j\beta d \sin(\theta_0 + n\Delta\theta)}$.

$$P[\theta] = \boldsymbol{a}[\theta]^H \boldsymbol{R}_x \boldsymbol{a}[\theta] \quad (2)$$

where:

θ = steering angle
$P[\theta]$ = Bartlett power spectrum, for given angle
$\boldsymbol{a}[\theta]$ = steering vector, for given angle
$\boldsymbol{R}_x$ = estimated correlation matrix
$[\ \ ]^H$ = complex transpose

Since the radar only provides a single snapshot for each range-doppler cells, the matrix multiplications in (2) can be simplified as (3) where x is the receiver input for N array elements:

$$P[\theta] = \left|\sum_{n=0}^{N-1} a[\theta, n] \cdot x[n]\right|^2 \quad (3)$$

4.3 Parallelization

The main parallelization opportunities lie within the sheer size of the data cube where each operation (range, velocity, and azimuth estimation) needs to be performed tens of thousands of times per radar frame. In the proposed algorithm, the final size of the processed cube is larger than the one for the existing solution. The CZT enables the use of all range samples, and the Bartlett DOA was configured to cover a FOV of $-90°$ to $90°$ with $1°$ increments. The proposed data cube has the following dimensions: $1024 \times 256 \times 180$ or ~47.2M complex samples. Due to the new size, the *Cube Compression* function will also take longer to execute.

Note that before any signal processing can be done, the raw data cube samples must be transferred from the CPU to the GPU. This step takes time and must be considered when measuring the acceleration.

4.3.1 Chirp Z Transform

As shown in Figure 5, the CZT can be done in five steps. Three of these steps are just element-by-element multiplication and the other two steps are the FFT and the IFFT. Algorithm 1 shows the pseudocode associated with the execution of the parallel CZT. Lines 1, 3, and 5 are dedicated kernels which perform the multiplication steps where each launched thread maps to a sample of the 3D data cube. Lines 2 and 4 are the FFT and IFFT steps of the algorithm which use cuFFT plans. The plans are executed across the entire data cube in the *range* direction.

Algorithm 1. Parallel CZT on GPU

```
1:  Multiply_Kernel_1( )
2:
3:  // Do FFT
4:  cufftExecZ2Z(FORWARD)
5:
6:  // Fast Convolution
7:  Multiply_Kernel_2( )
8:
9:  // Do IFFT
10: cufftExecZ2z(INVERSE)
11:
12: // Final Multiply
13: Multiply_Kernel_3( )
14:
15: // End of Parallel CZT Algorithm
```

4.3.2 Bartlett DOA

Since the Bartlett DOA can be expressed as (3) for the single snapshot case, parallelization becomes quite natural. For each angle of interest, two kernels must be created: one which performs multiplication of the input samples with the steering vector, and another which performs the sum across N samples for each range-doppler combination. Algorithm 2 shows the parallel GPU implementation of the Bartlett DOA. Line 1 indicates that the subsequent steps are performed for all angles of interest. For a FOV of 180 degrees with increments of 1 degree, the parallel *Multiply* and *Sum* kernels (line 4 and 7) are performed 180 times sequentially. Note that the *Multiply* kernel of Algorithm 2 is different than the kernels from the CZT.

Algorithm 2. Bartlett DOA on GPU

```
1:  for all angles of interest {
2:
3:        // Perform element by element multiplication
4:        Multiply_Kernel( )
5:
6:        // Perform Sum across channel direction
7:        Sum_Kernel( )
8:  }
9:
10: // End of Parallel Bartlett DOA
```

Figure 6 illustrates the execution of Algorithm 2 for a given angle θ. Note that the multiplication of the steering vectors is performed across the entire data cube. The cube is then compressed into a 2D sheet using the *Sum* kernel and stored in the output cube at the correct bearing location.

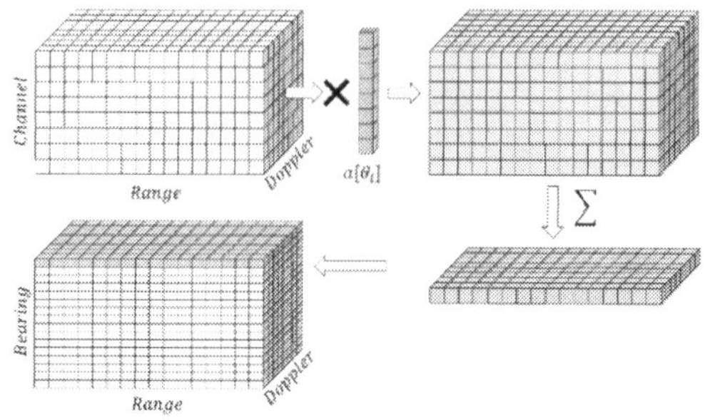

Fig. 6. GPU Implementation of Bartlett DOA

4.3.3 Cube Compression

As discussed in Section 2.1, the Cube Compression algorithm must generate the pixel values for the range-doppler and range-bearing images, which consists of performing many power sums. Since the calculation of a pixel is completely independent from the other pixels, each of them can be computed in parallel. Algorithm 3 shows the *Cube Compression* kernel, which is called form the CPU. In this kernel, each thread is mapped to a pixel of the range-doppler or range-bearing image. If the range-doppler image contains A pixels and the range-bearing image contains B pixels, a total of $A + B$ threads should be launched. Line 1 seeks to assign the thread ID to a variable. Lines 4 and 12 evaluate if the thread belongs to the range-doppler or the range-bearing image. Lines 7-8 is the power sum across all bearings, and lines 15-16 is the power sum across all doppler bins. The pixel outputs are stores in line 10 and 18 for the range-doppler and range-bearing images respectively.

Algorithm 3. Cube Compression Kernel

```
1:  id = thread ID within grid
2:
3:  // Do Range-Doppler Calculation
4:  if (id < A){
5:        find range and doppler bin indexes r and d
6:        temp = 0
7:        for all bearings b{
8:              temp = temp + |cube[r, d, b]|^2
9:        }
10:       out_d[r, d] = temp
11: }
12: if (id ≥ A and id < A + B){
13:       find range and bearing bin indexes r and b
14:       temp = 0
15:       for all doppler bins d{
16:             temp = temp + |cube[r, d, b]|^2
17:       }
18:       out_b[r, b] = temp
19: }
20:
21: // End of Cube Compression Kernel
```

4.3.4 GPU Implementation

Finally, the parallel GPU solution is inserted into the program. Once the radio samples have been collected, the raw data cube must be transferred to the GPU. From there, 3D signal processing is performed followed by Cube Compression. The range-doppler and range-bearing images are then transferred back to the CPU for display, completing the radar frame. Figure 7 shows the proposed implementation on GPU.

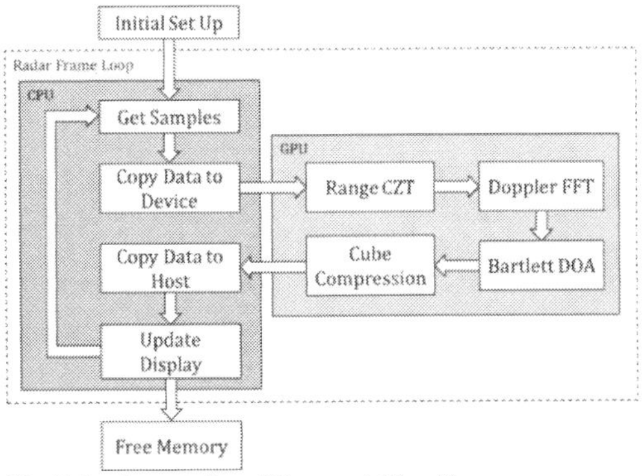

Fig. 7. Implementation of Proposed Algorithm

4.3.5 Parallel Work Efficiency

The parallel implementation of the proposed method is work efficient, where no additional computations on the GPU is required to execute the proposed algorithms. CuFFT is one of the most efficient FFT libraries available for the GPU. However, as with all GPU kernels, additional time is required to transfer the between the host and the device. To mitigate this overhead, the proposed solution limits the transfer of data to the beginning and the end of the radar frame calculations.

5 Simulation and Results

A simulated target echo environment was created to generate reflections for M targets with individual characteristics (i.e., range, speed, bearing, heading, RCS, etc.). To compare the image quality between the existing 3D FFT method and the proposed method, four targets with similar characteristics were simulated. The simulator outputs a raw data cube for the 8×8 MIMO radar with 256 chirps, each containing 1024 complex samples. This data cube is then processed to produce the radar images. As seen from Figure 8, the existing FFT method is not fine enough to resolve all four targets within the range-bearing image. The targets are clearly visible, however, when using the proposed algorithm.

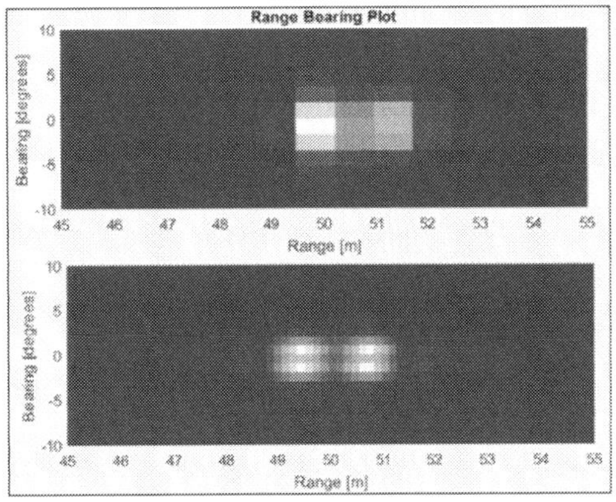

Fig. 8. Resolution Comparison: (above) 3D FFT Method (below) Proposed Algorithm

For speed up measurements, the proposed algorithm was performed sequentially, and in parallel on GPU. The workstation in fitted with two Intel Xeon Gold 5120 Processors and a NVIDIA GeForce RTX 2080 Ti GPU. The execution times for the proposed algorithm, the *Cube Compression*, and the memory transfer times are compared and displayed in Table 2. When implementing the proposed algorithm on GPU, an acceleration of 453.7x is achieved, compared to sequential execution on the CPU. The total execution time of the signal processing on GPU is 407.8 ms. For comparison, the existing 3D FFT method (including *Cube Compression*) takes approximately 3 seconds when done sequentially. In [3], GNU Radio somewhat leveraged the multiple cores of the CPU and was able to execute the 3D FFT and *Cube Compression* in 1.25 seconds.

Table 2. Execution Times and Speed Up

Process Name	**Execution Time (ms)**		**Speed Up**
	Sequential	GPU	
Proposed Algorithm	171,752	339.5	506x
Cube Compression	13,268	8.64	1536x
Memory Copy		59.7	
Overall	**185,020**	**407.8**	**453.7x**

6 Conclusion

In this paper, a parallel implementation of MIMO radar signal processing was proposed, which replaced the range and azimuth FFTs with the CZT and the Bartlett Beamformer respectively. By using a step-by-step approach which requires the computations to be performed simultaneously on the entire data cube, parallel execution on the GPU yielded a speed up of 453.7x when compared to the single-threaded sequential CPU implementation. The aim of producing high quality MIMO radar images in less than 1 second was achieved.

References

[1] J. Bergin and J. R. Guerci, *MIMO Radar Theory and Application*. Boston, MA, USA: Artech House, 2018.

[2] S. Sun, A. P. Petropulu, and H. V. Poor, "MIMO Radar for Advanced Driver-Assistance Systems and Autonomous Driving: Advantages and Challenges," *IEEE Signal Process. Mag.*, vol. 37, no. 4, pp. 98–117, Jul. 2020, doi: 10.1109/MSP.2020.2978507.

[3] R. T. Gilpin, "Real-Time Multiple Input Multiple Output (MIMO) Radar Using Sotware Defined Radio," Royal Military College of Canada, Kingston Ontario, 2021.

[4] W. Wang, D. Liang, Z. Wang, H. Yu, and Q. Liu, "Design and Implementation of a FPGA and DSP Based MIMO Radar Imaging System," vol. 24, no. 2, Art. no. 2, 2015.

[5] L. Sit, B. Nuss, S. Basak, M. Orzol, W. Wiesbeck, and T. Zwick, "Real-Time 2D+ velocity Localization Measurement of a Simultaneous- Transmit OFDM MIMO Radar using Software Defined Radios," Oct. 2016.

[6] A. Ganis, "Architectures and Algorithms for the Signal Processing of Advanced MIMO Radar Systems - CORE," Universita' Degli Studi Di Udine, 2018. Accessed: Mar. 09, 2022. [Online]. Available: https://core.ac.uk/display/195748925?recSetID=

[7] A. Figueroa, N. Joram, and F. Ellinger, "A fully modular, distributed FMCW MIMO radar system with a flexible baseband frequency," in *2021 IEEE Radar Conference (RadarConf21)*, May 2021, pp. 1–6. doi: 10.1109/RadarConf2147009.2021.9455174.

[8] D. B. Kirk and W. W. Hwu, *Programming Massively Parallel Processors*, Third. Cambridge, MA, USA: Elsevier, 2017.

[9] J. Nickolls, I. Buck, M. Garland, and K. Skadron, "Scalable parallel programming," in *2008 IEEE Hot Chips 20 Symposium (HCS)*, Aug. 2008, pp. 40–53. doi: 10.1109/HOTCHIPS.2008.7476525.

[10] G.-J. van den Braak, B. Mesman, and H. Corporaal, "Compile-time GPU memory access optimizations," in *Modeling and Simulation 2010 International Conference on Embedded Computer Systems: Architectures*, Jul. 2010, pp. 200–207. doi: 10.1109/ICSAMOS.2010.5642066.

[11] P. Rajmic, Z. Prusa, and C. Wiesmeyr, "Computational cost of Chirp Z-transform and Generalized Goertzel algorithm," in *2014 22nd European Signal Processing Conference (EUSIPCO)*, Sep. 2014, pp. 1004–1008.

[12] T. Yong and C. Wang, "Echo DOA Based High-resolution Target Location," in *2019 IEEE International Conference on Signal, Information and Data Processing (ICSIDP)*, Dec. 2019, pp. 1–4. doi: 10.1109/ICSIDP47821.2019.9173194.

Performance metrics for vehicular visible light communication systems

Fatima zahra Raissouni[1,2*], *Abdeljabbar* Cherkaoui[1], *José Luis* Lázaro Galilea[2], and *Alfredo* Gardel Vicente[2]

[1]Laboratory of Innovative Technologies (LTI), National School of Applied Sciences, Abdelmalek Essaâdi University, Tangier, Morocco.
[2]Department of Electronics, University of Alcalá, Alcalá de Henares, 28801 Madrid, Spain.

Abstract. Vehicular Visible Light communication (VVLC) presents a new paradigm for providing vehicle connectivity, increasing road safety, and achieving autonomous driving. It can be chosen as an alternative solution to radio frequency-based inter-vehicle communication systems and/or as a complementary solution to ensure redundancy. Nevertheless, there are still significant challenges in incorporating visible light communication systems into vehicular networks. The main purpose of this paper is to provide an outline of the performance metrics of a VVLC system. Furthermore, the study is performed based on a vehicle-to-vehicle dynamic model close to reality, considering the effect of geometrical changes in the LOS path and the variation of the inter-vehicular distance. The analysis of the proposed system is discussed in terms of Signal to noise (SNR), Field of view (FOV), mobility, and average capacity.

Keywords: Visible light communication (VLC), vehicular visible light communication (VVLC), vehicle-to-vehicle (V2V), mobility, performance metrics.

1 Introduction

Vehicular visible light communication is a new promising communication technology that sends data through light waves of the visible spectrum.It relies on the ability to modulate the light intensity of LEDs to communicate information and enables the dual use of headlamps and road traffic for both illumination and communication purposes [1]. Vehicular communication systems are essential in improving intelligent transportation systems (ITS). Specially, in the context of reducing the number of traffic accidents and related fatalities. Enabling real-time data exchange between vehicles and traffic infrastructures has been emerging as one of the key solutions for improving road safety, traffic efficiency, and driver comfort.

Compared to existing forms of wireless communications, VLC offers several advantages, such as huge and unregulated bandwidth, no interference with RF systems, better security in the physical layer, high energy efficiency, and low-cost implementation [2]. In addition, VLC brings a completely new field of possibilities to solve the problem of increasing mobile data traffic exponentially for decades, which has highlighted the limitations of RF-only mobile communications. In contrast, the visible light spectrum, which includes hundreds of terahertz of unlicensed bandwidth, is completely untapped for communication [3]. It can complement RF-based mobile communication systems to design high-capacity mobile data networks. It can also significantly increase the available wireless channel capacity [4],[5].

The VLC technology must enable reliable communications while maintaining the main purpose of the lighting device. Thus, the pattern of light radiation and the optical power output should be in accordance with the regulations, although their adjustment allows significant optimization.

Currently, the most extended version of VLC research is for the indoor environment Vehicular Visible light communications technology entails more research efforts.

To evaluate the performance of applications enabled by V2V communications, it is vital to determine the time distribution of the communication link between two vehicles. Furthermore, the conditional distribution of the link duration, such as vehicle speed, receiver field of view FOV [6], atmospheric effects, could be modeled to adjust its parameters as [7], so that a loss of connection is less likely to occur in the middle of an active transmission and affect the final performance.

In previous performance evaluation studies [8],[9] some basic performance metrics such as received optical power and signal-to-noise ratio (SNR) have been derived for different road surfaces for a Vehicle to infrastructure(V2I) communication system between a street/ traffic light and a vehicle. The experimental work in [10] has used commercial LEDs and photodiodes

* Corresponding author: Fatimazahra.raissouni@etu.uae.ac.ma

https://doi.org/10.1051/itmconf/20224801014

(PDs) to study the throughput and error rate of VLC-based V2V and V2I links. This work has proved the reliability of vehicular VLC.
Nevertheless, this technology remains in its beginning stage and requires additional research efforts in several areas, including channel modeling and physical layer design.
The study in [11] has evaluated the degradation of data rate, BER, and SNR performance of VLC systems in the presence of sunlight at different locations.
The VLC system performance on a LOS channel is examined in [12], [13].
The distribution of received power and signal-to-noise ratio (SNR) with intersymbol interference (ISI) of directed light is presented [14]. The bit error rate (BER) and outage performance of LOS VLC system is simulated for random orientation of the receiver in [15].

In order to design and implement a communication system, it is necessary to define and study the performance metrics of the overall system.
Up to now, there is no work that defines the metric performance of the outdoor systems based on visible light communication considering the key parameters of a realistic model. Most works do not consider the dynamic aspect of the outdoor model and the variation of angles or distance of the communicating cars.
The objective of this work is to fill this gap by presenting some key performance parameters based on a realistic outdoor VLC simulation model.

The remainder of this paper is organized as follows: first we present the system model for the vehicular visible light communication system, then we present and discuss the key performance metrics. In the third section, we present the simulation results. Finally, the conclusion is presented in the fourth section.

1.1 System model

The VLC system model proposed (presented in Fig. 1), consisting of a transmitter, a light-emitting diode is employed as the lighting source which converts the original electrical signal to the visible-light signal. Then, the optical signal is propagated in the VVLC channel. At the receiver, a photodiode (PD) is used to carry out the optical-electrical conversion.
To model, the point-to-point response of the VLC link, the three main blocks (VLC transmitter, optical wireless channel, and VLC receiver) are used.

.

Fig. 1. Illustration of a communication link between two vehicles.

The VLC channel is modeled using the classical path loss additive white Gaussian noise (AWGN) channel model [16].
In this model, the output current Y (t) generated by the photodiode (PD) is related to the transmitted optical power X(t) by:

$$Y(t) = X(t) \otimes h(t) + N(t) \quad (1)$$

where $h(t)$ denotes the channel impulse response.
Based on the Lambertian emission, and it can be modeled (fig.2 shows a simplified model-based LOS path between two vehicles) by [16] :

$$h(t) = \frac{(m+1)A}{2\pi(\mathrm{d}(t))^2}\cos^m(\varphi)T_s(\theta)g(\theta)\cos(\theta) \times \delta\left(t - \frac{\mathrm{d}(t)}{c}\right) \quad (2)$$

Where A is the physical area of the PD, d(t), φ, and θ are the time-varying distance, the angle of irradiance, and the angle of incidence from the transmitter to the receiver, respectively. θ is also known as the field of view of the PD, and $\delta(t - \frac{\mathrm{d}(t)}{c})$ is the propagation delay of the signal.
The inter-vehicular distance between the transmitter and receiver can be expressed by:

$$d = \sqrt{{d_L}^2 + {d_h}^2} \quad (3)$$

Where d_L and d_h are the lateral and the longitudinal distances between the transmitter and the receiver respectively. d_h is directly related to the speed as:

$$d_L(t) = tv(t)\cos(\varphi) \quad (4)$$

$T_s(\theta)$ is an optical filter gain used to block the natural and artificial light signals, g(θ) is the concentrator gain is given as:

$$g(\theta) = \begin{cases} \frac{n^2}{sin^2(\theta)}, 0 \le \theta \le FOV \\ 0,\ \theta > FOV \end{cases} \quad (5)$$

in which m is the order of Lambertian model specifying the directivity of the transmitter and obtained by: $m = -\frac{\ln(2)}{\ln(\cos(\varphi_{1/2}))}$, and $\varphi_{1/2}$ is the half-power angle of the radiation and n indicates lens refractive index of PD
The average received optical power P_r is computed by:

$$P_r = H.Pt \quad (6)$$

Where H is the VLC channel gain [17]:

$$H = \frac{(m+1)A}{2\pi({d_L}^2 + {d_h}^2)}\cos^m(\varphi)T_s(\theta)g(\theta)\cos(\theta) \quad (7)$$

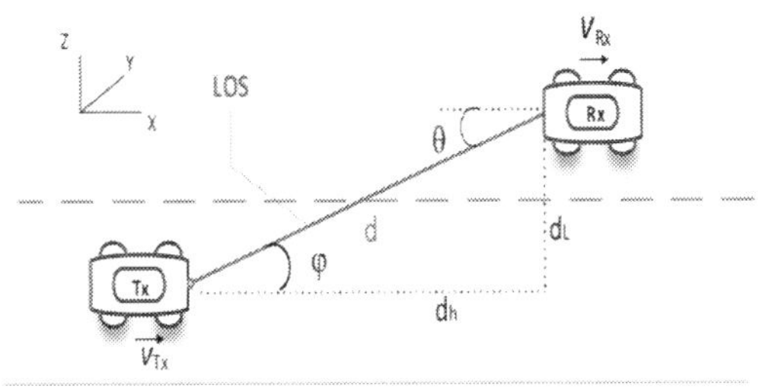

Fig. 2. The V2V system model

2 Performance metrics analysis

In this section, we will propose, evaluate and discuss the main performance measures and their impact in the VVLC system.

2.1 Signal to Noise Ratio Analysis (SNR)

A basic specification used to measure the quality of communication is the signal-to-noise ratio (SNR). We have analyzed the signal-to-noise ratio (SNR), at the detector of the VVLC system.

The quality of the transmission in optical channels is mostly dominated by the shot noise. The shot noises consist mainly of shot noise caused by signals and thermal noise induced by intense ambient light. We assume that the noises are additive white Gaussian noises (AWGN).

The received electrical power S and the SNR are given by:

$$S = \gamma^2 P_{rSignal}^2 \tag{8}$$

where $\boldsymbol{\gamma}$ is the responsivity of the PD. The desired signal power $P_{rSignal}$ is given by:

$$P_{rSignal} = \int_0^T \left(\sum_{i=1}^{LEDs} h_i(t) \otimes X(t)\right)dt \tag{9}$$

The equalizer output contains a Gaussian noise having a total variance N that is the sum of contributions from shot noise, thermal noise [17].

$$N = \sigma_{shot}^2 + \sigma_{thermal}^2 \tag{10}$$

According to [17] the shot and thermal-noise variances are given as:

$$\sigma_{shot}^2 = 2q\gamma(P_{rSignal})B + 2qI_{bg}I_2B \tag{11}$$

where q is the electronic charge (1.602×10−19 C), B is equivalent noise bandwidth, Ibg is the received background current.

And we have defined the noise bandwidth factors I2 = 0.562 [18]. The variance of the thermal noise is given by:

$$\sigma_{thermal}^2 = \frac{8\pi K T_K}{G}\eta A I_2 B^2 + \frac{16\pi^2 K T_K \Gamma}{g_m}\eta^2 A^2 I_3 B^3 \tag{12}$$

Where k is Boltzmann's constant, ,G is the voltage gain in open loop, TK is absolute temperature, η is the constant capacitance of photo detector per unit area, Γ is the FET channel noise factor, A is the detection area of the PD, g_m is the FET transconductance, and $I_3 = 0.0868$ [19]. the noise bandwidth factor. In our numerical examples, we choose the following parameter values [17]: T = 298 K, γ = 0.54 A/W, G = 10, gm = 30 mS, Γ = 1.5, η = 112 pF/cm^2 (Table1).

Lastly, the electrical SNR at the receiver, which is a key metric for measuring the quality of the communication link, can be determined by:

$$SNR = \frac{S}{N} = \frac{(\gamma P_{rSignal})^2}{\sigma_{shot}^2 + \sigma_{thermal}^2} \tag{13}$$

And we assume the background current from direct sum light. The distribution of SNR is shown in Fig. 6 and Fig.7.

2.2 Average Capacity

The average capacity can be used to effectively evaluate the performance of VVLC. For a communication system, it can be defined as the maximum data rate that can be achieved over the channel of the system within a certain bandwidth, considering the noise effect. Unlike conventional radio frequency (RF) communication, VLC adopts the intensity modulation/direct detection (IM/DD) scheme, in which the transmitted information is modulated on the intensity of light. Therefore, the transmitted signal must be real, not negative, and the Gaussian distribution that achieves the conventional channel capacity of RF cannot be used in VLC channels [20].

By taking the noise parameters of equation (9), the system capacity C in bits/s, can be evaluated directly by:

$$C = B\log_2(1 + SNR) \tag{14}$$

The average channel capacity, C_{av} . Mathematically, the average capacity of this system can be calculated as:

$$C_{av} = \int_0^\infty \log_2(1 + \mu) f_\mu(\mu) d\mu \tag{15}$$

where μ is the average SNR and $f_\mu(\mu)$ is the PDF of the perturbed system.

$\cos(\varphi)$ and $\cos(\theta)$ represents the effect of receiver vehicle location on channel gain. Based on our model presented in fig.2, Their expressions can be written as:

$$\cos(\varphi) = \cos(\theta) = \frac{d_h}{d} \tag{16}$$

Further in order to obtain a statistical model of channel gain (h), the vehicle receiver position is considered to be random and assuming d_L to be following a uniform distribution with PDF [21]:

$$f_{V_r}(d_r) = \frac{2d_r}{{d_r}^2} \tag{17}$$

The average capacity of the VLC link can be calculated from:

$$C_{av} = \int_{t_{min}}^{t_{max}} \log_2(1+\mu) f_\mu(\mu) d\mu \quad (18)$$

Where: $t_{min} = \frac{K^2}{(R^2+d_h{}^2)^{m+3}}$, and $t_{max} = \frac{K^2}{(d_h{}^2)^{m+3}}$.

By replacing (16) in (7), the channel gain is given by [21]:

$$H = \frac{(m+1)AT_s(\theta)g(\theta)d_h{}^{m+1}}{2\pi(d_L{}^2+d_h{}^2)^{\frac{m+3}{2}}} \quad (19)$$

The PDF of the channel gain can be expressed as [21], [22]:

$$f_H(h) = \frac{2}{(m+3)R^2} K^{\frac{2}{m+3}} h^{-\frac{2}{m+3}-1} \quad (20)$$

Where: $k = \frac{(m+1)AT_s(\theta)g(\theta)d_h{}^{m+1}}{2\pi}$, is assumed to simplify our analysis.
Consequently, the PDF of H^2, can be obtained as follows:

$$f_{H^2}(h) = \frac{-K^{\frac{2}{m+3}}}{(m+3)R^2} h^{-\frac{m+5}{m+3}} \quad (21)$$

Substituting Equation (21) in Equation (18), Equation (22) is obtained:

$$C_{av} = \frac{1}{2\log[2]} \int_{t_{min}}^{t_{max}} \log\left(1 + \frac{P_{r\text{Signal}}}{N^2} h\right) \left(\frac{-K^{\frac{2}{m+3}}}{(m+3)R^2} h^{-\frac{m+5}{m+3}}\right) \quad (22)$$

Figure 5 presents the average capacity of the system with respect to the source power.

2.3 Effect of mobility

The FOV is an important factor for defining the visible light range at the receiver.
In visible-light communication, the interferences depend on a data rate and the FOV of receiver since a transmitter should have a wide irradiation angle for the lighting function [23]. Different VLC receivers have various field-of-view angles.
Optical transmissions arrive outside of the field-of-view angle of the receiver are generally not received as the receiver power is negligible.

As the field of view of the VLC receiver has a significant influence on the communication parameters, the performance of the VLC was evaluated for a variable field of view from 20° to 80°.

In this section, we will discuss the relation between received power, FOV, and the variable distance between the two communicating cars.
Figures 2 and 3 show the relation between FOV and received power and distance.

3 Simulation results and discussion

In this section, we conducted simulations of the vehicle-to-vehicle key performance metrics.
The performance of the different V2V-VLC system configurations is discussed in this subsection in terms of key performance metrics.

The main simulation parameters are illustrated in Table1.

Table 1. System model parameters.

Parameters	**Symbol**	**Value**
FOV of the PD	θ	70
PD area	A_r	1 (cm^2)
Order of Lambertian diffuser	m	1
Luminous efficacy of radiation	LER	250.3 (lm/W)
Responsivity of PD	γ	0.54 (A/W)
Electronic charge	q	1.6×10^{-19} (C)
Received background noise current	I_{bg}	5100 (μA)
Noise bandwidth factor	I_2	0.562
Boltzmann's constant	K	1.38×10^{-23} (J/K)
Absolute temperature	T_k	298 (K)
Open-loop voltage gain	G	10
FET channel noise factor	Γ	1.5
FET transconductance	g_m	30 (mS)
System bandwidth	B	2 (MHz)
Fixed capacitance of PD per unit area	η	112 (pF/cm^2)

Figure 3 depicts Pr and distance for different values of FOVs.

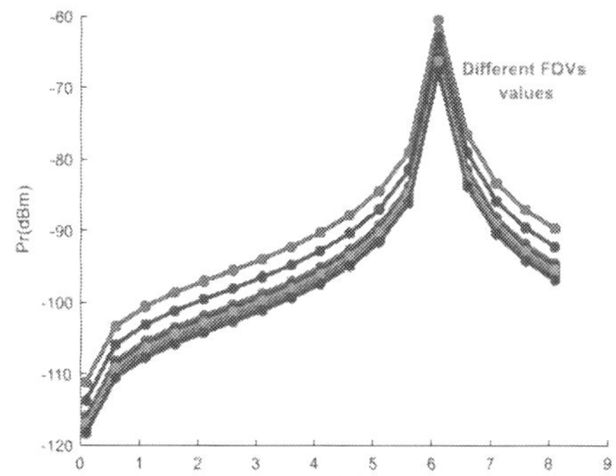

Fig.3. Power received vs. distance for different FOVs values

Figure 4 shows the effect FOV on the distribution of received optical power. Clearly that with increasing FOV, received optical power decrease.

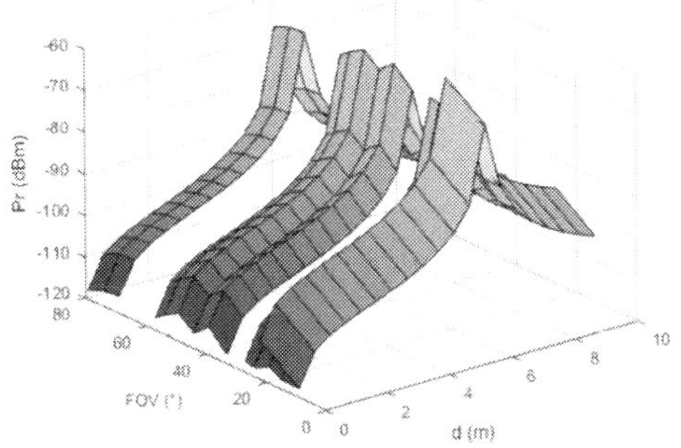

Fig. 4. The distribution of Power received vs. distance for different FOVs values

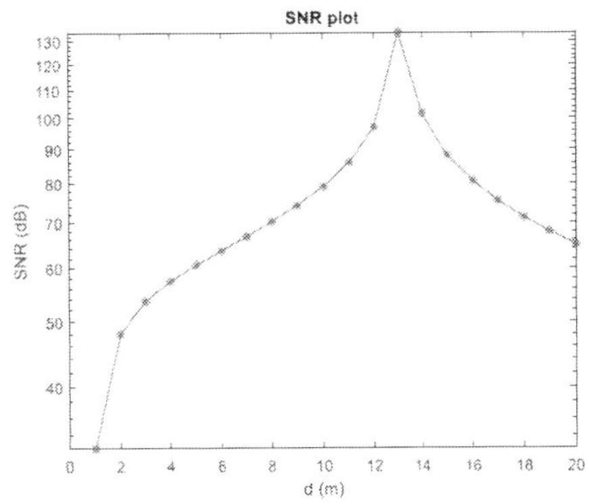

Fig. 6. The relation between the SNR and the distance

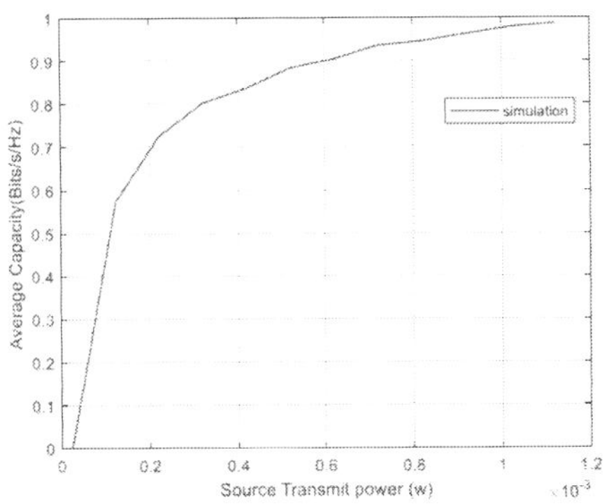

Fig. 5. Average capacity for VVLC based system with respect to source transmit power

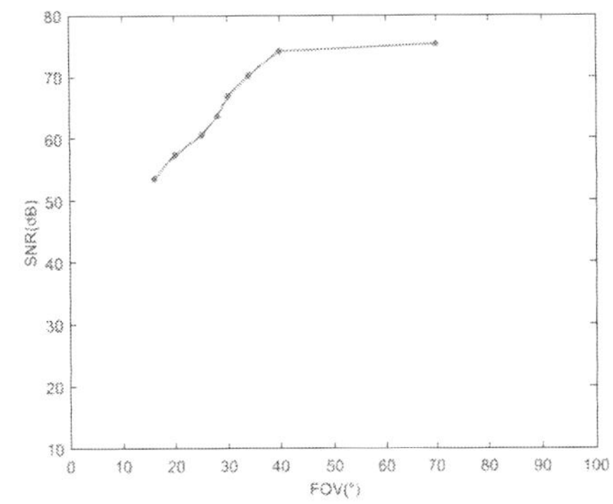

Fig. 7. The relation between the SNR and FOV

In Figure 5 the average capacity of the system is plotted with respect to the source power. As can be seen, with the increase of the source transmit power, the average capacity increases. This indicates that performance system becomes more evident due to the higher transmission power source.

On the other side, Figure 6 shows the SNR against the relative distance between two vehicles.

As the SNR decreases with the increase of perception rate. The SNR has very close behavioral curves at small distances. The maximum SNR that a vehicle can receive in the case of variable speed is 133.75 dB at 13 m ranging.

Ultimately, Fig 7 shows the dependence of SNR on FOV. So, when FOV is smaller than 20 deg., the blind area exists. However, up to 40 deg the SNR has increased until 70 deg. From this figure, with respect to SNR performance, it was found that for very narrow and larger FOVs, lower SNR values were found.

The effect of changes in the dynamic behavior of the system parameters is also included for all parameters studied. The discussion began with Figure 2, where the total received power performance at the receiver was analyzed by varying the FOV.

The performance of automotive VLC receivers can be significantly improved by making the FOV dynamic and adaptable by following the position of the sensor and the receiving vehicle. Because a narrow FOV also restricts the system's mobility, as useful light may also be prevented from reaching the VLC receiver.

However, the narrowing of the FOV also reduces the amount of optical noise (i.e., from artificial light or sunlight) that reaches the PD detector, and it can be considered among the simplest solutions to avoid the saturation of the photosensitive element is therefore improve the signal-to-noise ratio. Another advantage is that it can increase the dominance of the line of sight (LOS).

This means that an appropriate FOV adjustment is very critical to achieve good illumination and higher data rate in a V2V-VLC link. Therefore, the problem of maintaining a direct line of sight between vehicles is very challenging and needs further study.

Through figure 5, it is noticeable that the capacity improves as the input power increases. On the other hand, through fig.6 a performance degradation is clearly visible when the distance varying between the Tx and the Rx increases.

4 Conclusion

In this paper, we have investigated the key performance metrics of V2V-VLC link. The proposed system performance is measured by its power receiver (Pr), Signal to Noise Ratio (SNR), field-of-view (FOV), mobility and average capacity.
We discussed those requirements and investigated their impact in the system performance. The results simulations showed that the performance of the proposed system is affected by variations in: the field of view, mobility (variable distance inter-vehicles) and input power.
The study presented in this paper was performed based on one specific scenario. We will further investigate and develop the performance metrics considering other scenarios with the impact of disturbing elements such as atmospheric effects and re-specify and confirm the approach in future work.
Furthermore, we intend also to find the optimal values of different metrics in various situations specific to outdoor VLC in order to maximize the system performance.

References

1. A. Memedi and F. Dressler, "Vehicular Visible Light Communications: A Survey," IEEE Commun. Surv. Tutor., vol. 23, no. 1, Art. no. 1, 2021.
2. S. U. Rehman, S. Ullah, P. H. J. Chong, S. Yongchareon, and D. Komosny, "Visible Light Communication: A System Perspective—Overview and Challenges," Sensors, vol. 19, no. 5, Art. no. 5, Jan. 2019.
3. P. H. Pathak, X. Feng, P. Hu, and P. Mohapatra, "Visible Light Communication, Networking, and Sensing: A Survey, Potential and Challenges," IEEE Commun. Surv. Tutor., vol. 17, no. 4, Art. no. 4, Fourthquarter 2015.
4. A. Jovicic, J. Li, and T. Richardson, "Visible light communication: opportunities, challenges and the path to market," IEEE Commun. Mag., vol. 51, no. 12, Art. no. 12, Dec. 2013.
5. G. Pang, T. Kwan, C.-H. Chan, and H. Liu, "LED traffic light as a communications device," in Proceedings 199 IEEE/IEEJ/JSAI International Conference on Intelligent Transportation Systems (Cat. No.99TH8383), Oct. 1999, pp. 788–793.
6. H. B. Eldeeb and M. Uysal, "Vehicle-to-Vehicle Visible Light Communication: How to select receiver locations for optimal performance?," in 2019 11th International Conference on Electrical and Electronics Engineering (ELECO), Nov. 2019, pp. 402–405.
7. M. Karbalayghareh, F. Miramirkhani, H. B. Eldeeb, R. C. Kizilirmak, S. M. Sait, and M. Uysal, "Channel Modelling and Performance Limits of Vehicular Visible Light Communication Systems," IEEE Trans. Veh. Technol., vol. 69, no. 7, pp. 6891–6901, Jul. 2020.
8. M. Akanegawa, Y. Tanaka, and M. Nakagawa, "Basic study on traffic information system using LED traffic lights," IEEE Trans. Intell. Transp. Syst., vol. 2, no. 4, Art. no. 4, Dec. 2001.
9. S. Kitano, S. Haruyama, and M. Nakagawa, "LED road illumination communications system," in 2003 IEEE 58th Vehicular Technology Conference. VTC 2003-Fall (IEEE Cat. No.03CH37484), Oct. 2003, vol. 5, pp. 3346-3350 Vol.5.
10. C. B. Liu, B. Sadeghi, and E. W. Knightly, "Enabling vehicular visible light communication (V2LC) networks," in Proceedings of the Eighth ACM international workshop on Vehicular inter-networking, New York, NY, USA, Sep. 2011, pp. 41–50.
11. M. S. Islim and H. Haas, "An investigation of the solar irradiance effect on visible light communications," in 2017 IEEE 28th Annual International Symposium on Personal, Indoor, and Mobile Radio Communications (PIMRC), Oct. 2017, pp. 1–6.
12. P. Luo, Z. Ghassemlooy, H. Le Minh, E. Bentley, A. Burton, and X. Tang, "Fundamental analysis of a car to car visible light communication system," in 2014 9th International Symposium on Communication Systems, Networks & Digital Sign (CSNDSP), Manchester, UK, Jul. 2014, pp. 1011–1016.
13. P. Luo, Z. Ghassemlooy, H. L. Minh, E. Bentley, A. Burton, and X. Tang, "Performance analysis of a car-to-car visible light communication system," Appl. Opt., vol. 54, no. 7, Art. no. 7, Mar. 2015.
14. M. Uysal, Z. Ghassemlooy, A. Bekkali, A. Kadri, and H. Menouar, "Visible Light Communication for Vehicular Networking: Performance Study of a V2V System Using a Measured Headlamp Beam Pattern Model," IEEE Veh. Technol. Mag., vol. 10, no. 4, pp. 45–53, Dec. 2015.
15. M. Irfan et al., "Optical-Interference Mitigation in Visible Light Communication for Intelligent Transport Systems Applications," Energies, vol. 13, no. 19, Art. no. 19, Jan. 2020.
16. J. M. Kahn and J. R. Barry, "Wireless infrared communications," Proc. IEEE, vol. 85, no. 2, pp. 265–298, Feb. 1997.
17. T. Komine and M. Nakagawa, "Fundamental analysis for visible-light communication system using LED lights," IEEE Transactions on Consumer Electronics, vol. 50, no. 1, pp. 100–107, Feb. 2004.
18. T. Komine, J. H. Lee, S. Haruyama, and M. Nakagawa, "Adaptive equalization system for visible light wireless communication utilizing multiple white LED lighting equipment," IEEE Trans. Wirel. Commun., vol. 8, no. 6, pp. 2892–2900, Jun. 2009.
19. I. E. Lee, M. L. Sim, and F. W. L. Kung, "Performance enhancement of outdoor visible-light communication system using selective combining receiver," IET Optoelectron., vol. 3, no. 1, pp. 30–39, Feb. 2009.

20. A. Chaaban and S. Hranilovic, "Capacity of optical wireless communication channels," Philos. Trans. R. Soc. Math. Phys. Eng. Sci., vol. 378, no. 2169, p. 20190184, Apr. 2020.
21. A. Gupta, N. Sharma, P. Garg, and M.-S. Alouini, "Cascaded FSO-VLC Communication System," IEEE Wirel. Commun. Lett., vol. 6, no. 6, pp. 810–813, Dec. 2017.
22. J.-Y. Wang, Y. Qiu, S.-H. Lin, J.-B. Wang, M. Lin, and C. Liu, "On the Secrecy Performance of Random VLC Networks With Imperfect CSI and Protected Zone," IEEE Syst. J., vol. 14, no. 3, pp. 4176–4187, Sep. 2020.
23. S.-A. Avătămăniţei, C. Beguni, A.-M. Căilean, M. Dimian, and V. Popa, "Evaluation of Misalignment Effect in Vehicle-to-Vehicle Visible Light Communications: Experimental Demonstration of a 75 Meters Link," Sensors, vol. 21, no. 11, Art. no. 11, Jan. 2021.

Performance Evaluation of OTFS and OFDM for 6G Waveform

Abderrahim Mohammadi[1*], *Saad Chakkor*[1], *Ahmed El Oualkadi*[1], *Mohamed Moussaoui*[1], *Aziz Dkiouak*[2] and *Mostafa Baghouri*[3]

[1]LabTIC, ENSA of Tangier, University of Abdelmalek Essaâdi, Tangier, Morocco
[2]Laboratory of sciences and advanced technology, ENSA of Tetouan, University of Abdelmalek Essaâdi, Tetouan, Morocco
[3]Systems Team, Structural Engineering, Intelligent Systems and Electrical Energy Laboratory, ENSAM of Casablanca, University of Hassan II, Morocco

Abstract. Orthogonal time frequency space (OTFS) modulation is a coming technique addressing the challenges of 5G and 6G. It's appeared to bring about outstanding error performance in delay – Doppler (DeDo) channels. Still, a really fair comparison with its competitor's direct orthogonal frequency division multiplexing and is extensively used in this participation, we will make a comparison of the performance of OTFS and OFDM with the digital modulation format in terms of bit error rate (BER) versus signal-to-noise rate (SNR) for the 16- QAM technique. From the numerical simulations under MATLAB, the results showed that the OTFS admits a small BER to the OFDM in given cases

1 Introduction

New generations of wireless communications are expecting to deliver efficient service with lower latency in high-speed environments. Recently, OTFS modulation has been proposed an effective solution for achieving a trusted wireless communication service [1]. That is because of an optimum choice for vehicular communication at high speed caused by its rise susceptibility to inter-carrier interference, which is attributed to Doppler spreading [2]. OTFS is an extended generalization of classic time and frequency modulation techniques such as time division multiplexing (TDM) and frequency division multiplexing (FDM) and, from a stronger perspective, it conceptually links radar to communications. OTFS waveform technologies are combined with the wireless channel to directly capture hidden physics, leading to a microwave radar image with greater reliability of the constituent reflectors. Therefore, the selected time-frequency channel is regenerated in an associate degree unchanged, divisible, and orthogonal interaction, whither all admitted QAM symbol data undergo the identical specific deterioration, and all delay-Doppler variety parts are summed in a cohesive way. [3].
Basically, OTFS has emerged to overcome the drawbacks that OFDM could not cover. In short, OFDM is a digital modulation technique that uses digital data by encoding onto a number of orthogonal subcarriers. OFDM has become a popular high-bandwidth digital communication technology, applied in many fields for instance digital television, audio transmission service, digital subscriber line (DSL) web access, wireless network, electrical connection, fiber optic networks, and 4G/5G mobile telephony. [4].

Although both, OTFS and OFDM serve the same purpose for QAM method, but technically are different. Studies have demonstrated that the results of [5] show OTFS is stronger Technology against Doppler compared to OFDM. Also, OFDM update for inferior Doppler environments is not effective big for these applications [5]. Furthermore, OFDM is severely limited by synchronization errors, which is an ancient discovery. In comparison, we get that OTFS has great flexibility to such a large residual timing error though it is not totally impervious to it. Thus, it can be understood that the time domain channel estimation provides acceptable flexibility to OTFS against timing errors during compensation [6]. In addition to that, the proportion resource utilized for drivers and general losses in OTFS is lower than that of OFDM, and therefore the spectrum efficiency (SE) of OTFS is greater from of OFDM [6].
Besides, the user data of OTFS (constellation of data symbols) is arranged in the DeDo interval unlike the frequency-time (TF) interval in OFDM. Therefore, the data is propagated through the TF domain using a unitary transformation. It is pursued by the modulation OFDM [7]. The prefix of cyclic [7] or OFDM block [8] is provided to dampen the channel delay spread.
Not to mention that OFDM is not an optimal choice for high-rapidity vehicular communications due to its greater susceptibility to interference between inter-carriers, this is a consequence of Doppler propagation. However, the subcarrier bandwidth in OFDM can be increased to decrease the inter-carrier interference at the cost of spectral efficiency loss.
In the [6], authors show that reduced CP OTFS signals can be generated by interleaving the signal of a block OFDM system. The performance of OTFS in AWGN

* A Mohammadi: mohammadiabde@gmail.com

channel is shown, where it is found to be identical with a single carrier as well as OFDM. Moreover, OTFS gives diverse gains in different situations with high-frequency dispersion. Such situations are facing, For example, in millimeter wave systems, both due to the higher phase noise and higher Doppler propagation are encountered there. [7]

In this work, in the first part, we will talk about OTFS modulation. In the second section about ODFM modulation and in the third part, we will assess the yield of OTFS and OFDM codes modulation applied QAM symbols. This comparison is done in terms of BER vs SNR for 16-QAM, and finally, we will give a conclusion and a perspective to our future vision.

2 Modulation OTFS

Considering OTFS modulation when M is a number of subcarriers such that subcarrier bandwidth is Δf and number N of symbols with symbol duration is T. the total bandwidth is given by B = M Δf and the total duration Tf = NT. Additionally, if TΔf = 1 the OTFS technique in the case of a critical sample . [11]

According the [1, 2, 6] Quadrature Amplitude Modulation (QAM) modulated the data symbols x(k,l) on the NxM array are coordinated in the Doppler delay range. It is represented on an N × M grid, X(n, m) in the (TF) interval according to the use of the inverse Symplectic Fourier transform (ISFFT),

$$X(n,m) = \frac{1}{NM}\sum_{k=0}^{N-1}\sum_{m=0}^{M-1} x(k,l)e^{j2\pi(\frac{nk}{N}-\frac{ml}{M})} \quad (1)$$

Then the signal X(n, m) in the time interval is changed to x(t) from the Heisenberg transform as follows,

$$x(t) = \sum_{k=0}^{N-1}\sum_{m=0}^{M-1} X(n,m)g_{tx}(t-nT)e^{j2\pi m\Delta f(t-nT)} \quad (2)$$

Where, $g_{tx}(t)$ is transmitter pulse and Δf the subcarrier spacing and T symbol duration. We have a channel that has a response in the delay-Doppler domain h(τ,v), with the variables, τ and v are respectively the delay and the Doppler. Therefore The transmitted signal x(t) is transformed on this channel into a received signal in the time domain y(t) on the receiver in the following form,

$$y(t) = \iint h(\tau,v)\, x(t-\tau)e^{j2\pi v(t-\tau)}d\tau dv \quad (3)$$

After using the Wigner transform The signal y(t). We find a signal in the intervel TF, as follows

$$Y(n,m) = A_{g_{x,y}}(t,f)|t = nT, f = m\Delta f \quad (4)$$

$$A_{g_{x,y}}(t,f) = \int g^{*}_{rx}(t'-t)y(t)e^{-j2\pi f(t'-t)}\,dt' \quad (5)$$

With the receive pulse is of the form $g_{rx}(t)$. If the bi-orthogonality condition $g_{rx}(t)$ and $g_{tx}(t)$ is satisfied [2], the equation below represents the input-output function in the TF range

$$Y(n,m) = H(n,m)X(n,m) + V(n,m) \quad (6)$$

Where $V(n,m)$ represents the noise in TF and $H(n,m)$ is given by

$$H(n,m) = \int_{\tau}\int_{v} h(\tau,v)e^{j2\pi vnT}e^{-j2\pi vn(v+m\Delta f)\tau}\,dvd\tau \quad (7)$$

According to the symplectic finite Fourier transform (SFFT), the Y(n,m) is marked on a map of the DeDo domain signal y(k,l), so we have,

$$y(k,l) = \sum_{n=0}^{N-1}\sum_{m=1}^{M-1} Y(n,m)e^{-j2\pi(\frac{nk}{N}-\frac{ml}{M})} \quad (8)$$

Starting (1)-(7), the input-output equation is given by [2]

$$y(k,l) = \frac{1}{NM}\sum_{l'=0}^{N-1}\sum_{k'=0}^{M-1} x(k',l')h_w\left(\frac{k-k'}{NT},\frac{l-l'}{M\Delta f}\right) + v(k,l) \quad (9)$$

Where $h_w(v,\tau)$ is the channel response circular convolution with a windowing function $w(v,\tau)$ and

$$h_w\left(\frac{k-k'}{NT},\frac{l-l'}{M\Delta f}\right) = h_w(v,\tau)|v = \frac{k-k'}{NT}, \tau = \frac{l-l'}{M\Delta f} \quad (10)$$

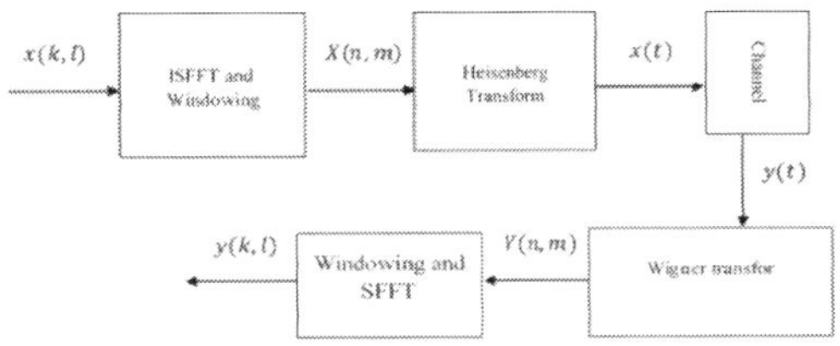

Figure 1 : OTFS modulation scheme.

3 Modulation OFDM

Assuming that a block of QAM modulated symbols $d = (d_0\, d_1 \ldots d_{N-1})^T$.

Either the data symbols are numerically independent and identical, i.e.., $[d_l d_l^*] = \sigma_d^2$; $\forall l$ and $E[d_l d_l^*] = 0$ while $l \neq q$. Let divide B total bandwidth into N number of subcarriers where the symbol duration $T = \frac{N}{B}$ second and $\frac{B}{N}$ Hz is the subcarrier bandwidth. The transmitted OFDM signal, $x(n)$, can be given as:

$$x(n) = \sum_{r=0}^{N-1} d_r\, e^{j2\pi\frac{rn}{N}} \quad (11)$$

the signal transmitted can be given as,

$$x = W_N d \quad (12)$$

where W_N is N-order inverse discrete Fourier transform (IDFT) matrix. Hence, transmitter can be implemented using N-point IDFT of complex data symbols, d. IDFT can computed using low complexity radix-2 fast Fourier transform (FFT) algorithm. When OFDM signal is passed through multipath channel, symbol spreads in time. To ignore inter symbol interference (ISI), a prefix of cyclic (CP) of length LCP is added. OFDM-CP transmitted signal can be donated as:

$$x_{cp} = [x(N + L_{cp} - 1): x(N-1); x] \tag{13}$$

Let, $h = (h_0, h_1, \dots h_L)^T$ L is the number of elements of the channel impulse response vector, and h_r, for $1 \leq r \leq L$, is the complex baseband channel factor of rth trail, which we suppose is a zero mean circular symmetric complex Gaussian (ZMCSC). We too suppose that the channel factors concerning the different trajectories are uncorrelated. We consider, $N_{cp} \geq L$.
at the level of the receiver the vector is of length

$$N_{cp} + NM + L - 1 \tag{14}$$

is given by,

$$z_{cp} = h * x_{cp} + v_{cp} \tag{15}$$

Where v_{cp} is AWGN vector of length $N_{cp} + NM + L - 1$ With element variance σ_v^2.

$$\sigma_v^2 = E(v^2) \tag{16}$$

At the receiver, we have:

$$y = [z_{cp}(N_{cp} + 1 : N_{cp} + MN)] \tag{17}$$

The received vector of length MN after removing the CP can be written as follows,

$$y = HW_N d + v \tag{18}$$

Where H represents circulant convolution matrix of size
$MN \times MN$, that we can give like this

$$H = \begin{bmatrix} h_1 & h_1 & \cdots 0 & h_L \cdots h_2 \\ h_2 & h_1 & \cdots 0 & 0 \cdots h_3 \\ \vdots & \ddots & & \\ h_L & h_{L-1} & \cdots\cdots\cdots & 0 \\ 0 & h_L & \cdots\cdots\cdots & 0 \\ \vdots & & \ddots \vdots \vdots \vdots & \vdots \\ 0 & 0 & h_L \cdots\cdots & h_1 \end{bmatrix} \tag{19}$$

Whereas v means WGN vector of length MN with elemental variance σ_v^2. H can be factorized as,

$$H = W\Lambda W^H . \Lambda = \sqrt{N} \times diag\{W_N^H h\} \tag{20}$$

where, h is the first column of H. This implies that Λ holds channel frequency samples which are sampled at the rate of $\frac{N}{B}$ samples per Hz in its diagonals.
Hence, received vector after CP removal may be represented as,

$$y = W_N \Lambda W_N^H W_N d + v = W_N \Lambda d + v \tag{21}$$

Further, N-point FFT of y is computed i.e.,

$$z = W_N^H y = \Lambda d + W_N^H v \tag{22}$$

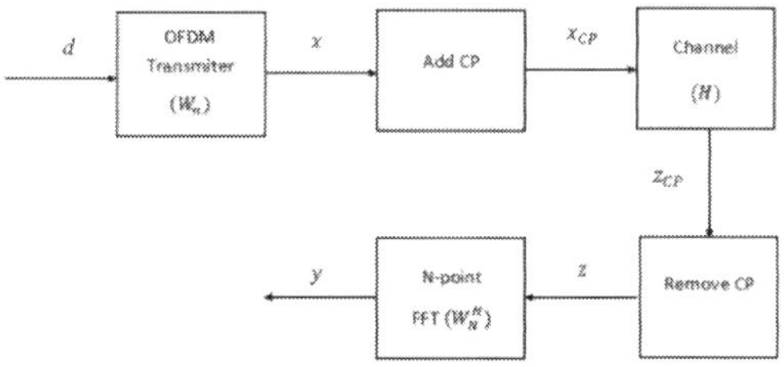

Figure 2: OFDM modulation scheme.

4 Results and discussions

In this part, we will compare the efficiency of OTFS and OFDM modulation technique using MATLAB. The comparison was made in term of BER versus SNR for number of subcarriers (OTFS, M=8) and (OFDM, M=64) with 16-QAM technique.

4.1 OTFS modulation results

The simulated results of the BER versus SNR performance for OTFS modulation with 16-QAM is plotted in figure 3. From this figure, the BER is higher for low SNR value and progressively decreases as the SNR increases. Then we can say that the BER performance is not good for a SNR ranges from 0 to 8 dB, but this parameter can be decreased for a high SNR (SNR greater than 8dB).
physically the BER decreases as SNR increases because as SNR increases the signal power becomes high relative to the noise power.

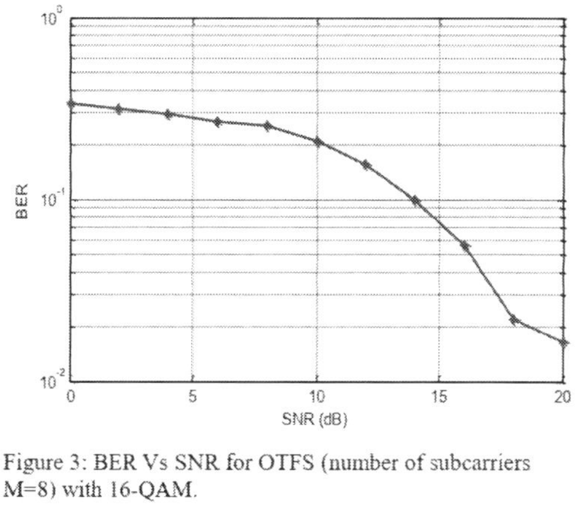

Figure 3: BER Vs SNR for OTFS (number of subcarriers M=8) with 16-QAM.

4.2 OFDM modulation results

Figure 4 shows the simulation results of BER versus SNR for OFDM modulation with 16-QAM. In this figure, we can observe that as the SNR increases, the BER decreases gradually. We can see also that the variation of the BER according to the SNR from 0 to 8 dB, is very slow and almost noticeable on this interval.

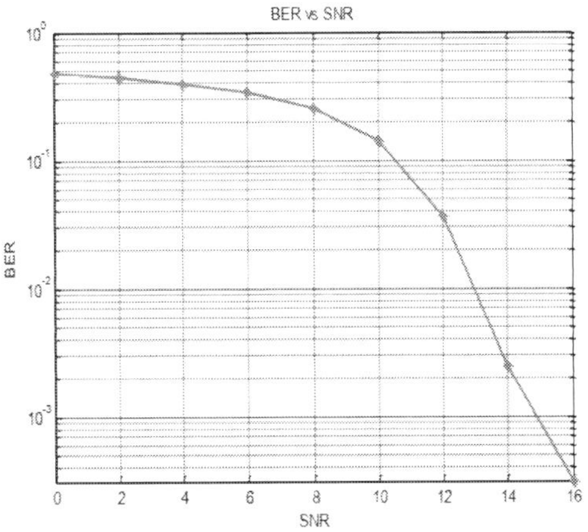

Figure 4: BER vs SNR for OFDM (number of subcarriers M=8) with 16-QAM.

4.3 Comparison results

Figure 5, plot BER from SNR variation for OTFS and OFDM with 16-QAM. From this figure, we can see that the BER of OFDM is higher than the OTFS one between 0 and 8dB. From 8dB and more, we observe that OTFS surpasses OFDM in the entire rest of the remaining range as SNR increases. We can say that the performance improvement is very low or null when the signal to noise ratio increases.
In SNR cases higher than 8 dB the BER of OFDM becomes lower than that of OTFS because we used a number of deferent subcarriers for OTFS (N=8) and for OFDM (M=64).

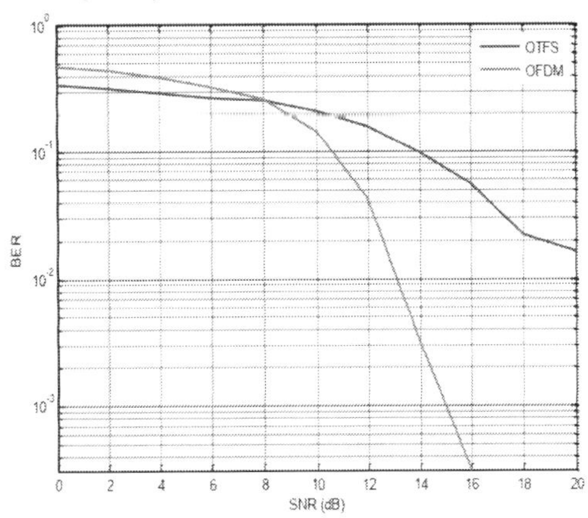

Figure 5: BER Vs SNR for OFDM (number of subcarriers M=64) and OTFS (number of subcarriers M=8) with 16-QAM.

The following table shows the results obtained by comparing OTFS and ODFM using Figure. 5

Table1. comparison OTFS and ODFM

BER / SNR (dB)	OTFS	OFDM
0	$3.5.10^{-1}$	$3.5.10^{-1}$
4	$3.5.10^{-1}$	5.10^{-1}
8	$2,8.10^{-1}$	$2,8.10^{-1}$
10	2.10^{-1}	$1,5.10^{-1}$
12	$1,6.10^{-1}$	4.10^{-2}
14	10^{-1}	3.10^{-3}
16	5.10^{-2}	0

5 Conclusion

According to previous results, the performance of both OFDM and OTFS waveforms are plotted and compared in terms of BER vs SNR for 16-QAM signals using numerical simulation. From the study findings, the OTFS modulation can be preferred for its improved reliability than OFDM one. Furthermore, OTFS is an excellent candidate for modulation and future 6G network requirements. As perspectives of this work, we propose to study the performance of OTFS using NOMA and OTFS combined with Filter Bank Multi Carrier (FBMC).

References

1. Fred Wiffen, Lawrence Sayer, Mohammud Z. Bocusy, Angela Doufexi, Andrew, "Comparison of OTFS and OFDM in Ray Launched sub-6 GHz and mmWave Line-of-Sight Mobility Channels", DOI: 10.1109/PIMRC.2018.8580850 (Publisher: IEEE) (20 December 2018)
2. Ashwitha Naikoti, A. Chockalingam, "A DNN-based OTFS Transceiver with Delay-Doppler Channel Training and IQI Compensation", DOI: 10.1109/PIMRC50174.2021.9569634 (Publisher: IEEE), (22 October 2021)
3. Ronny Hadani, Anton Monk Cohere, "OTFS: A New Generation of Modulation Addressing the Challenges of 5G" (arXiv.1802.02623) (7 Feb 2018)
4. Mustafa Ergen. "Mobile Broadband: including WiMAX and LTE". Springer Science+Business Media. doi:10.1007/978-0-387-68192-4. (2009)
5. Noriyuki HASHIMOTO, Noboru OSAWA, Kosuke YAMAZAKI, and Shinsuke IBI, "Channel Estimation and Equalization for CP-OFDM-based OTFS in Fractional Doppler Channels", DOI: 10.1109/ICCWorkshops50388.2021.9473532 (Publisher: IEEE) (09 July 2021)
6. Suvra Sekhar Das, Ramjee Prasad, "OTFS: Orthogonal Time Frequency Space Modulation A Waveform for 6G Series River Publishers", ISBN: 8770226563, (book) (2022)
7. R. Hadani, S. Rakib, A. F. Molisch, C. Ibars, A. Monk, M. Tsatsanis, J. Delfeld, A. Goldsmith, R. Calderbank, : "Orthogonal Time Frequency Space (OTFS) modulation for millimeter-wave communications systems", DOI:

10.1109/MWSYM.2017.8058662 Publisher: IEEE (05 October 2017)

8. P. Raviteja, Y. Hong, E. Viterbo, and E. Biglieri, "Practical Pulse-Shaping Waveforms for Reduced-Cyclic-Prefix OTFS", IEEE Transactions on Vehicular Technology, (Jan. 2019)
9. 3GPP, Service requirements for V2X services, TS 22.185, V 14.3, 3GPP, http://www.etsi.org/standards-search, (17-3-2017)
10. Suvra Sekhar Das, Ramjee Prasad, "Evolution of Air Interface Towards 5G Radio Access Technology and Performance Analysis", ISBN: 978-87-93609-81-5, River Publishers (2018)
11. Shashank Tiwari, Suvra Sekhar Das, Member, IEEE and Vivek Rangamgari, "Low complexity LMMSE receiver for OTFS", DOI: 10.1109/LCOMM.2019.2945564 Publisher: IEEE (2019)

Development of Compactness of Reconfigurable LPF/BSF Using a Simple Technique for Radar Applications

Elmahjouby Sghir[1]*

[1]MIET Laboratory, FST of Settat, Hassan 1st University, Settat, Morocco.

Abstract. In this paper a new compact reconfigurable low pass/band stop filter using a miniature omega resonator placed in a small DGS in the ground. The reconfiguration in frequency response can be achieved with the same topology and the same dimensions by changing just the position of channel slot which connect the omega resonator to the metallic ground without any other requirements. Several advantages obtained with this structure based on the use of one omega resonator cell such as simplicity of fabrication, very low insertion loss in the pass bands, a high rejection in the stop band with a sharp transition. This filter is suited for many areas of modern communication applications and microwave technology since it shows a simple frequency reconfiguration with a several benefits such a good wide rejection of out of bands signals, high selectivity, compactness and small losses and seize, with total area of 9.08×10mm^2.

1 Introduction

Reconfigurable and compactness filters have received significant attention for their applications in wireless communications, electronic surveillance and countermeasures by adapting their properties to achieve selectivity in frequency and bandwidth to obtained more surface and improving performance to replace the standard filters, [1]. These filters can only furnish a progressive transition from pass-band to stop-band with narrow bands and large size [2]. Such as stepped-impedance [3-5], semi-lumped element filters [6], open-circuited stubs [7], [8] and others. Many effort has gone to overcome this problem into designs of compactness and simple structure of filters using micro-strip ground structure (DMS) [9-11], electromagnetic band gap (EBG) and detected ground structure (DGS) [12-14], periodic and non-periodic DGS [15], [16], cascaded resonators [17-20], Though all these previous techniques, structures have confines on the minimization of the effective size reduction of circuits, because their resonators based absolutely on the half-wavelength resonance [21],[22].

In other, complementary cell resonators with high quality factor have gotten much interest for their simple integration with planar topology, low-cost and easy fabrication. Parallel to this previous device, a complementary split ring resonator formed omega topology is better than the conventional microwave shaped resonators, since it has very high-quality factor, and it can resonate with very small size, which offers low radiation losses, [1].

The basic structure is an omega complementary resonator with Ω-topology, as shown in Fig. 1. A circuit model has been extracted to characterize the proposed structure, [1]. All the dimensions of the proposed Ω-omega resonator are depicted in Table 1.)

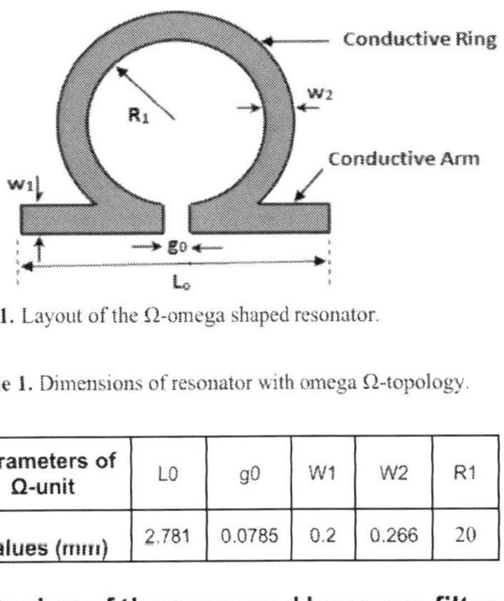

Fig. 1. Layout of the Ω-omega shaped resonator.

Table 1. Dimensions of resonator with omega Ω-topology.

Parameters of Ω-unit	L0	g0	W1	W2	R1
Values (mm)	2.781	0.0785	0.2	0.266	20

2 Design of the proposed low pass filter

The present structure is a new coplanar waveguide (CPW) low pass filter (LPF) as schematic shown in Fig. 2. The basic unit of the proposed LPF layout is a single omega resonator which is inserted in a small DGS in the metallic ground plane of coplanar line and connected vertically to the ground plane via slot-channel (Sc=0.2x0.05mm^2) between the top arm heads of omega resonator and the ground of structure. While the omega resonator speared horizontally with a disconnecting gap (g=0.05x0.2mm^2) between the two arms and the ground

* Corresponding author: elmahjouby.sfhir@gmail.com

plane. The new low pass filter has been simulated on a FR4 substrate with a relative dielectric constant of εr = 4.4 and a thickness of 0.635 mm.
The input-output characteristic impedance of the CPW transmission line is 50Ω, with a signal line width of W=0.83 mm and the gap of G=0.125 mm. While the width of the rectangular DGS in the ground is 0.696 mm and length of 2.881 mm. The surface area of the whole structure is 9.08 mm in width and length of 10 mm (9.08×10mm^2). The design procedure is optimized and validated by using Momentum based on moments method.

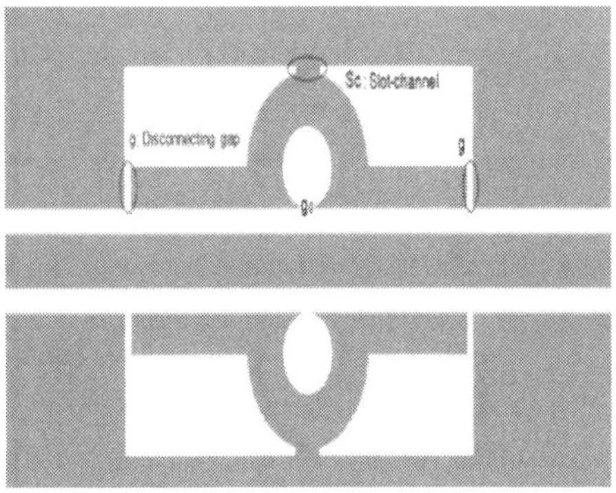

Fig. 2. The schematic of the proposed LPF.

The proposed structure has been designed, optimized and simulated. The results of EM simulation are shown in Fig. 3(a). The proposed LPF have a cut off frequency at 16.02 GHz. The first attenuation pole at 11.5 GHz with level more than -30dB. The insertion loss in the passband around 0.5dB. While the suppression level is greater than -20 dB from 20 GHz up to 30 GHz. Thus, the filter exhibit sharp roll-off at transition knee of 32 dB/GHz. While the stopband bandwidth is equal to 14 GHz. Fig. 3(b), depicts the phase simulation in term of return loss (S11) and insertion loss (S21) of the proposed LPF, which is accepted as a linear for wide band such as here from 0.1 GHz up to 30 GHz.

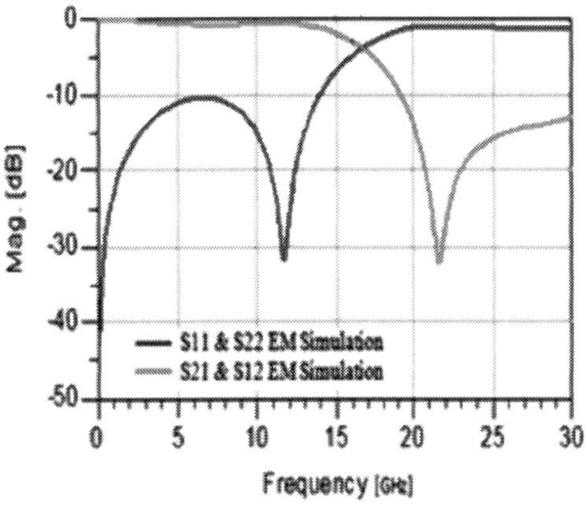

(a)

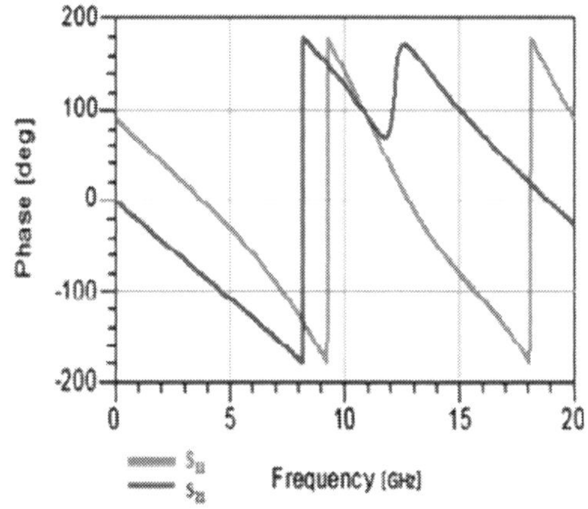

(b)

Fig. 3. EM-Simulation results of the proposed CPW LPF (a), and its simulation phase responses (b).

3 EM-field distribution of the proposed LPF filter

To demonstrate the level accord between the EM-simulation results and the current distribution of this proposed structure, we propose the study of surface energies as depicts in Figs. 4(a), and 4(b). The topology characteristic was revalued in two different frequencies, at 10 GHz and 25 GHz, which corresponds to pass- and stop-bands state, respectively.
Firstly, at frequency 10 GHz, as observed in the results of the Fig 6(a), it is clearly that the current energies distributed from the input to the output along the filter, which confirms the passband state of structure. Secondly, in the Fig 4(b) at frequency of 25 GHz, we can see clearly that the energy distribution remains blocked in the first part of the filter and mainly concentrated around the omega resonator, hence no current energies at the output port, which signifies that the structure is in the stopband response. Finally, a good agreement between EM-Simulation results and EM-Field distribution.

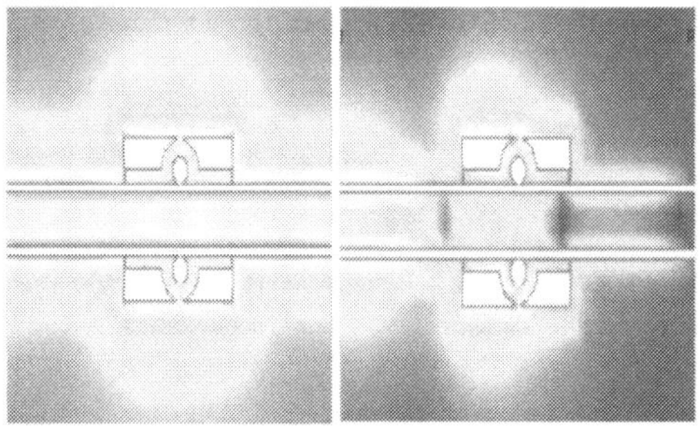

(a) (b)

Fig. 4. The current distribution of the simulated structure: (a) at 10 GHz & (b) at 25 GHz.

4 Reconfiguration of low pass (LPF) to band stop filter (BSF) based on switches technique

The goal of this study is to reconfigure the proposed low pass filter to the band stop filter, in order to save the same dimensions of the global structure with the same topology, as shown in Fig.5(a). Consequently, limiting the occupied volume on the surface area of this structure then minimize the level of complexity of the proposed filter in the transformation bands. However, the trick in this work is a sample technique based on switches mechanism, compared to the conventional transformation and other complicated method that is need to change the topology and adding other parameters which occupied an overall volume of structure. The band stop filter is composed of a single omega resonator that is placed in DGS in the metallic ground plane and directly connected with the left arm heads to the ground of coplanar line with a small slot channel with of $0.05x0.2mm^2$. The extracted circuit of the proposed filter is compatible to the layout structure as shown in Fig.5(b).

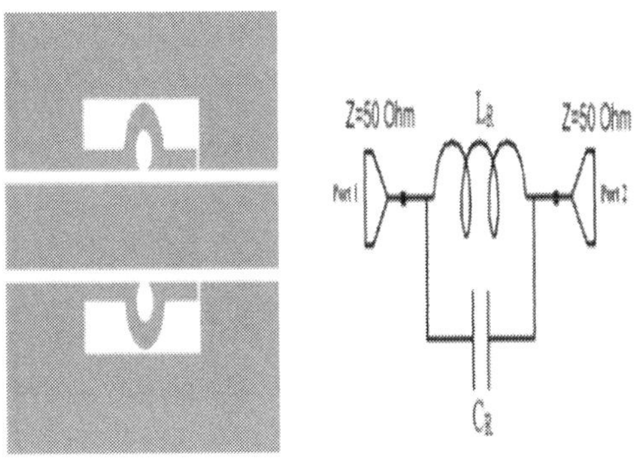

(a) (b)

Fig. 5. The schematic of the BSF (a) & Its equivalent circuit (b).

The EM and phase simulation results of the reconfigurable BSF filter, are shown in Fig. 6(a) and (b), respectively. As can be seen from the results, the band stop filter has resonance frequency, lower and higher cut-off frequencies of f0=12.1GHz, fl=9.9 GHz and fh=14.6 GHz, respectively. The stop band rejection level is extended over −20 dB, while it is greater than -15 dB from 11 GHz to 13 GHz. Thus, the stop band filter exhibits a sharp cutoff within a range of 2 GHz, from -3 dB to - 23 dB. The insertion loss is less than -1dB, the return loss is more than -23 dB in the stop band. While the stop band bandwidth is equal to 4.6 GHz. While the phase is accepted as a linear in term of return loss (S11) and insertion loss (S21) for the BSF for wide band from 0.1 GHz to 20 GHz.

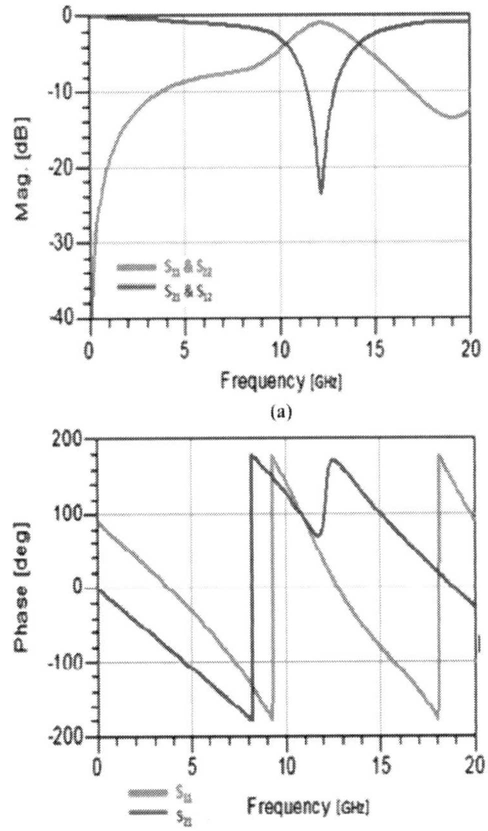

(b)

Fig. 6. Simulation results of the reconfigurable SBF (a), and its simulation phase responses (b).

5 Conclusion

This paper presents, a novel design and development of a compact coplanar wave guide low pass filter and band stop filter with good performances as sharp-rejections, low loss and high rejection using a single omega resonator cell has been designed. The reconfiguration of frequency response can be achieved easily by using four switches. Parallel to this previous several advantages, a small size and wide rejection band of reconfigurable LPF/BSF is highly demanded in modern microwave communication systems, especially in radar applications in attempt to suppress spurious and harmonics signals. There are several general requirements for a LPF, such as low loss.

Acknowledgment

We thank Mr. Mohamed Latrach from ESEO Angers in France, for allowing us to use and to perform simulations by using electromagnetic solvers and measurement instruments.

References

1. Elmahjouby Sghir, Ahmed Errkik, Jamal Zbitou, Otman Oulhaj, Ahmed Lakhssassi, Mohamed Latrach, "A Novel Concept of a Miniature CPW Band Stop Filter Using Omega Resonators for Radar Applications", *International Journal of Microwave And Optical Technology*, Vol 15, Issue 05, pp. 433-439, Sep. 2020.
2. M.J. chao, Y.W. hua, Sun Hou-jun, Lv Xin, "A compact low pass filter with good stopband performance using an open-loop-arms defected ground structure unit", *3rd IEEE International Symposium on Microwave, Antenna, Propagation and EMC Technologies for Wireless Communications*, pp.462-465, 27-29 Oct. 2009.
3. Islam Mansour, Hadia Elhennawy, Adly S. Tag El-dein, "Design of Stepped Impedance Microstrip Low Pass Filter with DGS", *International Journal of Engineering Research and Development*, Vol.10, Issue 7, pp.58-67, July 2014.
4. Sheetal. Mitra, D.K. Kumuda, "Stepped Impedance Microstrip Low-Pass Filter Implementation for S-band Application", *International Journal of Latest Trends in Engineering and Technology*, Vol. 5 Issue 3, pp 248-255, May 2015.
5. Budhadeb Maity, "Stepped impedance low pass filter using microstrip line for C-band wireless communication", *IEEE Conference Paper, 2016 International Conference on Computer Communication and Informatics (ICCCI)*, Coimbatore, INDIA, 07-09, Jan. 2016.
6. J.W. Sheen, "A compact semi-lumped low-pass filter for harmonics and spurious suppression", *IEEE Microwave and Guided Wave Letters*, Vol 10, Issue 3, pp. 92-93, April 2000.
7. Keyel Kim, Sinai Kim, Haosen Han, Ikmo Park, Hunhwa Lim, "Compact microstrip lowpass filter using shunt open stubs and coupled slots on ground plane", *Electronics Letters*, Vol 40, Issue 5, pp.313-314, April 2004.
8. Mohammed A. Aseeri, Meshaal A. Alyahya, Hatim A. Bukhari, and Hussein N. Shaman, "Compact Microstrip Lowpass Filter with Low Insertion Loss for UWB Medical Applications", *Hindawi, Wireless Communications and Mobile Computing*, Vol. 2018, 5 pages, June 2018.
9. Wua, H.-W, C. C. Lina, Y.F. Chenb, and H.Y. Leeb, "Compact triple-passband bandpass filter based on new modified stepped impedance resonators", Microelectronics Journal., Vol. 46, pp.490-495, 2015.
10. Boutejdar A, A. Omar, and E. Burte, "High performance wide stop band low pass filter using a vertically coupled DGS-DMS-resonators and interdigital capacitor", *MOTL Microwave Opt. Technol. Lett.*, Vol. 56, pp.87-92, November 2014.
11. Kazerooni M, N. P. Gandji, A. Cheldavi, and M. Kamarei, "A new microwave bandstop filter using defected microstrip structure (DMS)", *PIERS Proceedings*, pp.697-700, Moscow, Russia, 18- 21, August 2009.
12. A. Boutejdar, A. A. Ibrahim, and E. P. Burte, "DGS Resonators Form Compact Filters", *Microwave & RF*, March 2015.
13. M. Challal, A. Boutejdar, A. Azrar and D. Vanhoenacker Janvier, "Design of compact low-pass filter with large reject band using open circuit stubs and two cascaded DGS quasitriangular resonator", *Microwave and Optical Technology Letters*, Vol.55, Issue 1, pp. 122-127, January 2013.
14. A. Faraghi, M. Azarmanesh, M. Ojaroudi, "Small Microstrip LowPass Filter by using Novel Defected Ground Structure for UWB Applications", *ACES Journal the Applied Computational Electromagnetics*, vol. 28, no.4, April 2013.
15. A. Boutejdar, ABatmanov, A. Omar, E. Burte "A compact bandstop/band-pass filter using a narrow gap capacitive as J-inverter and a new octagonal defected ground structure (DGS)", *EuCAP 2009. 3rd European Conference on Antennas and Propagation*, pp.3210-3213, 23-27 March 2009.
16. Y.M. Huang, S. Zhenhai, C.J. You, W. Guoan, "Size-reduced bandpass filters using quarter-mode substrate integrated waveguide loaded with different defected ground structure patterns", *Microwave Symposium (IMS), 2015 IEEE MTT-S International*, pp.1-4, 17-22 May 2015.
17. Boutejdar. A, G. Nadim, S. Amari, and A. S. Omar, "Control of bandstop response of cascaded microstrip low-pass-bandstop filters using arrowhead slots in backside metallic ground plane", *Proc. IEEE AP-Symp.* Washington DC, July 2005.
18. Ahn, D, J. S. Park, C. S. Kim, J. Kim, Y. Qian, and T. Toh, "A design of the low-pass filter using the novel microstrip defected ground structure", *IEEE Trans. Microwave Theory Tech*, Vol. 49, pp.86-93, Jua.2001.
19. Nosrati, M, T. Faraji, and Z. Atlasbaf, "A compact composite broad stop-band elliptic function low-pass filter for ultra-wideband applications using interdigital capacitors", *Progress In Electromagnetics Research Letters*, Vol. 7, pp.87-95, June 2009.
20. G. Jang and S. Kahng, "Design Of A Dual-Band Metamaterial Bandpass Filter Using Zeroth Order Resonance", *Progress In Electromagnetics Research C*, Vol. 12, pp.149-162, January 2010.
21. A. Boutejdar, "Design of Compact Reconfigurable Broadband Band-Stop Filter Based on a Low-Pass Filter Using Half Circle DGS Resonator and Multi-Layer Technique", *Progress In Electromagnetics Research C*, Vol. 71, pp.91-100, February 2017.

22. Khelil Fertas, Farid Ghanem, Mouloud Challal, and Rabia Aksas, "Design and Development of Compact Reconfigurable Tri-Stopband Bandstop Filter Using Hexagonal Metamaterial Cellsfor Wireless Applications", *Progress In Electromagnetics Research M*, Vol. 80, pp.93-102, April 2019.

The effect of tuning screws of the S-parameters of a 5G bandpass Filter

Amina Aghanim, *Otman* Oulhaj, and *Ismail* Ouboudrar

LSTA.FPL, Abdelmalek Essaadi University, Tetouan, Morocco

Abstract. In this paper, the design and simulation of a symmetrical waveguide filter are presented based on the insertion loss method. The designed WR-34 waveguide filter is operating at 28 GHz, where its dimensions are a=4.32mm and b=8.64mm, a fractional bandwidth of 2.15% a return loss level better than 16 dB, a frequency range from 22Ghz to 33 GHz. The structure is a 5th-order bandpass filter with five resonators and six symmetrical circular inductive irises. In order to optimize the frequency response, five tuning screws were added to the center of each resonator, while two other screws were positioned before the first iris and the last iris from the wave ports. The simulation model of the filter is developed in two different simulators ANSYS high-frequency Structure Simulator (HFSS) and CST Studio Suite so as to validate the obtained results and to depict the effects of tuning screws on the S-parameters of the filter while generating scenarios depending on the depth of penetration of each screw.

1 Introduction

The development of wireless communication has led to the emergence of two major trends, especially in microwave applications[1]. The first trend could be described as the increase in operational frequencies since microwave applications are on a rapid rise, therefore more bandwidths are available. While the second trend depicts the huge increase in bandwidth and data rate, where the bandwidth of new wireless devices doubles every 18 months, conforming to Edholm's law[2]. Otherwise, this increasing demand in wireless technology has commanded the saturation and limitation of the frequency spectrum Therefore, developing filtering devices has become an essential issue to avoid channel interferences and take advantage of the spectrum frequency's range.

In this context, microwave filters are one of the most significant devices in all wireless transmissions, since they can split between the desired and none desired frequencies. Especially mobile communications which are based on the transfer of millimeter waves ranging from 3 to 300Ghz [3]. As already known, the quantity of insertion loss in the passband is usually inversely proportional to the bandwidth of the filter[4], which causes a problem in the case of designing a narrow bandpass filter. To remedy this, an excessively high Q resonator factor should be used to design the filter so that it could respond to the required narrow bandwidth and the low insertion loss quantity[5]. For this reason, waveguide filter technology is considered to be the most qualified for this purpose[6] as it offers low passband insertion loss, extremely high rejection in the stopband, and a high Q factor[7].

For many years, this domain was facing challenges concerning the cost, size, and time of developing such efficient devices, especially when the response of a given filter after fabrication does not meet the required specifications. This mismatch is generally due to the manufacturing tolerances and the type of material adopted in production. In order to remedy this, a tuning procedure after fabrication has become an overriding issue, this process could be done either manually or automatically[8–10].

Unfortunately, the regulation has been limited to the level of technologists and experts in the microwave domain, since there is no direct or unique relationship between the depth of the screws inside the filter and the S-parameters. At this point, it should be mentioned that screws play a major role in adapting the filter response such as the other discontinuities: irises and posts that could take the symmetrical aspect or an asymmetrical one[11].

This paper presents the design of a bandpass filter dedicated to 5G applications with the insertion loss method, where we will vary the depth of penetration of screws to see their impact on the frequency response of the concerned paper. This work will be organized as follows. A discussion about the adopted design method is elaborated in section2, the synthesis and simulation process are presented in section 3. While section 4 presents a discussion about the obtained results in the four scenarios after varying the depth of penetration of the screws to prove the effect of their depth of penetration on the frequency response of the filter. Finally, the conclusions and some short terms perspectives are drawn in section 5.

2 Insertion loss Method

Generally, two famous methods are used to design microwave filter. The first method was helpful in the case of low frequency filters in radio, telephony, Called image parameter method. While, the modern recent procedure, named insertion loss method uses networks synthesis techniques to design an effectively specific response[12]. It allows high degree of control over the passband and stopband amplitude. This method takes as a first step a low-pass filter prototype, normalized in impedance and frequency. The design process is illustrated in Fig.1

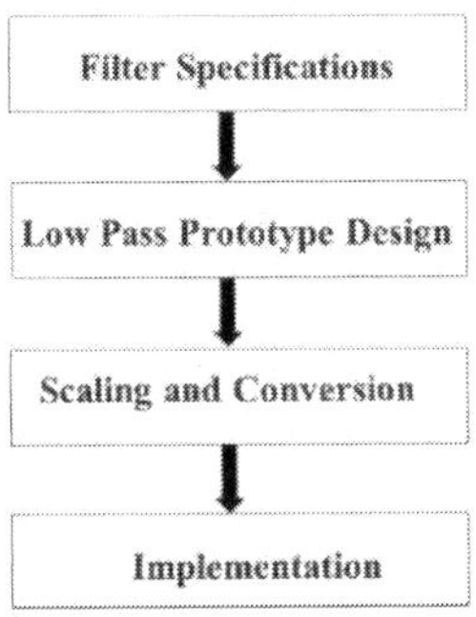

Fig 1: Process of filter design by insertion loss method

The detailed steps of this method are elaborated in the references[13–15].

3 Synthesis and simulation process

To deepen the theory presented in the previous section, we have chosen a bandpass filter with inductive irises of fifth-order designed in waveguide technology. This filter operates in the extremely high-frequency band (EHF), ranging from 22 GHz to 33 GHz, with a resonance frequency of 28 GHz, a relative bandwidth of 2.13%, and a return loss quantity around 15 dB. The illustration in Fig.2 depicts the geometrical structure of the filter, which is characterized by six inductive circular irises separating between five cavities, within a screw in the center of each cavity with two other screws positioned before the first iris and the last iris from the wave ports, which make them seven in total. It should mention that the screws were added with the aim of tuning the frequency response or the S-parameters of the waveguide filter whose internal dimensions are a = 8.64 mm, b = 4.32 mm.

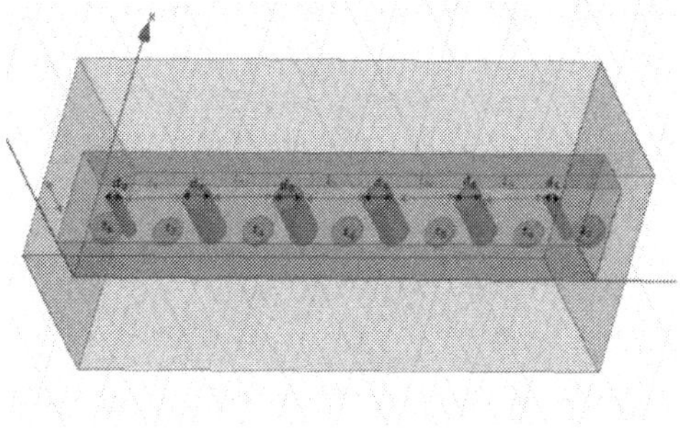

Fig 2: Interior topology of the waveguide filter structure

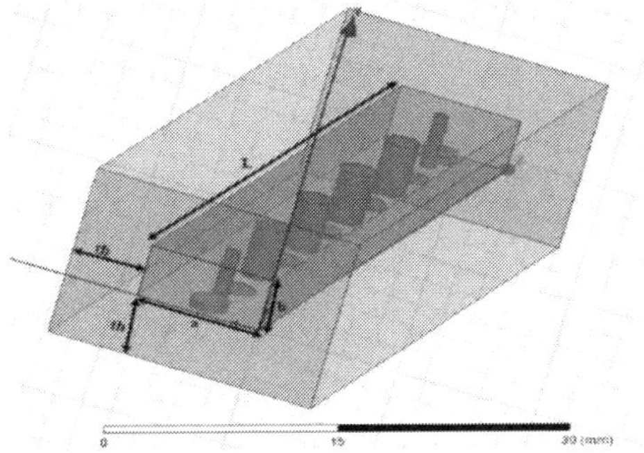

Fig 3: Exterior topology of the waveguide filter structure

In order to calculate the irises dimensions and the lengths of the five resonators, a Matlab code was used, and Table 1 illustrates the numerical values in mm of the parameters of these irises, where d_i is the diameter of each iris and b is their common height. Whereas, Table 2 presents the lengths l_i of each resonator.

d_0	d_1	d_2	d_3	d_4	d_5
1.096	2.315	2.436	2.436	2.315	1.096
b =4.32					

Table 1: The parameters of the five irises in mm

l_1	l_2	l_3	l_4	l_5
5.557	5.616	5.572	5.616	5.557

Table 2: Lengths of the resonators in mm

Knowing that the tuning screws were added such as discontinuities in the form of inductive circular insertions, they allow adjusting the S-parameters, in this purpose Table.3 summarizes the numerical values of their dimensions, where t_i is the height and D is the common diameter of all the tuning screws.

t_1	t_2	t_3	t_4	t_5	t_6	t_7
0.658	0.015	0.262	0.332	0.262	0.015	0.658
D = 2.571						

Table 3: The dimensions of the tuning screws in mm

Other dimensions worthen to be noted are the length of the waveguide L = 45.72 mm, the thickness of the waveguide was chosen to be th =5 mm and R = 3.054 mm referring to the distance between the wave ports and the first and last posts, these previous dimensions are depicted in Fig.3.

4 Results and discussion

4.1 Parametric study on the screw's properties

A parametric study was carried out, to determine the optimal value of D the diameter of the screws where Fig.4 shows the geometry of the considered structure. The range of the parametrisation goes from 2.5mm to 3mm. This step was effectuated under the CST environment.

After running the simulation, the obtained graph depicts that the optimal value of the diameter of the five screws is D= 2.72mm this value presents a good quantity of return loss and insertion loss around the resonance frequency. The Fig.5 and Fig.6 show clearly this aspect.

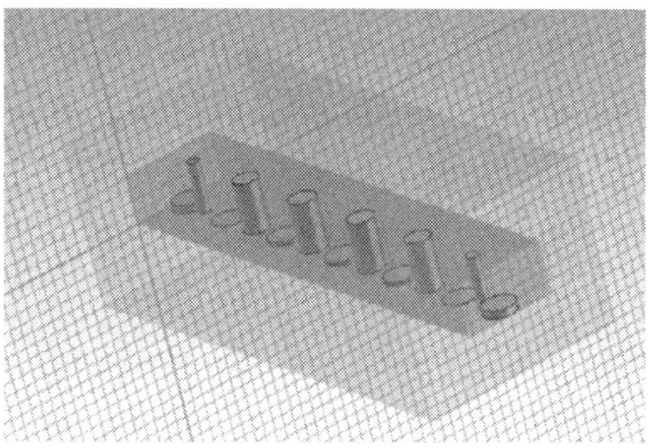

Fig 4: Geometry of the structure under CST

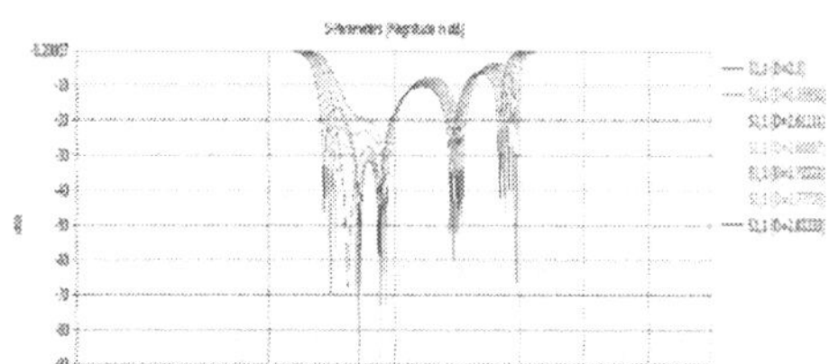

Fig 5: The plot of S_{11} after the parametric study on D

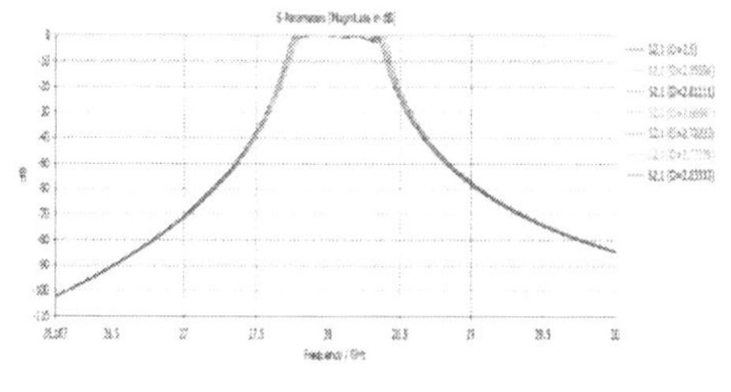

Fig 6: The plot of S21 after the parametric study on D

4.2 The variation of the depth of screws

The wavelength is around 10 mm at the frequency of 29Ghz, in this case, the performances of our filter may be affected by the variation of the depth of penetration of each one of the screws. With this aim, we structured the experiment to analyse how this mechanical penetration affects the S-parameters of our filter. After the creating the structure geometry, we will take into consideration four scenarios, to see if the depth of penetration of these discontinuities do influence in a way or another the response of the filter. For this Table.4 summarizes the depth of the screws in the four cases.

Scenario	t_1	t_2	t_3	t_4
1	0.658	0.04	0.262	0.332
2	0.326	0.015	0.262	0.166
3	0.658	0.015	0.3	0.332
4	0.658	0.015	0.262	0.332

Table 4: Corresponding screw depth combinations for the four scenarios

Taking into consideration that the structure geometry of the filter is symmetrical compared to the central screw (4th screw), where $t_1 = t_7$, $t_2 = t_6$, and $t_3 = t_5$. Only four geometrical variables will be varied, that are as follows: t_1, t_2, t_3, and t_4.

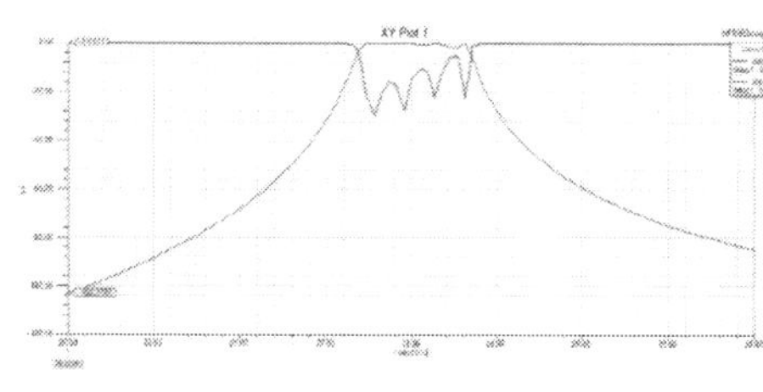

Fig 7: First scenario

As could be seen, the Fig.7 presents the S-parameters of our structure, based on the combination 1 of the depth of the screws. From the illustration, the bandwidth and center frequency had nearly met the specifications. We can also see that the amount of the return loss is around 9dB. Besides, an appearance of four zero poles, and a lower and upper stopband insertion loss quantities of 103.25 dB and 84dB. While, the illustrations of both Fig.8 and Fig.9 depicting the second and third scenarios does not show a good response for both S_{11} or S_{21} since there is a ripple in S_{21} in the bandwidth. A disappearing of the zero poles in the bandwidth, and the performance of the insertion loss had decreased in both cases.

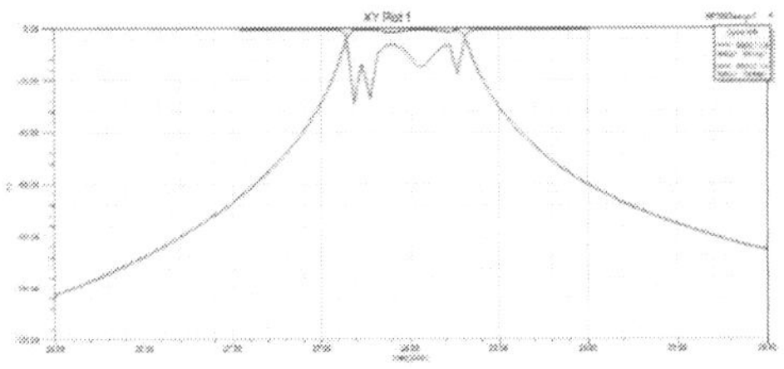

Fig 8: Second scenario

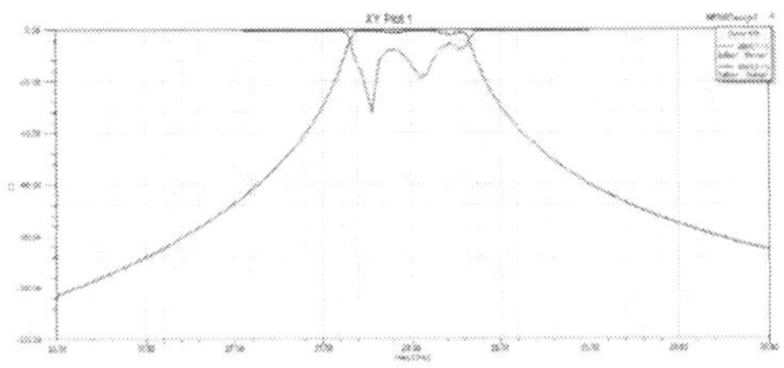

Fig 9: Third scenario

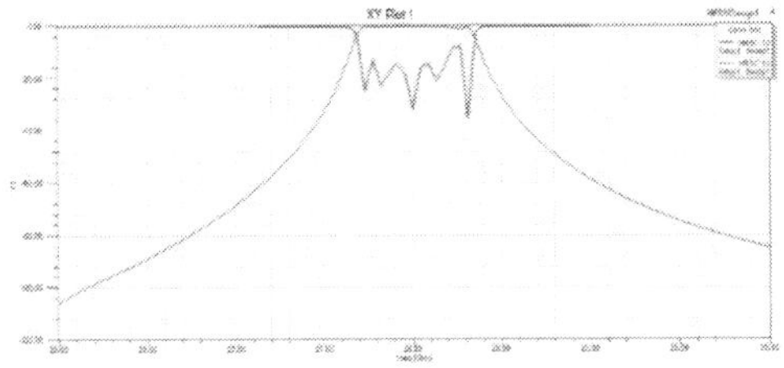

Fig 10: Fourth scenario

The 4th and last scenario have shown good agreement in terms of the fractional bandwidth, the center frequency which is 28GHz. The return loss quantity is around 10dB, and we can clearly see that the number of zero poles has increased to five in the bandpass. Considering the lower and upper stopband insertion loss measurements we can extracts the quantity of 106 dB at 26GHz and 85 dB at 30GHz.This scenario had showed better performances in view of all the properties of our filter, as could be seen in Fig.10.

The results of the paper [5] presented the aspect of the same specifications of our geometry, and we can clearly see that the results are similar. They used HFSS to elaborate the results compared to the mathematical model obtained from Matlab, while we used both simulators HFSS and CST.

Reference	***[9]***	***[5]***	***Proposed design***
Return Loss(dB)	17dB	15dB	16dB
Resonance frequency (GHz)	3 GHz	28 GHz	28 GHz
Number of screws	5	5	5
Screws Diameters(mm)	3	2.68	2.72

5 Conclusion

In this work, the design and simulation of a bandpass waveguide filter, with inductive discontinuities in the form of symmetrical circular irises, along with the effect of the depth of penetration of the seven screws to improve the frequency response of our filter. It was clear that the S-parameters are sensitive to the depth of our discontinuities, since the operating wavelength is around 10mm in the range of frequencies going from 26GHz to 29Ghz. The obtained results have showed a good agreement with the previously fixed specification. Where these results depict a low passband insertion loss and a good return loss level for the entire passband. Therefore, the main addition of this work is that this type of design is a reliable filtering device structure especially of millimeter-wave band, that presents a large bandwidth and a fast data transfer rate to overcome with the limitations of the spectrum range of frequency.

A short-term prospect was settled, which is developing an automated tuning process to control the response of the filter based on an intelligent controller such as FLC, which duplicates the human reasoning. This control will be effectuated on the depth of screws, to overcome the obligation of using only experts in the domain.

References

1. J. S. Parish, N. Somjit, and I. C. Hunter, IET Microwaves, Antennas & Propagation **12**, 2332 (2018)
2. BT, M. Kern, R. Ainslie, BT, D. Williams, BT, A. Jacquet, BT, N. Wong, and BT, in *Proceedings of SW21 The OR Society Simulation Workshop* (Operational Research Society, 2021)
3. Y. Al-Yasir, R. A. Abd-Alhameed, J. M. Noras, A. M. Abdulkhaleq, and N. O. Parchin, in *Loughborough Antennas & Propagation Conference 2018 (LAPC 2018)* (Institution of

Engineering and Technology, Loughborough, UK, 2018), p. 61 (4 pp.)-61 (4 pp.)

4. M. S. Anwar and H. R. Dhanyal, in *2018 15th International Bhurban Conference on Applied Sciences and Technology (IBCAST)* (IEEE, Islamabad, 2018), pp. 866–869
5. Y. P. Lim, C. Sovuthy, and P. W. Wong, 3 (n.d.)
6. S. Alotaibi and J.-S. Hong, Int J RF and Microwave Comp Aid Eng **18**, 1 (2008)
7. I. C. Hunter, L. Billonet, B. Jarry, and P. Guillon, IEEE Trans. Microwave Theory Techn. **50**, 794 (2002)
8. S. Wu, W. Cao, M. Wu, and C. Liu, International Journal of Electronics and Communication Engineering **12**, 7 (2018)
9. Z. Wang, J. Yang, J. Hu, W. Feng, and Y. Ou, in *2015 IEEE International Conference on Robotics and Biomimetics (ROBIO)* (IEEE, Zhuhai, 2015), pp. 2145–2150
10. C. Kwak, M. Uhm, I. Yom, and H. J. Eom, IEEE Microw. Wireless Compon. Lett. **22**, 539 (2012)
11. D. Sun and J. Xu, IEEE Microw. Wireless Compon. Lett. **26**, 475 (2016)
12. G. F. Craven and C. K. Mok, IEEE Trans. Microwave Theory Techn. **19**, 295 (1971)
13. R. Levy, R. V. Snyder, and G. Matthaei, IEEE Trans. Microwave Theory Techn. **50**, 783 (2002)
14. R. J. Cameron, C. M. Kudsia, and R. R. Mansour, *Microwave Filters for Communication Systems: Fundamentals, Design and Applications*, 2nd ed (Wiley, Hoboken, 2018)
15. D. C. Youla, Proc. IEEE **59**, 760 (1971)

 https://doi.org/10.1051/itmconf/20224802003

A GaAs Low Noise Amplifier for 2.4 GHz ISM band Applications

Mohamed Ribate[1*], *Rachid* Mandry[1], *Larbi* Elabdellaoui[1], and *Mohammed* Benbrahim[2]

[1]MIET Laboratory, Faculty of Sciences and Techniques of Settat, Hassan 1st University, Morocco
[2]ENSA of Kenitra, Ibn Tofail University, Morocco

Abstract. In this paper, a Low Noise Amplifier (LNA) operating in the Industrial Scientific and Medical (ISM) band of 2.45 GHz is proposed. The designed LNA is based on GaAs ATF21170 active device. To improve gain and matching parameters, a hybrid matching technique based on smith chart and optimization is used. The proposed biasing network as well as the stabilization circuit allow achieving a minimum noise figure with an unconditionally stability over the whole operating frequency ranging from 2.25 GHz to 2.55 GHz. Finally, some tradeoffs were done to obtain the minimum noise figure with the greatest available power gain. In terms of performances, the proposed structure shows an excellent noise figure of ~1Db.The simulated results show an excellent input/output return loss under -15 dB over the operation frequency band with an excellent small signal gain ranging between 17 dB and 15 dB. In terms of large signal performance, the proposed LNA achieves an output power of 17 dBm (52 mW) with a PAE of 22% and an output intercept IP3 around 27 dBm.

1 Introduction

A Low Noise Amplifier (LNA) is an RF device that amplifies a very weak signal without significantly degrading its SNR (signal-to-noise ratio) [1-3]. In fact, an RF amplifier will not only amplifies the power of both the signal and the noise available at its input, but it will introduces certain additional noise due to its internal noise sources [4-8]. As a result, the LNAs are designed so that its internal noise sources minimally contribute to output noise power [9-10].

The LNAs are generally placed at the front-end wireless receivers, and plays the role of amplifying the very weak received signals to the point that they can correctly processed by the next stage signal-processing devices [11-13]. In addition, the LNAs set the system total noise figure or sensitivity [14].

The additional noise introduced by the LNA internal noise sources can reduced by choosing low-noise components, the bias operating points and the circuit topology. In addition, the input matching circuit for an LNA is designed to be low loss to reaches the minimum possible noise figure [15-17].

Suitable characteristics of an LNA are unconditionally stability, low noise figure and high gain, high third-order intercept point (IP3), high dynamic range, low input/output reflexion coefficients, broad bandwidth, low power dissipation and compact size. However, there are tradeoffs in these characteristics during designing [13].

The LNA can in a single stage or multistage configuration. In a multistage topology, the input matching circuit is designed for minimum noise figure, the interstage matching is synthesised for flat gain, and the output matching circuit is designed for maximum output power and gain [18-19].

The present paper exposes the design and implementation of a Low Noise Amplifier operation around the 2.4 GHz ISM band. The paper is organised as follows. In section 2, the proposed LNA design flow is presented. Section 3 is devoted to the obtained results and discussions, and the concluding remarks are summarized in section 4.

2 The proposed LNA design flow

2.1 Overview

The design of an LNA, as that of any other electronic circuit, is typically subdivided into a series of systematic steps, from the LNA design specifications up to the final characterization of the fabricated circuit. In fact, starting from the LNA design specifications, the first pass consists of selecting the design process and the subsequent selection of the active device type. The latter is chosen according to the required operating frequency and the targeted application. After the proper active device selection, the next step resides in DC analysis in order to select the proper bias operating point that fulfil the electrical specifications.

On the other hand, the overall LNA stability is very crucial and has to be enforced not only in the operating frequency band but for the whole active device operating frequency band: especially, for the low frequency range, where the higher active device gain could introduce oscillations because of the feedback components intrinsically present in active device and the

* Corresponding author: ribate.mohamed@gmail.com

LNA circuit. Thus, the subsequent step of DC analysis is stability analysis.

The remaining steps consist of designing the biasing, input and output matching networks. The latter can be designed using both lumped or distributed elements. In this work, the distributed transmission lines are adopted.

In terms of characterization and validation process, the proposed LNA pass first by small signal analysis, then, the obtained results are consolidated by nonlinear analysis and finally validated by the EM Co-Simulation.

2.2 DC analysis

In this work, the adopted active device is ATF21170. The proposed LNA is designed to operate around the class A operating point, which allows the highest possible linearity, the $Id_{ss}/2$ bias operating point is selected.

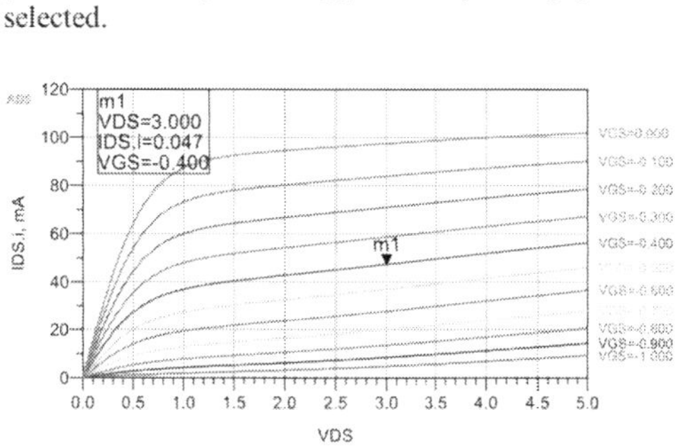

Fig. 1. DC analysis curves (active device ATF21170)

From the figure 1, the selected biasing point (corresponding to $Id_{ss}/2$) is: $\{V_{DS} = 3v; V_{GS} = -0.4v\}$.

2.3 Stability analysis

The operating frequency of the selected active device starts from 0.5 GHz to 6 GHz, therefore, the main goal during the stability analysis is the achieve the unconditionally stability and tradeoffs over the whole active device operating frequency band.

On the other hand, if we have to stabilize the active device, the easiest way consists of introducing some resistive components at the input and the output of the active device respectively. For example, at the input side, we can simply introduce a series resistor, this latter can significantly improve the active device stability. However, in case of LNA design where the main goal is to obtain minimum noise figure possible, by introducing a series resistor at the active device input side, we can achieve the unconditionally stability but at the same time, we increase the noise figure and decrease the small signal gain.

Therefore, as a good practical tip, we should avoid any lossy components at the input side of an LNA as much as possible so the noise figure is not distorted beyond a point. Hence, if we have to introduce any lossy component, especially a series resistor, we should always include it at the output side, because referring to the Friis formula (used to calculate the total noise figure in case of a cascade architecture), the noise introduced by the output series resistor will be divided by the active device gain. The whole idea here is to balance the active device stability, the noise figure and the gain.

Basing on the above discussion, the proposed stability components are shown in figure 2.

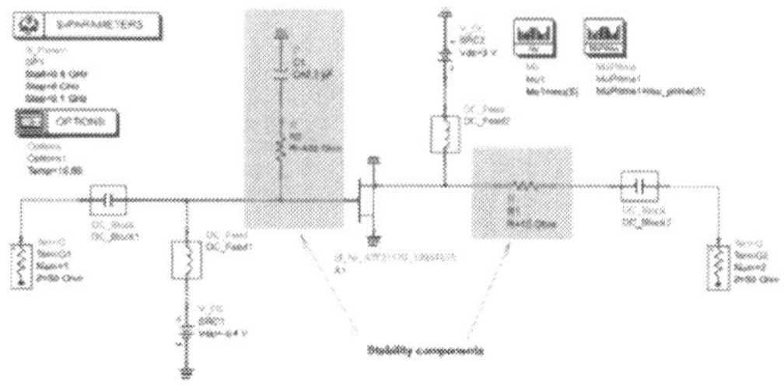

Fig. 2. The proposed stability components

At the active device output, a series resistor of 15 Ω is selected, while at the input side, a shunt RC network is introduced. Here a proper attention is taken while selecting the resistor value; this latter is selected in such a way that is the highest possible value to achieve the stability plus not letting the gain and the noise figure decrease so much around the desired operating frequency band. The role of the capacitor is simply blocking the DC going to the ground, because we want the DC going directly to the active device and not anywhere else. Using these stability networks, the obtained results are shown in figure 3.

As we can remark, the stability analysis was performed using the Edward-Sinsky factors :$\{\mu ; \mu'\}$, which are geometrical factors.

From figure3, using the proposed stability networks, the active device is unconditionally stable across the entire operating frequency band ranging from 0.5 GHz to 6.0 GHz. It is worth noting that in terms of R1, R2 and C1 values, this is the best possible combination we could come up with that provides an unconditionally stability over the whole active device operating frequency as well as a good noise figure and small signal gain at the desired operating frequency (at 2.4 GHz).

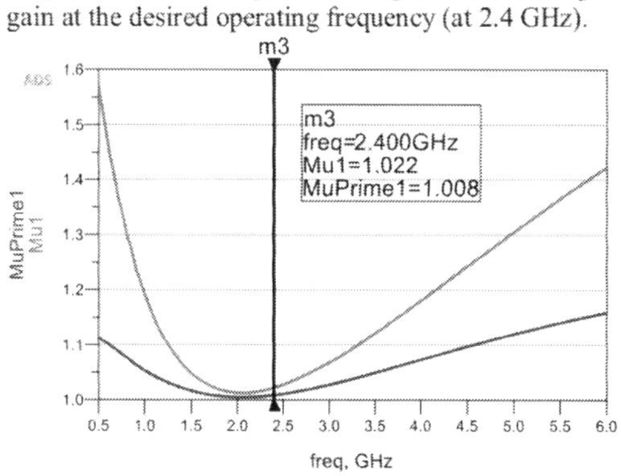

Fig. 3. Stability factors curves from 0.5 GHz to 6.0 GHz.

2.4 Bias network design

Biasing of active devices comprises two parts: the selection of a bias Q-point for an optimum performance

https://doi.org/10.1051/itmconf/20224802003

in terms of noise figure, gain, output power, etc. and the biasing network itself. Here, the first part was done above, and we will handle the second part at this section.

In fact, a biasing network consists of a DC block and an RF chock. It is up to the designer, the chock can be realized using an RFC, especially for low frequency applications, or using a $\lambda/4$ high impedance transformer. In this work, we opt for the $\lambda/4$ bias network. In terms of substrate, the Rogers RO4350B was used. The proposed biasing network is shown in figure 4.

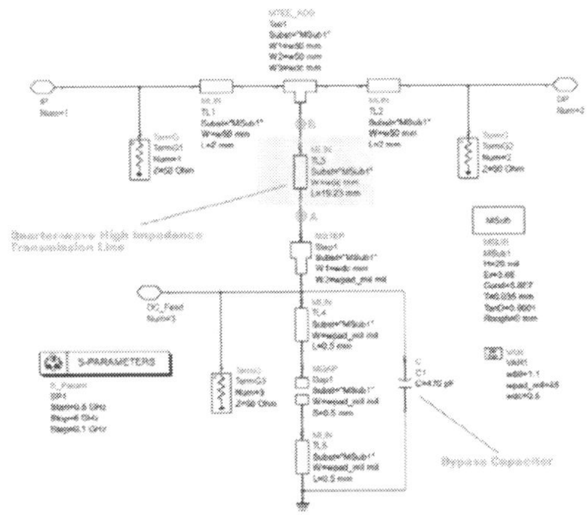

Fig. 4. The proposed biasing network

The length of the 90° quarter-wave high impedance transmission line (TL3) is calculated at the frequency center of our frequency band (19.23 mm @ 2.4 GHz).

For the connection purposes, two 50Ω transmission lines segments are added in the input and the output of the proposed biasing networks as shown in figure 4. It is worth noting that it is very important to model all possible discontinuities to account for their effect including the decoupling capacitors etc. In addition to the high impedance quarter-wave microstrip line, a bypass capacitor of 470 pF is used to connect this latter to the ground. However, a good thing to remember is that capacitor will have a footprint on the PCB, and this footprint will have certain physical size. In fact, the concept on which the quarter-wave microstrip line is working it is assuming a ground at the A point (figure 4), consequently, the $\lambda/4$ microstrip line will be considered as an open circuit at the B point shown in figure 4. Therefore, an RF signal from the input port (IP) will see a perfect open circuit at the B point, and the RF energy will transmitted to the output port (OP).

The obtained S-Parameters simulation results are shown in figure 5 and figure 6 respectively. From the S_{11} and S_{22} responses, we obtain a perfect matching at the frequency center of our desired frequency band as well as we obtain a good enough bandwidth to play with (the red dashed square represent the bandwidth under -20 dB).

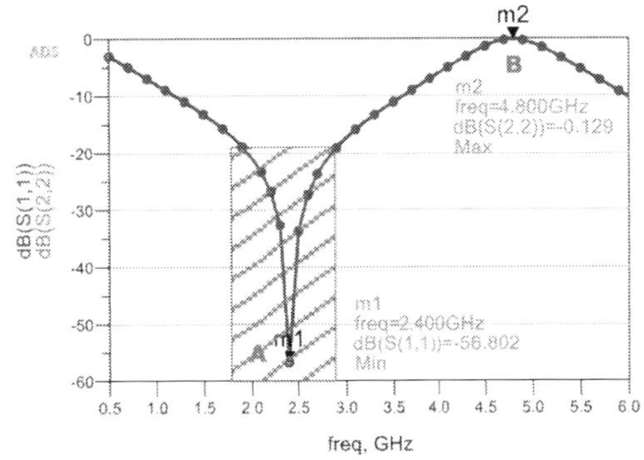

Fig. 5. S-parameters results of the proposed biasing network (S_{11} and S_{22})

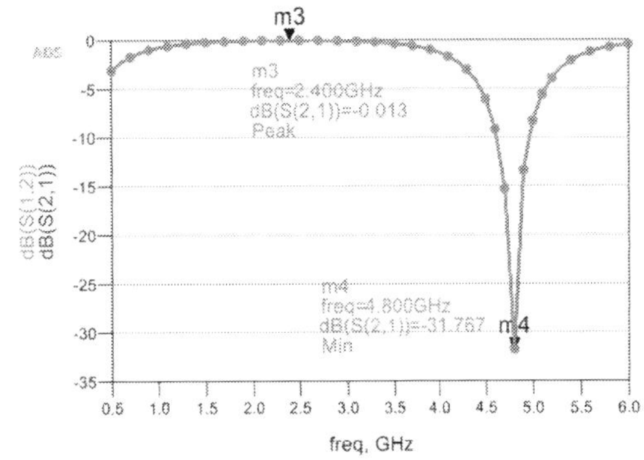

Fig. 6. S-parameters results of the proposed biasing network (S_{12} and S_{21})

In addition, from the S_{21} curve, we obtain a perfect transmission at the 2.4 GHz. On the other hand, the λ/4 line which we have at the 2.4 GHz (the red A point) will be λ/2 at the 4.8 GHz (the red B point), consequently, we obtain a perfect short-circuit at the 4.8 GHz. Therefore, if we have any harmonic coming into the amplifier, it will be grounded. Similarly, from the S_{21} response at the 4.8 GHz, we obtain a very high rejection from the input to the output

2.5 Impedance matching requirement

Before we start looking at impedance for matching purposes, it is recommended to visualize and model as many transmission lines, discontinuities etc. as possible. This will help to avoid extra rework during matching network design phase, which will make the circuit design suitable for layout generation eventually.

On the other hand, before starting the impedance matching design process, we need information about the input and output impedance values of the resulted circuit including the active device, the stability components and the biasing networks.

For this reason, we perform a preliminary S-parameters simulation. The obtained results are illustrated in figure 7, 8 and 9 respectively.

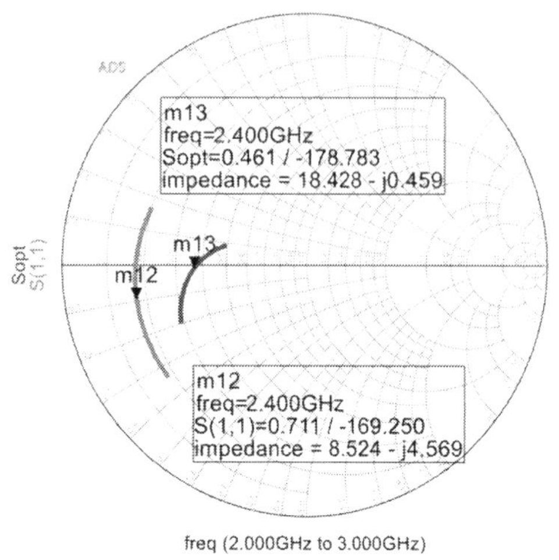

Fig. 7. Input impedance value for Γ_{opt} and S_{11}

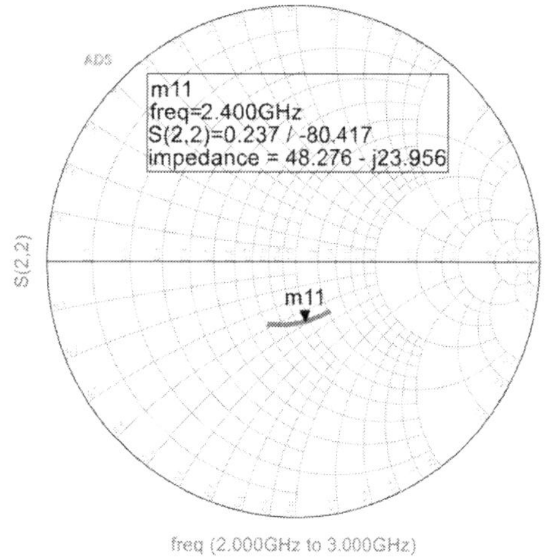

Fig. 8. Output impedance value

From figure 8, the output impedance value is around {49 – 24*j Ω}. However, for the input side, three approaches can be adopted during the determination of the input impedance value. From figure 7, if we opt for a noise figure minimum (NF_{min}), then we can use 18.42 – 0.45*j Ω as an input impedance, however, if we opt for the best input reflexion coefficient, then 8.52 – 4.5*j Ω can be adopted as an input impedance value.

Moreover, from the figure 9, the third approach consists of choosing between the input impedance that produce a minimum noise figure around 1.1 dB {9 + 5.7*j Ω} and the other that allows achieving a high gain around the 16 dB {7.73 – 0.87*j Ω}.

However, by comparing the input impedance that produce the best possible input reflexion coefficient and the other one that produce a noise figure around 1.1 dB, we can remark that are very close. Therefore, if we choose for example 8.5 – 4.5*j Ω as an input impedance, we will obtain the best possible input reflexion coefficient with a noise figure around 1.1 dB.

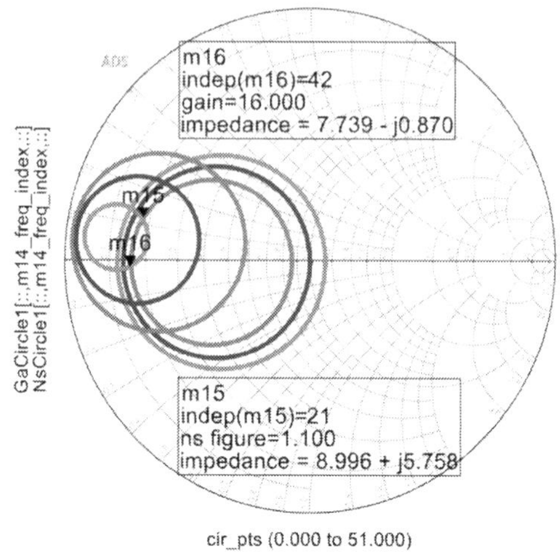

Fig. 9. Input impedance value for different noise figure and gain values using gain and noise figure circles

Basing on the above discussion, the selected source and load impedances that we have to match are summarized in the table below:

Table 1. Source and load target impedances.

Source target impedance (Z_{in})	49 – 24*j Ω
Load target impedance (Z_{out})	8.5 – 4.5*j Ω

2.6 The proposed matching networks

From the discussion above, we have now the information about the load and source impedances that we have to match. So, the next step consists of designing the input matching network that match 50 Ω to 8.5 – 4.5*j Ω as well as the output matching circuit that transform the output 50Ω to 49 – 24*j Ω. The preliminary proposed input and output matching networks are illustrated in figure 10 and 11.

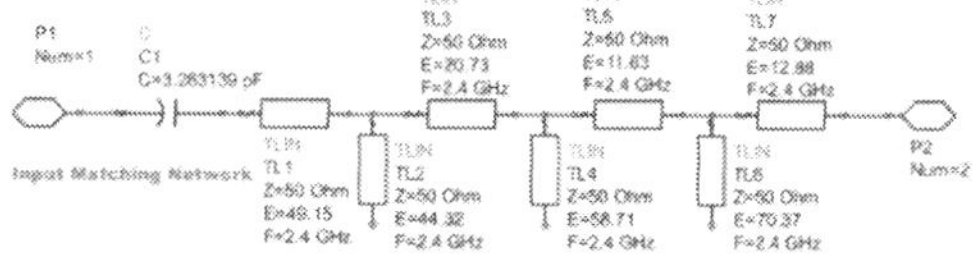

Fig. 10. Preliminary input matching network

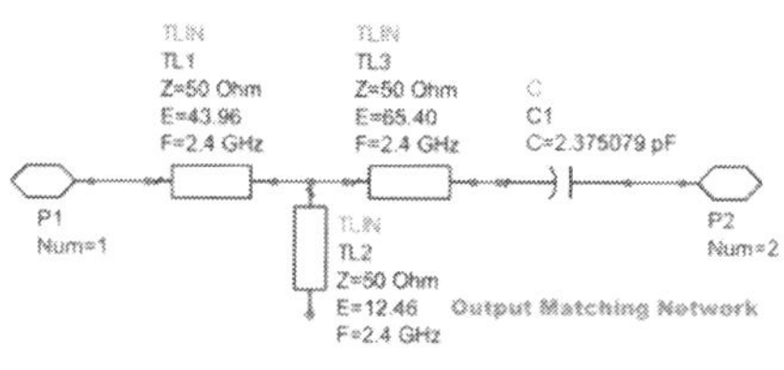

Fig. 11. Preliminary output matching network

As we can see, the preliminary proposed matching networks are based on transmission lines that are in terms of impedance and electrical length. Therefore, a further transformation is needed in order to obtain the practical matching networks.

2.7 The proposed LNA

The proposed LNA layout is illustrated in figure 12. The amplifier is implemented on Rogers RO4350B substrate and based on ATF21170 active device. The latter is a GaAs transistor. As we can remark, the input and output matching networks as well as the biasing circuitry all are designed using microstrip transmission lines. The proposed LNA is designed to operate around the ISM band 2.4 GHz.

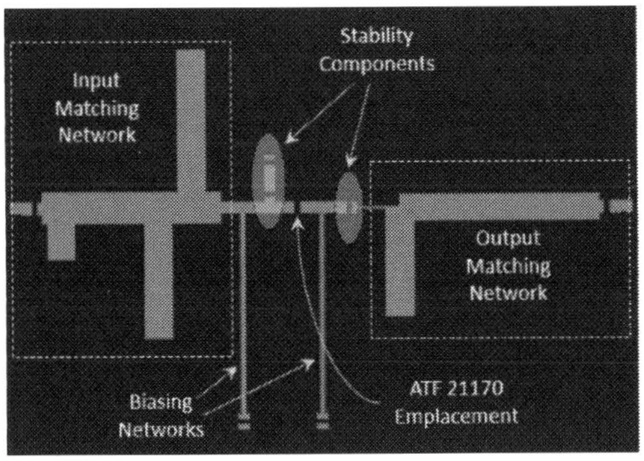

Fig. 12. The proposed LNA layout

3 Results and discussions

In this section, the obtained results are presented and discussed. The characterization and validation of the proposed LNA is subdivided into several steps. The first one consists of performing the small signal simulation. The latter allows checking small signal performances in terms of reflexion coefficients, the small signal gain, the reverse isolation coefficient, the stability and the noise figure. The next step consists of performing the EM Co-Simulation. In fact, the EM Co-Simulation provides the closest result to the real performances. Thereafter, the nonlinear simulation is performed. The aim of this simulation is to validate the nonlinear performances in terms of output power, the third order intercept point (IP3) and inherently the linearity, and the power added efficiency (PAE). Finally, the EM Circuit Excitation is performed as a final step in the characterisation and validation process.

3.1 Small-Signal & EM Co-Simulation results

The small-signal results versus the EM Co-Simulation performances are presented in this subsection. Figure 13 and 14 show the input and output reflexion coefficients performances. As we can see, the proposed LNA achieves an excellent impedance matching with an input/output return loss below than $-15\,dB$ over the operating frequency band ranging from 2.25 GHz to 2.55 GHz. On the other hand, the EM Co-Simulation shows a slight degradation of the input return loss and a bit improvement of the output return loss, but in overall, the obtained results still show excellent impedance matching performance.

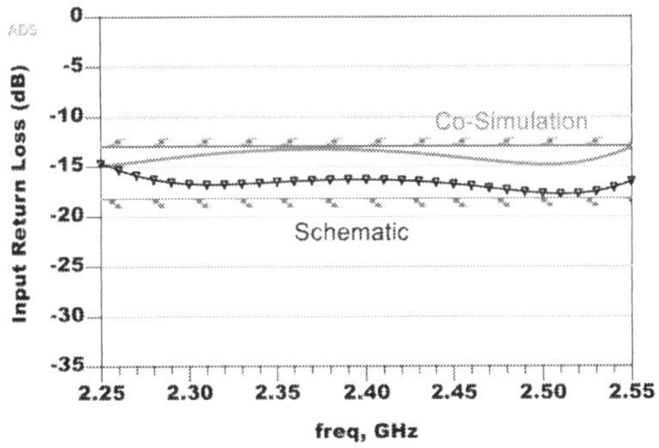

Fig. 13. Input return loss curves (schematic and EM Co-Simulation) versus frequency

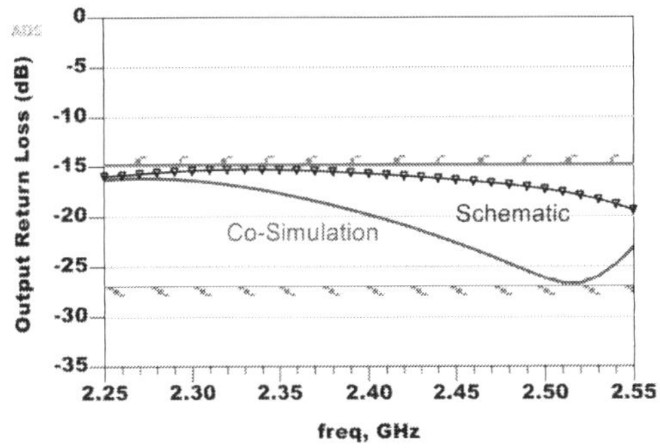

Fig. 14. Output return loss curves (schematic and EM Co-Simulation) versus frequency

On the other hand, the proposed LNA shows an excellent small signal gain changes between 15 dB and 17.5 dB over the completely operating frequency band. Again, the EM Co-Simulation shows a slight degradation of 0.5 dB but the proposed LNA still shows an excellent small signal gain as illustrated in figure 15.

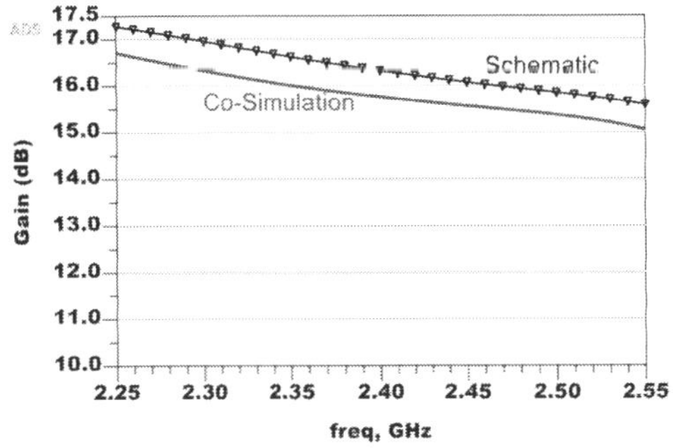

Fig. 15. Small signal gain versus frequency

In terms of stability performances, the proposed LNA is unconditionally stable over the entire operating frequency band. From figure 16, the Edward-Sinsky stability factors μ and μ prime, are both above the unity over the frequency band ranging from 2.25 GHz to 2.55 GHz. On the other hand, the proposed LNA shows an excellent noise figure (NF) around 1 dB over the entire operating frequency band as illustrated in figure 17.

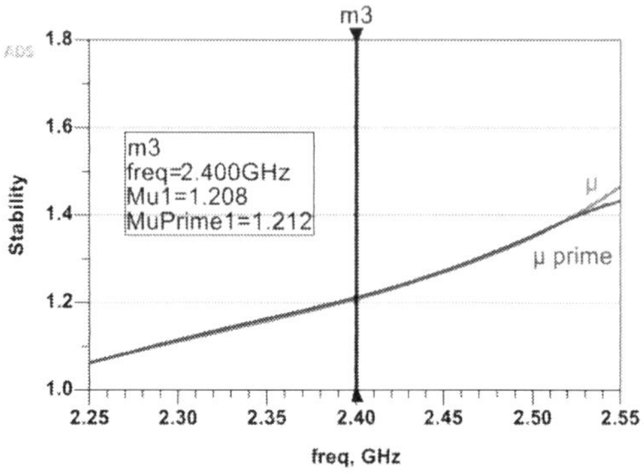

Fig. 16. Small signal gain versus frequency

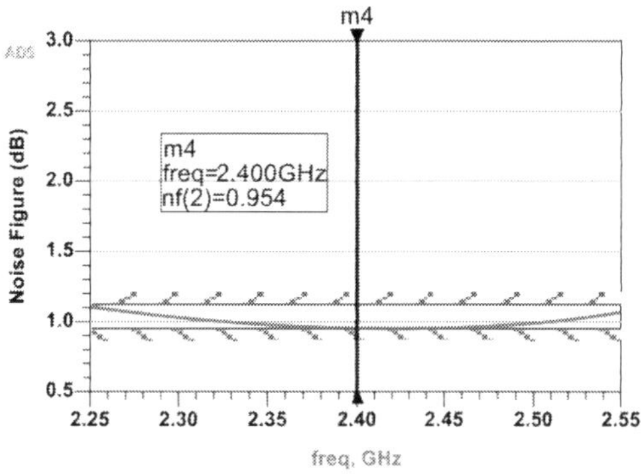

Fig. 17. Noise Figure response versus frequency

3.2 Nonlinear performances

In this subsection, the nonlinear performances in terms of output power, PAE, the gain compression, the intermodulation products and the IP3 are presented and discussed.

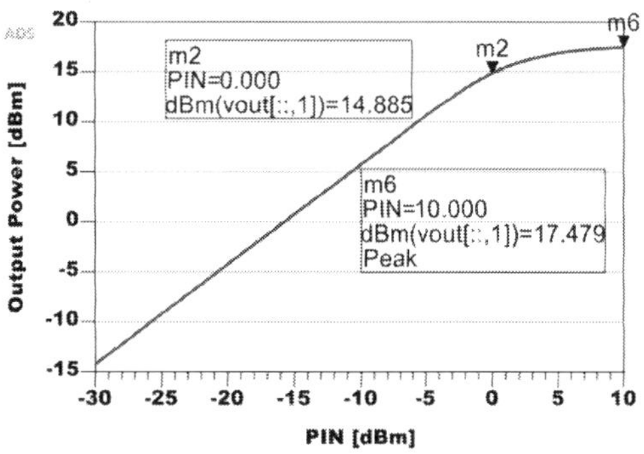

Fig. 18. Output power versus input power @ 2.4 GHz

From figure 18, the proposed LNA achieves a saturated output power of 17.5 dBm, which leading to about 56.2 mW. In addition, the proposed amplifier reaches a large signal gain of 15.8 dB around the fundamental frequency as shown in figure 19. On the other hand, from those figures, the required input power that tends the proposed LNA into 1-dB compression is around zero dBm. Also, the proposed LNA reaches a maximum PAE of 22.5% as depicted in figure 20.

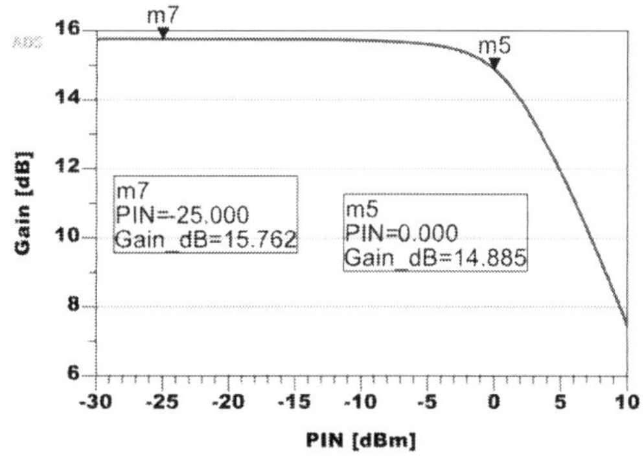

Fig. 19. Large signal gain @ 2.4 GHz

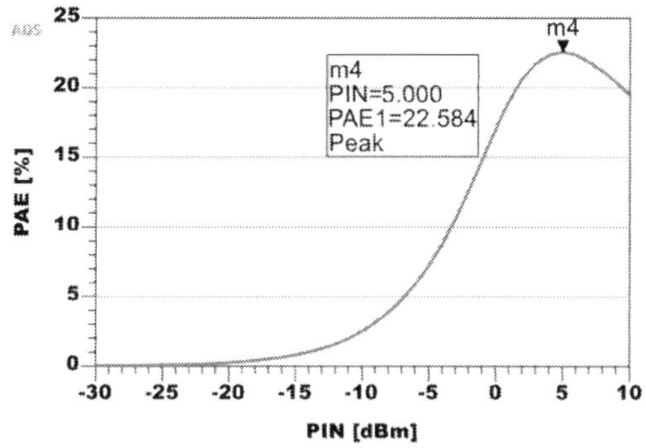

Fig. 20. Power Added Efficiency @ 2.4 GHz

On the other hand, harmonics and intermodulation products are undesired signals. They create leakage into adjacent channels, noise and distortion, etc. The typical way of dealing with the troublesome products is through filtering, but this becomes difficult when the products are very close in frequency to the desired (fundamental) signals. In fact, in two-tone simulation, the LNA output, in addition to the DC, fundamental frequencies, second harmonics and third harmonics, also contains second-order intermodulation products at $f1 \pm f2$ and third-order intermodulation products at $2f1 \pm f2$ and $2f2 \pm f1$. In a narrowband amplifier and when f2 and f1 are very close to each other, all frequencies except f1, f2, $2f1 - f2$, and $2f2 - f1$ can be filtered out. However, the third-order products, which are very close to the fundamental frequency f1 or f2, are very difficult to be filtered out and will distort the desired signal.

Referring to the third-order products, the intercept IP3 is defined as the point where the power in the fundamental tone and the third order product are equal if the amplifier is assumed to be linear. The two-tone simulation results of the proposed LNA are illustrated in figure 21.

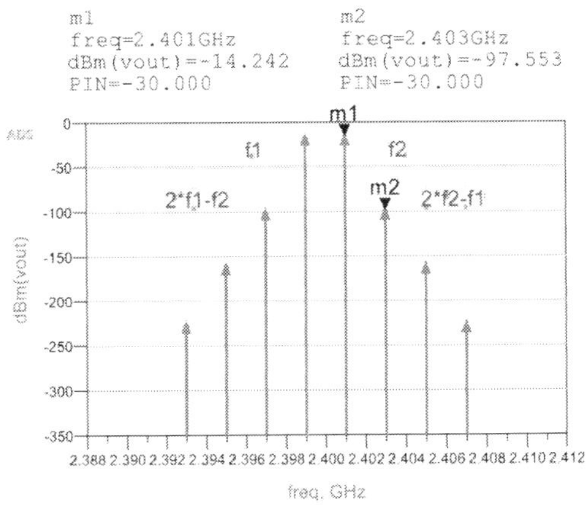

IIP3	OIP3
11.113	27.413

Fig. 21. Two-tone simulation results {f_1: 2.399 GHz, f_2: 2.401 GHz}

From figure 21, the third order product of the proposed LNA has a very weak power around -97 dBm at -30 dBm as input power, which is very weak, while the fundamental power is around -14 dBm at the same input power.

On the other hand, the input point where the power in the fundamental tone and the third-order product are equal (if the LNA assumed to be linear) is around 11 dBm as illustrated in figure 21 (IIP3=11.113 dBm). Moreover, the proposed LNA shows an output intercept OIP3 around 27.5 dBm.

3.3 EM Circuit Excitation results

The EM-Circuit Excitation simulation allows visualizing field data such as surface currents and far field of the EM Model/EM Co-simulation used with any circuit elements contribution in the schematic. The obtained results are presented below.

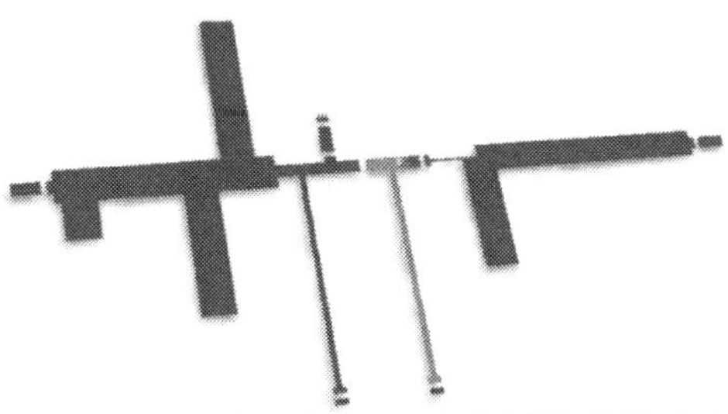

Fig. 22. EM-circuit excitation result @ 0 GHz (DC point)

At the DC frequency, the only current visualized on the structure is the drain current, the gate current is not visualized because it is very small with respect to the drain current as illustrated in figure 22.

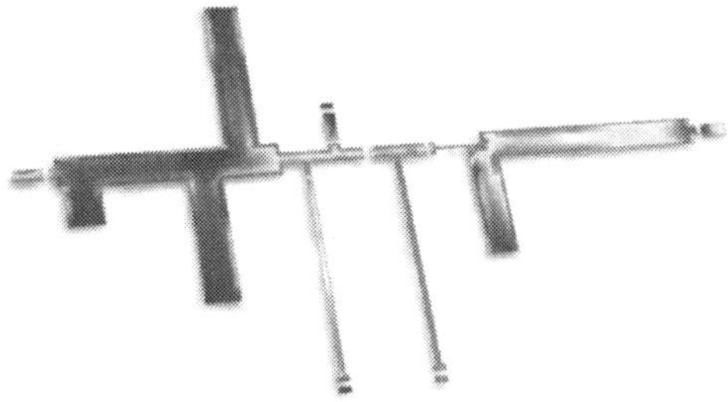

Fig. 23. EM-circuit excitation result @ the fundamental tone

Fig. 24. EM-circuit excitation result @ 2nd harmonic

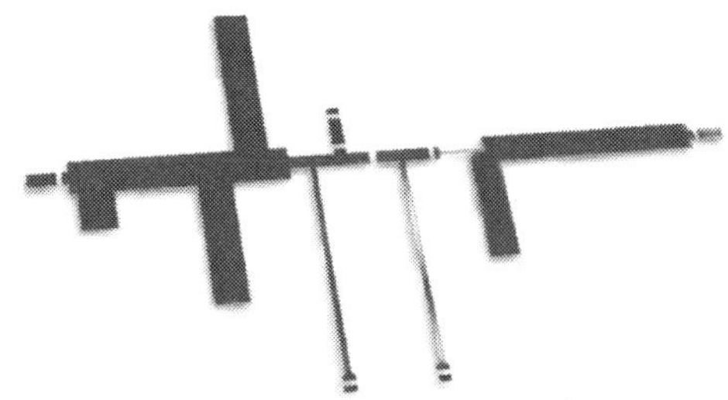

Fig. 25. EM-circuit excitation result @ 3rd harmonic

On the other hand, at the fundamental frequency (figure 23), the output part of the proposed structure shows more current propagation than the input part. This because of the active device gain. By feeding -20 dBm for example at the input port, we will obtain about – 5 dBm at the output side and inherently more current propagation.

However, as for the second harmonic, all current are absorbed by the bias network (figure 24), which confirms the above statement about the proposed biasing circuits. They are regarded as $\lambda/4$ at the fundamental frequency and as $\lambda/2$ at the second harmonic component.

For the third, fourth and fifth harmonics, the amount of the propagated current is very small (for example figure 25).

3.4 Benchmarking

Table below compares the proposed LNA performances with recently published works.

From table 2, the proposed LNA is well positioned compared to recently published works. Indeed, the proposed impedance matching techniques allow achieving excellent performance in terms of reflection coefficients (input and output) and small signal gain (S21).

Table 2. Performances comparison of the proposed LNA with contemporary published works.

	[10]	[12]	[14]	[16]	This work
Freq. [GHz]	2.4 & 5.2	2.45	2.4	2.4	2.25 – 2.55
S_{11} [dB]	-14|-12.8	-5	-15	-11|-9	-17|-15
S_{22} [dB]	-	-18.82	-7	-21|-15	-20|-15
S_{21} [dB]	14.3|15.9	26.48	23.9	15	15.5|17.2
Pout [dBm]	2.25	6.52	0	-10	17.47
NF [dB]	1.8|2.7	1.044	5.6	2.6	0.95|1.06
IIP3 [dBm]	-9	5	-5.8	-13	11.11
OIP3 [dBm]	0	4	18	0	27.41

On the other hand, thanks to the proposed bias and stability circuits, the proposed LNA shows excellent results in terms of noise figure and stability compared to the references listed in table 2. Moreover, the results obtained in terms of output power, gain compression and intermodulation distortion promote the positioning of the proposed LNA compared to other LNAs.

4 Conclusion

In this paper, a low noise amplifier (LNA) operating around the 2.4 GHz ISM band is presented. The proposed LNA is based on GaAs ATF 21170 active device and implemented on Rogers RO4350B substrate.

With the neatly designed stability, biasing and matching networks, the proposed LNA reaches an unconditionally stability over the whole operating frequency band ranging from 2.25 GHz to 2.55 GHz with an excellent noise figure around 1 dB. In terms of matching performances, the proposed LNA shows an excellent input/output return loss, both below that -15dB over the entire operating frequency band. On the other hand, the proposed LNA shows excellent nonlinear performances. In fact, the introduced LNA reaches a saturated output power around 17.5 dBm (56.2 mW) with a PAE of 22.5%. In terms of linearity, the proposed LNA achieve an excellent input third-order intercept point (IP3) of 11 dBm with and output IP3 around 27.5 dBm. An LNA with these performances is very suitable for many applications operating around the 2.4 ISM band such as medical microwave imaging, the 802.11 small radiation application, etc.

References

1. Aditi and M. Bansal, *A High Linearity and Moderate Gain LNA for Receiver Front-End Applications in 2.4GHz ISM Band,* 2017 International Conference on Innovations in Control, Communication and Information Systems (ICICCI), pp. 1-5, (2017)
2. V. K. Chalka and N. Z. Rizvi, *CMOS low-noise amplifier for 2.4-GHz wireless receivers*, 2016 International Conference on Electrical, Electronics, and Optimization Techniques (ICEEOT), Chennai, pp. 4235-4239, (2016)
3. P. Fish, *Electronic Noise and Low Noise Design*, Macmillan New Electronics Series, Macmillan Press, (1993)
4. A. C. Lowaton and E. G. O. Tabacon, *A 2.4GHz ISM Band Application Low Noise Amplifier with Active Inductor Load Implemented on 65nm CMOS Process*, 2019 19th International Symposium on Communications and Information Technologies (ISCIT), pp. 590-594, (2019)
5. T.T Nga, *Ultra Low-Power Low-Noise Amplifier Designs for 2.4 GHz ISM Band Applications*, Nanyang Technological University, July (2012)
6. S. Murad, S. Mohyar, et.al. "Design of a 2.4GHz CMOS Low Noise Amplifier for Wireless Sensor Network Applications", ARPN J. Eng. Appl. Sci., ***11**, no. 8, (2016)*
7. G. B. Rao, *A high gain and high linear LNA for low power receiver front-end applications*, 2016 International Conference on Communication and Signal Processing (ICCSP), pp. 1964-1967, Melmaruvathur,(2016)
8. M. Daoud, H. Mnif and M. Ghorbel, *A low power low noise amplifier for 2.45GHz ISM receiver for Body Area Network,* 2016 11th International Design & Test Symposium (IDT), pp. 296-301, (2016)
9. C.-C. Meng, et.al, *2.4/5.7-GHz CMOS Dual-Band Low-IF Architecture Using Weaver Hartley Image Rejection Techniques,* IEEE Transactions on Mic. The. Tech., **57**, no. 3, pp. 552–561, (2009).
10. H. Khosravi, et.al, *A Low Power, High Gain 2.4/5.2 GHz Concurrent Dual-Band Low Noise Amplifier,* IEEE 9th Annual Computing and Communication Workshop and Conference (CCWC), pp. 0788-0792, (2019)
11. D.M Pozar, *Microwave Engineering".* Hoboken, NJ: Wiley, APA. (2012).
12. P. K. Verma et.al, *A low power high gain low noise amplifier for wireless applications,* 2015 Communication, Control and Intelligent Systems (CCIS), *pp. 363-367, (2015)*
13. J. Bahl, *Fundamentals of RF and Microwave Transistor Amplifiers*, New Jersey, John Wiley & Sons, (2009).
14. S. A. Z. Murad, et.al, *High gain 2.4 GHz CMOS low noise amplifier for Wireless Sensor Network Applications,* 2013 IEEE International RF and Microwave Conference (RFM), pp. 39-41, (2013)
15. B. Guo, J. Gong, and Y. Wang, *A wideband differential linear lownoise transconductance amplifier with active-combiner feedback in complementary MGTR configurations,* IEEE Trans. Circuits Syst. I, Reg. Papers, **68**, no. 1, pp. 224–237, (2021).
16. J. -Y. Hsieh and H. -C. Kuo, *A 0.4-V High-Gain Low-Noise Amplifier Using a Variable-Frequency Image-Rejection Technology*, in IEEE Mic. Wir. Com. Letters, **32**, no. 4, pp. 324-326, (2022)
17. K.W Yeom, *Microwave Circuit Design A Practical Approach Using ADS,* Prentice Hall, Pearson Education Inc., (2015)
18. Akhchaf, S. Khoulji & et.al. *A Novel and Single Chip Tri-Band Low Noise Amplifier for WLAN, WiFi and WiMax Receivers*, Inter. J. Comp. Sci. & Inf. Tech. (IJCSIT), **4**, no. 6, (2012)
19. S. Joo, T. Choi and B. Jung, *A 2.4-GHz Resistive Feedback LNA in 0.13-μm CMOS,* IEEE J. Solid-State Circuits, **44**, no. 11, pp. 3019-3028, (2009).

Design of High-Power High-Efficiency Broadband GaN HEMT Power Amplifier for S Band Applications using Load-Pull Technique

Mohamed Ribate[1*], *Rachid* Mandry[1], *Larbi* Elabdellaoui[1], *Fouad* Aytouna[2], *Mohammed* Benbrahim[3]

[1]MIET Laboratory, Faculty of Sciences and Techniques of Settat, Hassan 1st University, Morocco
[2]LABTIC, ENSATE, Tetuan, UAE, Morocco
[3] ENSA of Kenitra, Ibn Tofail University, Morocco

Abstract. This work reveals the design for broadband high-power high-efficiency GaN HEMT power amplifier, operating in the frequency band ranging from 2.15 GHz to 2.65 GHz. The proposed structure is implemented using Cree GaN HEMT CGH40010 transistor. The high-power and high-efficiency performances over the broadband bandwidth are achieved using the load-pull technique. The proposed power amplifier is unconditionally stable over the entire operating frequency band. With the neatly designed matching circuits, the introduced power amplifier shows an excellent input/output matching. The simulated results show a flat power gain of 15 dB with an output 1-dB compression point of 14 dB. In terms of large-signal performance, the proposed amplifier reaches a saturated output power of 41.3 dBm (~13.6 Watts) with a PAE of 64% and a drain efficiency of 72%. The proposed design achieves an excellent linearity with an output third order two-tone intercept point TOI of 48 dBm.

1 Introduction

A power amplifier (PA) is usually located just before the antenna, and drives the transmitting power of the communication system [1-3]. It is the final amplification stage before the transmission of the RF signal, and as a result, it must provide enough output power to overcome channel losses between the transmitter and the receiver [4-5]. In a transmitter, the PA is the primary consumer of power; therefore, a major design constraint is how efficiently the PA can convert DC power into RF power [6-8]. The efficiency it can be translated either as longer battery life for example in case of wireless handheld or as lower operation cost, for example in case of cellular base station [9-11]. Another key requirement in today's microwave and radio frequency (RF) PA is the linearity; the PA input/output relationship must be linear in order to preserve the RF signal integrity [12-15]. However, the PA design often involves the trade-off of linearity and efficiency.

Referring to PA design, the output matching network that is always designed to deliver the maximum possible power, must be at the same time low loss to reaches the maximum possible PAE. In terms of PA design sequence, the output matching is designed first and the input matching is synthesised last [16-18].

Various techniques can be used to precisely determine an active device's input/output impedances at large-signal conditions, a crucial requirement for successful PA circuitry design. Among such techniques, the source-pull and load-pull techniques, which are used to determine source and load conditions for an RF active device that allow achieving an optimum solution in terms of PAE, output power, IP3 and ACPR [19-21].

In this paper, a high power high efficiency broadband GaN power amplifier operating in the frequency ranges from 2.15 GHz to 2.65 GHz is resented.

The paper is organized as follows. In section 2, the PA design flow is presented and described. Section 3 is devoted for the obtained results and discussions, while the concluding remarks are summarized in section 4.

2 The proposed PA design flow

2.1 Overview

The design of a PAs for a specific frequency range and application is a complex task in the sense that they have to fulfil several requirements, which are in many cases tradeoffs.

On the other hand, the PA performance requirements in terms of operating frequency band, output power, gain, noise figure, linearity, PAE, stability, input/output VSWR etc. are determined by the design process, active device technology/type, matching networks, biasing schemes, the operating class, the number of gain stages and the circuit size and topology.

* Corresponding author: ribate.mohamed@gmail.com

However, regardless of its target application, and up to the designer, the PA design may grouped into the following steps:

1. Design specifications
2. The Active device type/technology selection
3. DC analysis
4. Stability analysis
5. Biasing Networks design
6. Matching networks analysis/design
7. Circuit analysis and optimization

In this work, starting from the design specifications, the first pass resides in determining the design process and the subsequent selection of the active device. The latter is selected according the design specification goals in terms of the operating frequency band, output power, IP3 and PAE. After the proper active device selection, the subsequent step resides in DC analysis. The latter aims is to determine the proper biasing point according to the targeted operation class.

On the other hand, the stability analysis is an important part in the design process. The proposed PA is to be designed for an unconditionally stability over the entire operating frequency band. The easy way to achieve unconditionally stability is by using series resistors at the active device input/output. However, in case of PA design, a special care should be taken during the stability components selection/places to avoid any degradation of the gain and output power performances.

Whereas, the remaining steps consist of the biasing and matching networks design as well as the circuit analysis and validation. Basing on the DC analysis results, the biasing circuits are designed using microstrip line while the matching networks are synthesised using the load-pull technique.

In this work, the proposed matching networks are designed to provide the required bandwidth with an appropriate guard band. With the aim to reduce the size, the biasing networks are incorporated as part of the matching networks.

2.2 DC analysis and Q-point selection

According to the design specifications, the adopted active device in this work is CREE CGH40010. The latter is an unmatched, gallium nitride (GaN) high electron mobility transistor (HEMT). It operates from DC up to 6 GHz. The DC I-V characteristics curves of the used active device are illustarted in figure 1.

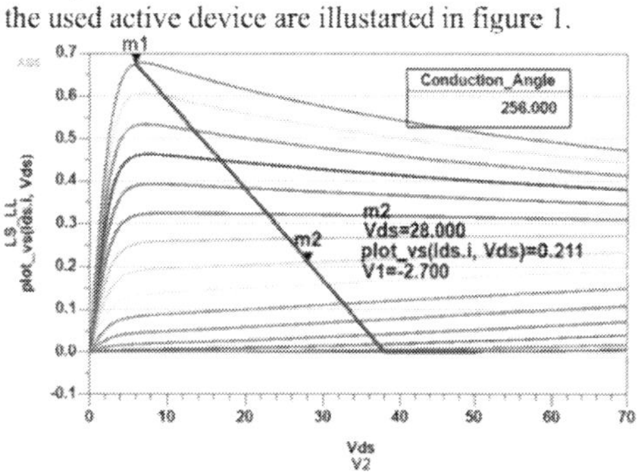

Fig. 1. DC I-V characteristics of the adopted active device.

With the aim to gain good linearity as well as high efficiency while balancing with the other design specification requirements, the proposed PA is designed to operate in class AB. In such class, the amplifier have no DC power consumption in the absence of the RF input (which is an advantage) and the drain current I_{ds} flows for a conduction angle between π and 2π.

As we can remark from figure 1, the proposed PA is biased around the class AB, and the selected biasing point is: $\{V_{ds} = 28v; V_{gs} = -2.7v\}$.

2.3 Stability analysis

By default, the adopted active device is not stable over its entire operating frequency ranging from DC to 6 GHz. As a result, a stabilization action is needed in order to achieve the unconditionally stability. The proposed stability circuit is illustrated in figure 2. Here, the emplacement of the stability series resistor is very important, it should be placed before the active device to avoid any degradation of de gain and output power. In addition, the selected stability resistor value represents the best value that combines the unconditionally stability with a good gain.

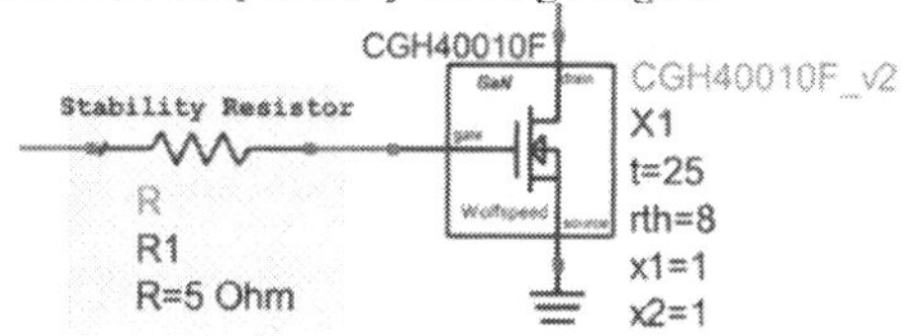

Fig. 2. The proposed stability network

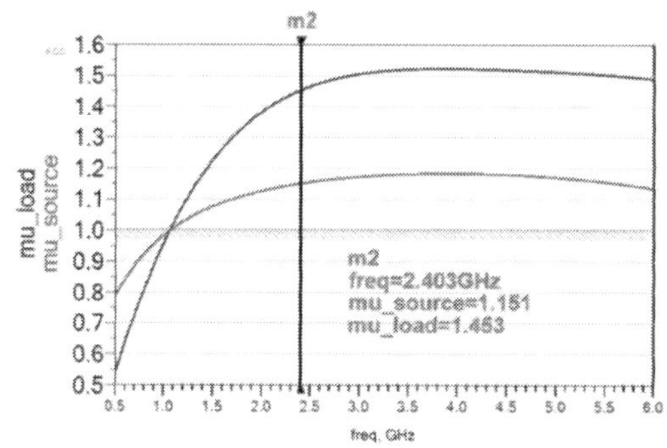

Fig. 3. Stability analysis results using Edwards-Sinsky stability factors μ and μ prime.

On the other hand, the stability analysis was performed using Edwards-Sinsky stability factors $\boldsymbol{\mu}$ and $\boldsymbol{\mu}$ prime, which are given by the following equations:

$$\mu = \frac{1 - |S_{11}|^2}{|S_{22} - \Delta S_{11}^*| + |S_{12}S_{21}|} > 1 \quad (1)$$

$$\mu' = \frac{1 - |S_{22}|^2}{|S_{11} - \Delta S_{22}^*| + |S_{12}S_{21}|} > 1 \quad (2)$$

Where $\boldsymbol{\mu}$ measures the distance between the center of the Smith chart and the nearest unstable point in the Γ_L–plan, while $\boldsymbol{\mu}'$ measures the same distance, but in the Γ_S–plan. According to Edwards-Sinsky stability criterion, the unconditionally stability is achieved where μ and μ', both are greater than unity.

Using the proposed stability circuit depicted in figure 2, the obtained results are shown in figure 3. As we can remark, because of the proposed stability circuit, the selected active device is unconditionally stable over the frequency band ranging from 1 GHz to 6 GHz.

2.4 Load-pull simulations

In this work, the proposed matching networks are designed using the load-pull matching technique. The initial load-pull results are illustrated in figure 4.

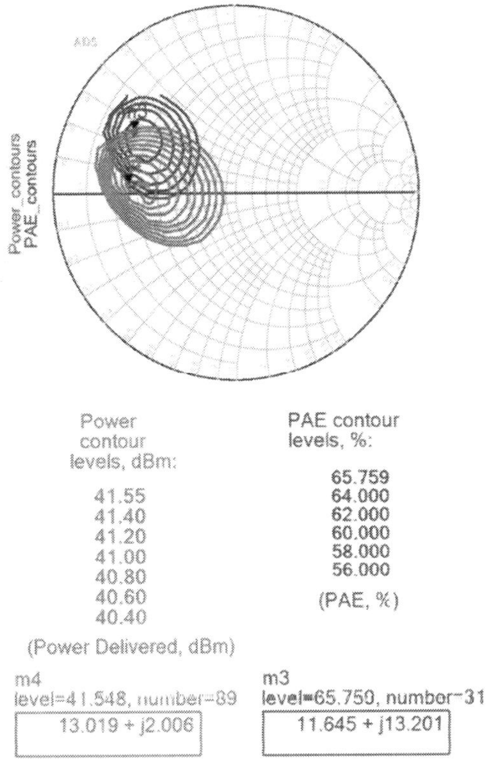

Fig. 4. Initial load-pull simulation results: PAE and delivered power contours with the output impedance for whether maximum PAE or maximum output power

Before starting the matching networks design, we need information about the active device input/output impedances that permit to achieve the design specification goals.

In fact, in order to achieve the high output power with the maximum possible efficiency, the output matching network for a power amplifier is designed to be low loss. However, to reaches this goal, the right active device output matching that can balance the high output power with the maximum possible PAE must be determined before. For that reason, the load-pull simulation is used in this work.

The initial load-pull simulation results are illustrated in figure 4, in which the delivered output power and the PAE contours, the appropriate output impedance that allows the high output power and the one that permits the maximum possible PAE are presented.

From that figure, and using the selected active device (Cree CGH 40010), the maximum PAE that can be reached is around 66%, and it can be achieved using the active device output impedance of 11.6 + j*13.2 Ω. On the other hand, the maximum output power that the active device can deliver is around 41.55 dBm, and it can be reached using an output impedance of 13 + j*2 Ω. The further information obtained from the initial load-pull simulation are presented in figure 5 and 6 respectively.

PAE_at_MaxPower	Gain_at_MaxPower	Pdel_dBm_Max
57.023	12.589	41.589

Zload_at_MaxPower	Z_In_at_MaxPower
13.189 + j1.883	5.696 + j4.211

Fig. 5. Input/output impedances for maximum output power (from the initial load-pull simulation results)

Pdel_dBm_at_MaxPAE	Gain_at_MaxPAE	PAEmax
39.965	10.965	65.825

Z_In_at_MaxPAE	Zload_at_MaxPAE
5.745 + j4.924	11.632 + j13.092

Fig. 6. Input/output impedances for maximum PAE (from the initial load-pull simulation results)

From figure 5, using 13.2 + j*19Ω as an input impedance and 5.7 + j*4.2Ω as an output impedance, we can reaches the maximum possible gain and output power, while using 5.7+j*5Ω (@ input) and 11.6+j*13 Ω (@output) instead, we can achieve the maximum possible PAE.

On the other hand, if information about the gain compression and the IP3 are required, the initial load-pull simulation is not enough. In this case, the 3-dB load-pull simulation is used.

In this work, after performing the 3-dB load-pull simulation, the adopted input/output impedances are the following:

- @ input side: $Z_S = 6.5 - j * 3.8\ \Omega$
- @ Output side: $Z_L = 28 + j * 0.5\ \Omega$

The adopted impedances allow achieving a maximum output power of 41.45 dBm with a PAE of 57% and large signal gain around 12.7 dB.

2.5 The proposed biasing and input/output matching networks

In this work, the proposed biasing and matching networks are designed using microstrip sections and implemented on Rogers RO4350B. The biasing circuits are designed to be an integral part of the matching networks. Referring to the biasing circuits, the RF choke is designed using λ/4 transmission line. The later dimensions are calculated around the frequency center of the operating frequency band. The proposed biasing and input/output matching circuits are illustrated in figure 7 and 8 respectively.

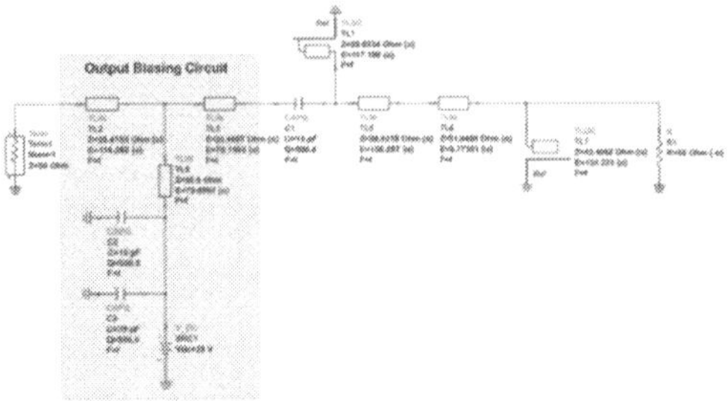

Fig. 7. The proposed output matching network.

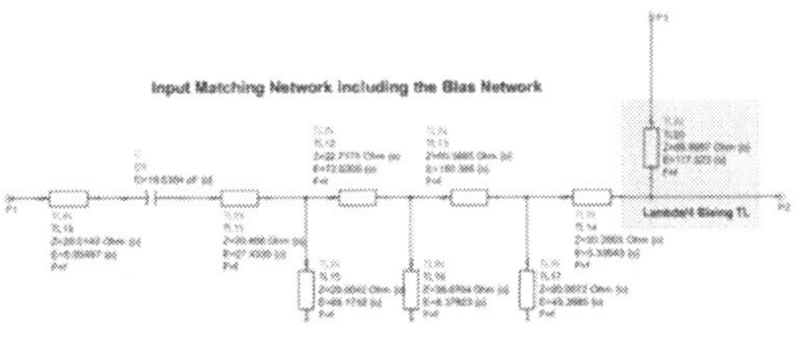

Fig. 8. The proposed input matching network

The output matching circuit depicted in figure 7 was designed using a special matching technique. In fact, the adopted active device output impedance obtained from the load-pull simulation is 28 + j*0.5 Ω. So, the output matching network aims is to match the output port 50Ω with the adopted active device output impedance. However, this latter is in Z format. Therefore, in order to use it as a target impedance, we need first to transform the S_{11} parameter of the proposed output matching network into Z format, then we perform an optimization process on the proposed output matching network with the aim to achieve the target output impedance obtained from the 3-dB load-pull simulation. On the other hand, the proposed input matching circuit illustrated in figure 8 is synthesised using the smith chart utility basing on the input matching impedance obtained from the 3-dB load-pull simulation.

2.6 The proposed PA circuit

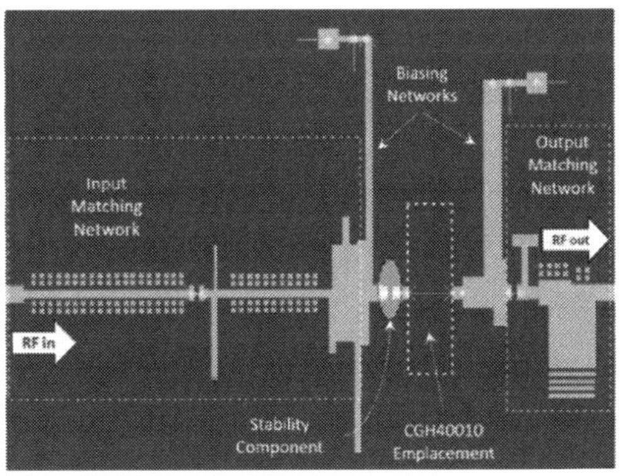

Fig. 9. The proposed PA circuit

The proposed PA circuit is shown in figure 9 in which all proposed networks in terms of biasing, input/output matching and stability networks are placed. The active device used in this work has a screw-down package, so it emplacement left free. The small square microstrip sections are added for after fabrication tuning purposes.

3 Results and discussion

In this section, the obtained results are presented and discussed. The characterization and validation of the proposed PA is partitioned into several steps. The first one consists of performing the small signal simulation. The latter allows checking small signal performances in terms of reflexion coefficients, the small signal gain, the reverse isolation coefficient and stability. The next step consists of performing the EM CoSimulation. This latter provides the closest result to the real performances. Thereafter, we will conclude with the nonlinear simulation. The aim of this simulation is to validate the nonlinear performances in terms of output power, the third order intercept point (IP3) and inherently the linearity, and the drain efficiency and power added efficiency (PAE).

3.1 Small-signal simulation results

As mentioned above, the Edwards-Sinsky stability factors (μ and μ prime) are used in this work for the stability analysis. The obtained results are depicted in figure 10. From that figure, we can remark that both μ and μ prime are greater than unity, therefore, the proposed PA is unconditionally stable over the entire operating frequency band ranging from 2.15 GHz to 2.65 GHz.

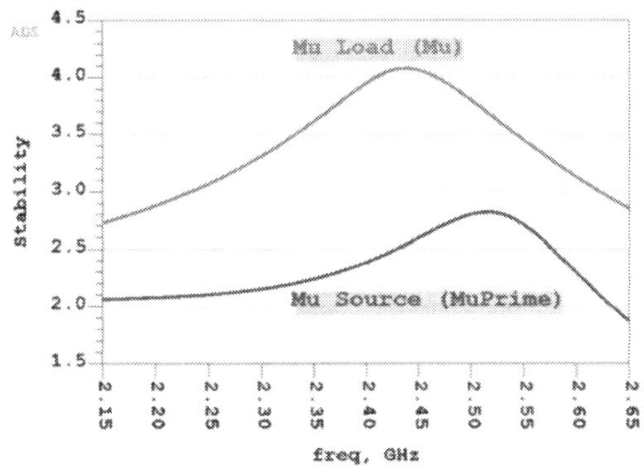

Fig. 10. Stability analysis of the entire PA circuitry over the frequency band ranging from 2.15 GHz to 2.65 GHz.

On the other hand, because of the neatly designed matching networks, the proposed PA achieves excellent matching performances. As illustrated in figure 11 and 12, from the schematic simulation, the proposed PA shows excellent input/output reflexion coefficients performances with an input return loss (under -16 dB) and an output return loss (under -18 dB) over the entire operating frequency band. However, comparing the schematic simulation results with the co-simulation, the latter shows a slight degradation of the input/output return loss performances, but we still achieving excellent matching performances.

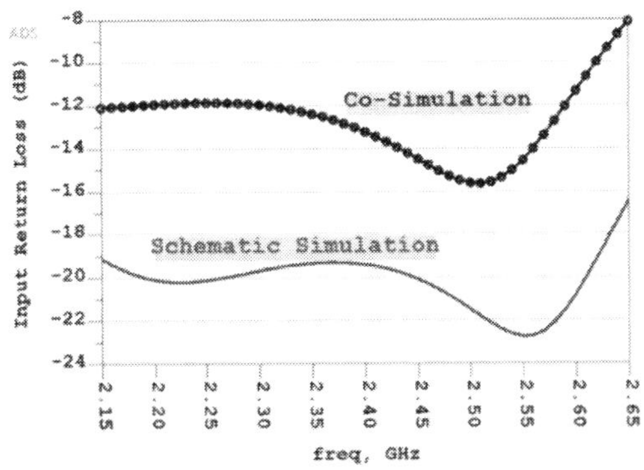

Fig. 11. Input return loss versus frequency.

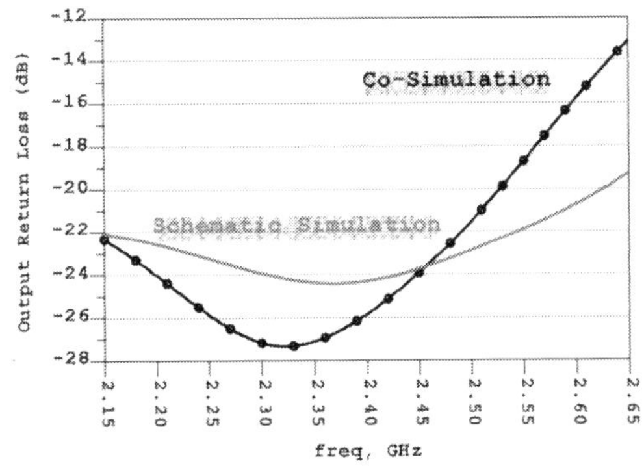

Fig. 12. Output return loss versus frequency.

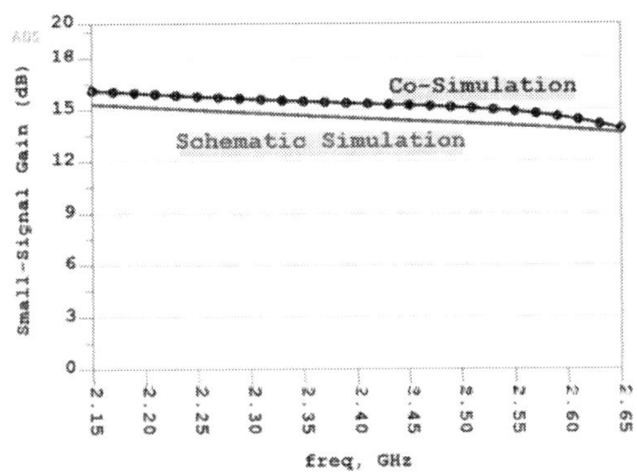

Fig. 13. Small-signal gain versus frequency.

From figure 13, the proposed PA reaches an excellent flat small signal gain around 15 dB over the entire operating frequency band from 2.15 GHz to 2.65 GHz. In addition, the schematic simulation result is very close to co-simulation performance.

3.2 Nonlinear performances

In this subsection, the nonlinear performances in terms of gain compression (figure 14), output power (figure 15), PAE (figure 16), drain efficiency (figure 16) and the IP3 (figure 17) are presented and discussed.

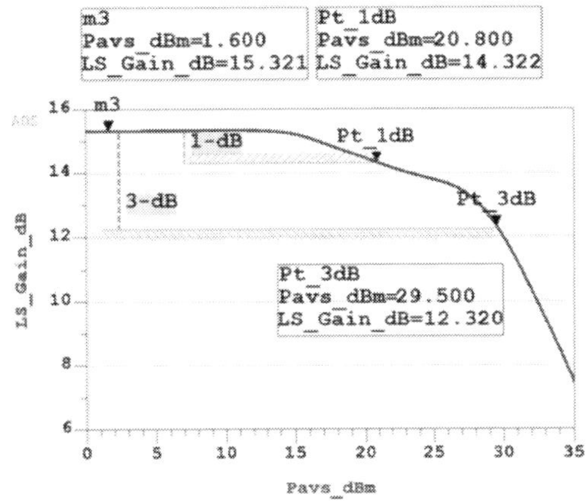

Fig. 14. Large signal gain versus input power

From figure 14, the proposed PA reaches an excellent large signal gain around 15 dB. In addition, the required input power to achieve the 1-dB compression point is about 20.8 dBm, while we need a 29.5 dBm as input power to reach the 3-dB compression point.

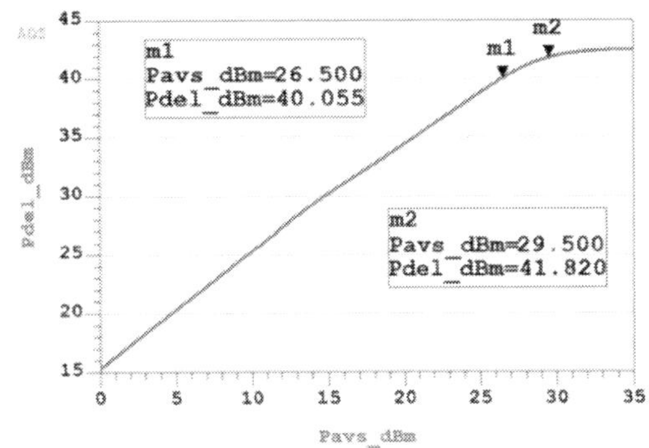

Fig. 15. Output power versus input power

On the other hand, the proposed PA achieves a saturated output power around 42 dBm (15 watts) as depicted in figure 15. In addition, at 1-dB compression point, the output power achieved is around 40 dBm, while the output power at 3-dB compression point is near than 42 dBm.

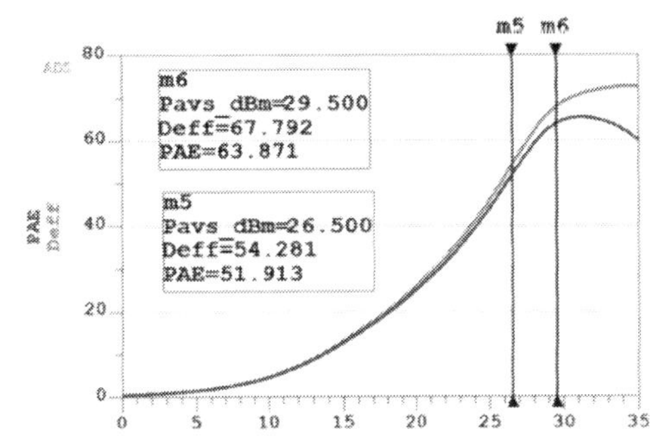

Fig. 16. PAE and drain efficiency versus input power

In terms of efficiency, the proposed PA touches a maximum PAE around 65% with a maximum possible drain efficiency of 75% as illustrated in figure 16.

https://doi.org/10.1051/itmconf/20224802004

Besides, at 1-dB compression point, the proposed PA achieves a PAE of 51% with a drain efficiency of 54%. However, at 3-dB compression point, the PA provides a PAE around 64% with a drain efficiency of 67%.

In terms of linearity, the third order intermodulation products as well as the intercept IP3 results are illustrated in figure 17. Referring to the third-order intermodulation products, the intercept IP3 is defined as the point where the power in the fundamental tone and the third order product are equal if the amplifier is assumed to be linear.

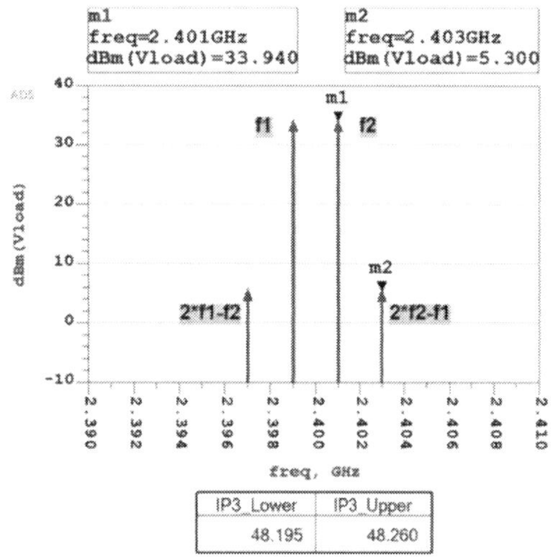

Fig. 17. Two-tone simulation analysis {f_1: 2.399 GHz, f_2: 2.401 GHz}.

From the two-tone simulation results illustrated in figure 17, the proposed PA shows an excellent output intercept IP3 around 48 dBm, which is an excellent linearity performance.

3.3 Benchmarking

Table 1 compares the proposed PA performances with the contemporary published works.

From table 1, the proposed PA is well positioned compared to recently published works. In fact, the proposed output impedance matching network as well as the selected bias point allow reaching excellent nonlinear performance in terms of output power output, PAE and OIP3. Compared to the contemporary published works, the proposed PA provides a wide bandwidth around the 2.4 GHz ISM band with an excellent output power and OIP3 using only one stage.

Table 1. Performances comparison of the proposed PA with contemporary published works

	[15]	[17]	[18]	[20]	This work
Freq. [GHz]	2.19	2.4	2.14	2.6	2.15 – 2.65
# Stage	2	2	2	2	1
Gain [dB]	23	15	16.6	7.5	15.35
Pout @ 3-dB [dBm]	38	37.3	36.9	41.63	41.34
PAE @ 3-dB [%]	70	63.3	55.7	49.48	64.64
DE [%]	75	68	57.0	58.2	71.94
OIP3 [dBm]	NA	NA	NA	NA	48

4 Conclusion

In this paper, a high output power and high efficiency power amplifier (PA) operating in the frequency band ranging from 2.15 GHz to 2.65 GHz is presented. The proposed PA is based on Cree CGH40010 GaN active device and implemented on Rogers RO4350B substrate.

The proposed input/output matching networks are designed based on the load-pull matching technique. The biasing circuits are designed to be an integral part of the matching networks. The RF choke used in the biasing circuitry was designed using λ/4 microstrip section. The latter dimensions were calculated around the 2.4 GHz ISM band.

In terms of performances, the proposed PA achieves an unconditionally stability over the entire operating frequency band ranging from 2.15 GHz to 2.65 GHz with excellent impedance matching performances. If fact, with the neatly designed input/output matching networks, the proposed PA reaches excellent input/output reflexion coefficients, both are under -16 dB over the entire operating bandwidth, with an excellent gain of 15 dB.

In terms of nonlinear results, the proposed PA provides a saturated output power around 42 dBm, which leading to about 15 watts, with a PAE around 63% and a drain efficiency nearby to 68% at 3-dB compression point. In terms of linearity, the proposed PA shows an excellent output intercept IP3 of 48 dBm.

A PA with the introduced performances would be very suitable for many applications operating around the 2.4 GHz ISM band that are require a high output power along a high efficiency such as radars, WCDMA, the modern and next generation wireless communication systems.

References

1. K.W Yeom, *Microwave Circuit Design A Practical Approach Using ADS*, Prentice Hall, Pearson Education Inc., (2015)
2. Y. Lee and Y. Jeong, *A High-Efficiency Class-E GaN HEMT Power Amplifier for WCDMA Applications*, in IEEE Microwave and Wireless Components Letters, **17**, 8, pp. 622-624, (2007)
3. S. C. Cripps, *Advanced Techniques in RF Power Amplifiers Design*. Norwood, MA: Artech House, (2002)
4. S. Dellier, T. Dehaene and E. Peragin, *GaN high-efficiency S-band power amplifier with power flexibility from 1 to 10 watts*, 2014 IEEE Topical Conference on Power Amplifiers for Wireless and Radio Applications (PAWR), pp. 28-30, (2014)
5. C. Poole, I. Darwazeh, *Microwave Active Circuit Analysis and Design*, Academic Press, (2016)
6. G. Hau, T. B. Nishimura and N. Iwata, *High efficiency, wide dynamic range variable gain and power amplifier MMICs for wideband CDMA handsets*, in IEEE Microwave and Wireless Components Letters, **11**, 1, pp. 13-15, (2001)
7. C. Shen, S. He, X. Zhu, J. Peng and T. Cao, *A 3.3–4.3-GHz High-Efficiency Broadband Doherty Power Amplifier*, in IEEE Microwave and Wireless Components Letters, **30**, 11, pp. 1081-1084, (2020)
8. S. Wong, S. Maisurah, M. N. Osman, F. Kung and J. See, *High Efficiency CMOS Power Amplifier for 3 to 5 GHz Ultra-Wideband (UWB) Application*, in IEEE Transactions on Consumer Electronics, **55**, 3, pp. 1546-1550, (2009)
9. H. Xu, S. Gao, S. Heikman, S. I. Long, U. K. Mishra and R. A. York, *A high-efficiency class-E GaN HEMT power amplifier at 1.9 GHz*, in IEEE Microwave and Wireless Components Letters, **16**, no. 1, pp. 22-24, (2006)
10. S. Gao, P. Butterworth, S. Ooi and A. Sambell, *High-efficiency power amplifier design including input harmonic termination*, in IEEE Microwave and Wireless Components Letters, **16**, 2, pp. 81-83, (2006)
11. J. Chen, S. He, F. You, R. Tong and R. Peng, *Design of Broadband High-Efficiency Power Amplifiers Based on a Series of Continuous Modes*, in IEEE Microwave and Wireless Components Letters, **24**, 9, pp. 631-633, (2014)
12. W. Nagy, J. Brown, R. Borges, and S. Singhal, *Linearity characteristics of microwave-power GaN HEMTs*, IEEE Trans. Microw. Theory Tech., **51**, 2, pp. 660–664, (2003)
13. A. Khan, H. Sarbishaei and S. Boumaiza, *High efficiency two-stage GaN power amplifier with improved linearity*, 2014 IEEE Topical Conference on Power Amplifiers for Wireless and Radio Applications (PAWR), pp. 4-6, (2014)
14. I. Khalil, M. Rudolph, A. Liero, M. Neumann and W. Heinrich, *High Power, High Linearity and Low-Noise Hybrid RF Amplifiers Based on GaN HEMTs*, 2009 German Microwave Conference, pp. 1-4, (2009)
15. Y. Takagi, et al., *High-Efficiency Asymmetric Doherty Power Amplifier with Spurious Suppression Circuit*, 2020 50th European Microwave Conference (EuMC), pp. 308-311, (2021)
16. M. Zhao, L. Sun, J. Wen, Z. Yu and J. Kang, *A 2.45GHz CMOS power amplifier with high linearity*, 2009 IEEE 8th International Conference on ASIC, pp. 383-386, (2009)
17. J. Kim, *Highly Efficient Asymmetric Class-F-1/F GaN Doherty Amplifier, in IEEE Transactions on Microwave Theory and Techniques*, **66**, 9, pp. 4070-4077, (2018)
18. H. Oh et al., *Doherty Power Amplifier Based on the Fundamental Current Ratio for Asymmetric cells*, in IEEE Transactions on Microwave Theory and Techniques, **65**, 11, pp. 4190-4197, (2017)
19. J. Bahl, *Fundamentals of RF and Microwave Transistor Amplifiers*, New Jersey, John Wiley & Sons, (2009).
20. J. Son, et al., *A highly efficient asymmetric Doherty Power Amplifier with a new output combining circuit*, 2011 IEEE International Conference on Microwaves, Communications, Antennas and Electronic Systems (COMCAS 2011), pp. 1-4, (2011)
21. D.M Pozar, *Microwave Engineering*. Hoboken, NJ: Wiley, APA. (2012).

Analysis of transmission lines considering frequency-dependent parameters with linear and nonlinear loads

Zahra Bouzidi[1], Hicham Rouijaa[2], Abdelaziz El Idrissi [1], and Mohamed Saih[3]*

[1]Laboratory of Electrical Systems, Energy Efficiency and Telecommunications, Faculty of Science and Technology, Cadi Ayyad University, Marrakesh, Morocco
[2]Laboratory of Systems Analysis and Information Processing, Faculty of Science and Technology, HASSAN 1 University, Settat, Morocco
[3]Laboratory of Automatic, Energy Conversion and Microelectronics, Faculty of Science and Technology, SULTAN MOULAY SLIMAN University, Beni Mellal, Morocco

Abstract. The primary objective of this paper is to model transmission lines with frequency-dependent parameters loaded with linear and nonlinear terminations using an approach based on Pade approximations. To validate our algorithm in the case of lines with constant losses and frequency-dependent parameters, examples of applications in the time and frequency domains are presented. The simulations and obtained results are reported and analysed.

1 Introduction

A transmission line is constructed using a conductor and a return (single transmission lines) or several conductors with a return (Multiconductor transmission lines), and its role is to transport energy from one place to another.

The modelling of transmission lines is more complicated than modelling discrete elements since the parameters of a line are distributed and generally dependent on the frequency. The analysis of transmission lines with frequency-dependent parameters is relevant for the design of high-speed applications. The measurements in [1] and [2] show that ignoring the effect of the dependence of the transmission line parameters leads to considerable errors in the simulation results.

The question therefore, is how to model a transmission line to simulate it on a computer. Many researchers have extensively researched This topic, especially since the 1960s, and even today, considerable work has been devoted to the development of precise and fast models (in terms of calculations). Note, for example, the work of Semlyen & Dabuleanu [3], Marti [4] and Marti [5].

Transmission lines have been modelled and analysed using several techniques and models[6].; the traditional lumped circuit segmentation macro-model allows the transmission line to be represented as a circuit model equivalent to the resistance, inductance, capacitance, and conductance RLCG-equivalent circuit model. Typical examples of such circuits are lumped-p or lumped-T circuits, which can be used for either time- or frequency-domain analyses of the line. These circuits can also be used in the analysis of multi--conductor transmission lines (MTL) [6], which complicates the equivalent circuit considerably. The method of characteristics or Branin's method [7], [8] is able to analyse an ideal transmission line by representing it in the form of a quadruple. The advantage of this quadruple form is that it does not presuppose the load conditions applied at its ends; however, this model is limited to only lossless lines.

An original method [9], [10] based on the Padé approximation makes it possible to model transmission lines in the time and frequency domains using the Padé macro-model.

This macro-model is decomposed into a set of first- and second-order cells. The cells that make up the macro-model are represented by admittance matrices. The number of these cells depends on the order of the Pade approximation of the matrix exponential, solution of the telegrapher's equation.

The aim of this paper is to integrate frequency-dependent parameters into the Padé approximation macro-model.

The paper is organised as follows. In Section 2, the fundamental aspects of the Pade approximation method are reviewed. Section 3 provides a numerical example to illustrate the validity of the Pade approximation macro-model and its application to a variety of interconnected structures.

2 Transmission lines with frequency-dependent parameters

When the frequency increases, an electric field is created, which causes the current distribution of conduction to change. As the frequency increases, the

* Corresponding author: zahra.bouzidi@edu.uca.ma

resistance and inductance per unit length change. These changes are divided into two categories: the effect of proximity and the effect of the skin. Due to the skin effect, the conductor is subjected to a variable frequency current with no applied external magnetic field due to the penetration of the internal magnetic field. In contrast to the skin effect, which is unique to a single conductor, the proximity effect occurs when multiple conductors are sufficiently close together.

2.1 Modelling of transmission line parameters

For simplicity of presentation, the discussion will concentrate on single transmission line structures.
The skin effect of transmission lines has been presented in the circuit configurations shown in Fig.1 [11], [12].

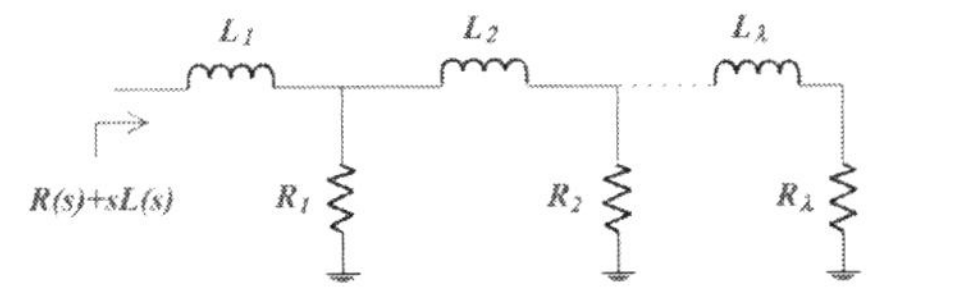

Fig. 1. Circuit configurations to model the frequency-dependency of R(s) + L(s)

Consider the transmission line represented by the following set of telegrapher's equations:

$$\begin{cases} \frac{\partial}{\partial z}\mathrm{V}(z,s) = -(\mathrm{R}+s\mathrm{L})\mathrm{I}(z,s) \\ \frac{\partial}{\partial z}\mathrm{I}(z,s) = -(\mathrm{G}+s\mathrm{C})\mathrm{V}(z,s) \end{cases} \quad (1)$$

Where R, L, C and G are the per unit length parameters (non-negative definite symmetric matrices).
The solution of the system in Eq. (1) can be written in the Laplace domain using the exponential function as follows:

$$\begin{bmatrix}\mathrm{V}(l,s)\\ \mathrm{I}(l,s)\end{bmatrix} = e^{z}\begin{bmatrix}\mathrm{V}(0,s)\\ \mathrm{I}(0,s)\end{bmatrix} \quad (2)$$

Where l is the length of the line.

We will use a matrix-rational function to approximate the exponential matrix, as follows:

$$e^{Z} = B_{kh}^{-1}(Z)A_{kh}(Z) \quad (3)$$

Where A_{kh} (Z) and B_{kh} (Z) are polynomial matrices written as closed-form Pade rational functions [13] as shown below.

$$B_{kh}(Z) = \sum_{j=0}^{k}\frac{(h+k-j)!\,k!}{(h+k)!\,j!\,(k-j)!}(-Z)^{j}$$
$$A_{kh}(Z) = \sum_{j=0}^{h}\frac{(h+k-j)!\,h!}{(h+k)!\,j!\,(h-j)!}Z^{j} \quad (4)$$

The Pade approximation order is represented by k and h.

The product of polynomials $B_{kh}^{-1}(Z)A_{kh}(Z)$ can be represented in terms of poles and zeros using the diagonal Pade approximation (h=k=q).

For even values of q:

$$B_{qq}^{-1}(Z)A_{qq}(Z) = \prod_{i=1}^{q/2}[(u_i I - Z)(u_i^{*} I - Z)]^{-1}[(u_i I + Z)(u_i^{*} I + Z)] \quad (5a)$$

For odd values of q:

$$B_{qq}^{-1}(Z)A_{qq}(Z) = (u_0 I - Z)^{-1}(u_0^{*} I + Z)\prod_{i=1}^{(q-1)/2}[(u_i I - Z)(u_i^{*} I - Z)]^{-1}[(u_i I + Z)(u_i^{*} I + Z)] \quad (5b)$$

$u_i = x + jy_i$: Complex pole of polynomial $B_{qq}(Z)$
u_0 : Real pole of polynomial $B_{qq}(Z)$

From the pole-zero pairs represented by Eq. (4) and Eq. (5), the transmission line can be modelled using a Pade macro-model. This macro-model consists of a set of first and second order cells, in which each subsection is composed of predetermined constants u_0 (real-pole), u_i (complex-pole) and the per unit length parameters, whose number depends on the order of the Pade approximation.

$$\begin{bmatrix}V_{i+1}\\ I_{i+1}\end{bmatrix} = [B_{qq}^{-1}]_i [A_{qq}]_i \begin{bmatrix}V_i\\ I_i\end{bmatrix} \quad (6)$$

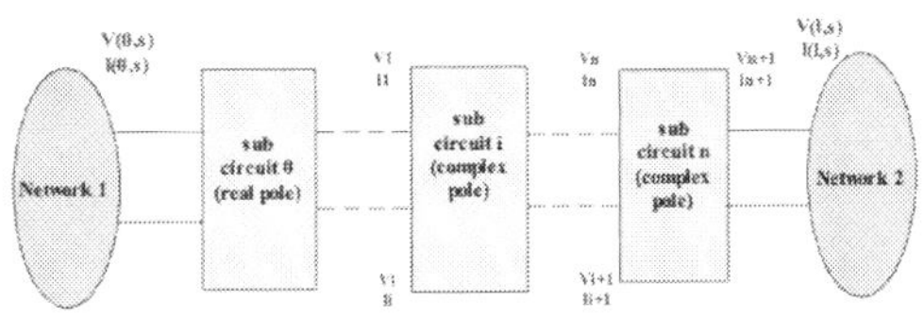

Fig. 2. Equivalent circuit of the line in the sense of Padé: represented by pole-zero section.

Where V(0,s) , I(0,s), V(l,s) and I(l,s) represent the terminal voltages and currents. Vi+1, Ii+1, Vi and Ii represent the internal voltages and currents created by the Pade macro-model.

2.2 Modelling with linear elements

Each subsection in Eq. (6) is converted into the Y-parameter to represent the circuit for transmission lines with frequency-dependent parameters in terms of resistor, inductors, capacitors and ideal transformers.

- Complex pole-zero subsection

$$Y_{11} = Y_{22} = \frac{(x_i^2+y_i^2)}{4x_i l}(a(s))^{-1} + \frac{a(s)l}{4x_i} + x_i\left(a(s)l + \frac{(x_i^2+y_i^2)}{l}\left(b(s)\right)^{-1}\right)^{-1}$$

$$Y_{12} = Y_{21} = \frac{(x_i^2+y_i^2)}{4x_i l}(a(s))^{-1} + \frac{b(s)l}{4x_i} - x_i\left(a(s)l + \frac{(x_i^2+y_i^2)}{l}\left(b(s)\right)^{-1}\right)^{-1}$$

Real pole-zero subsection

$$Y_{11} = Y_{22} = \frac{a_0}{2l}(a(s))^{-1} + \frac{l}{2a_0}(b(s))$$

$$Y_{12} = Y_{21} = -\frac{a_0}{2l}(a(s))^{-1} + \frac{l}{2a_0}(b(s))$$

https://doi.org/10.1051/itmconf/20224802005

The circuit topologies of the complex pole-zero subsection and the real pole-zero subsection are shown in Fig. 3 and Fig. 4, respectively.

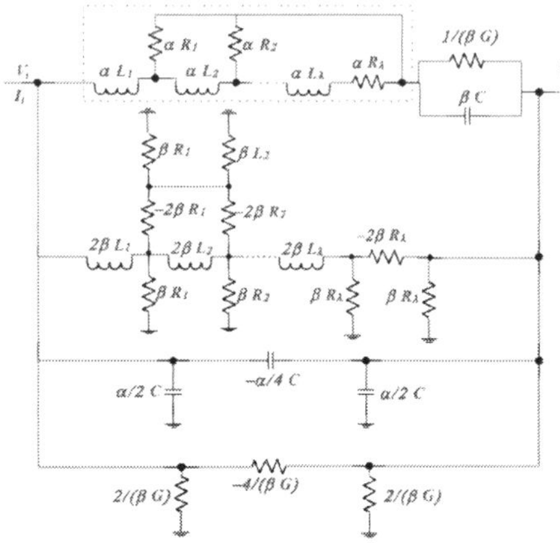

Fig. 3. Circuit topology of the complex pole-zero subsection

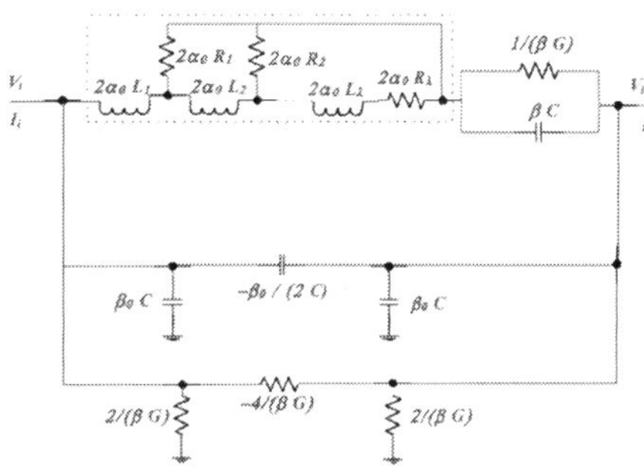

Fig. 4. Circuit topology of the real pole-zero subsection

The Pade approximation macromodel decomposes, as we have already seen, into a set of cells with real poles and with complex poles. The application of the MNA method [14] to each type of cell leads to matrices G and C.

3 Simulation and results

To demonstrate the points raised in the preceding section, we used the example shown in Fig. 5. We compared the simulation results obtained using Pade approximation with the results from the traditional lumped-circuit segmentation macro-model and Branin's method in Emcap2000-Esacap circuit simulator [16]. Pade approximation model is valid in both the time and frequency domains.

We considered a single transmission line of 1cm length see Fig. 5. The in and out charges equalled 50 ohms, as indicated in Fig. 6.

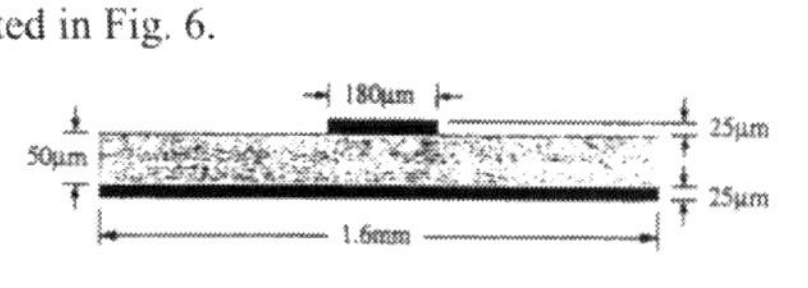

Fig. 5. Conductor over a ground plane.

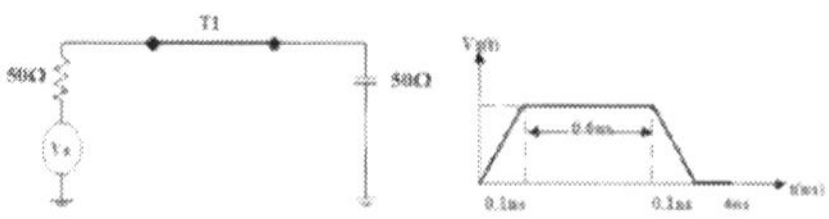

Fig. 6. Electrical representation of the line in Fig.5

3.1 Frequency–independent parameters

We first assumed the line to be lossless and the line parameters to be:

R=50Ω, G=0, C= 90PF et L= 2.65E-7 H.

The circuit in Fig. 6 was analysed in the time and frequency domains using three different approaches: the Branin method, the traditional lumped-circuit segmentation macro-model (decomposed into 100-cascaded cells) and the Pade approximation method (the order of Pade approximation used in this case was 20).

The waveform evolution of the line as a function of time is shown in Fig. 7, and Fig. 8 presents the frequency response.

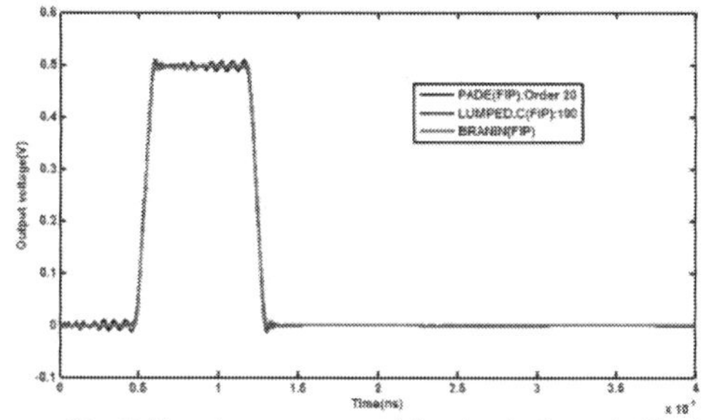

Fig. 7. Transient response of the circuit shown in Fig. 2

Note that in both comparisons, there was excellent agreement between the Pade approximation macro-model on the one hand and the lumped-circuit segmentation macro-model or Branin method on the other.

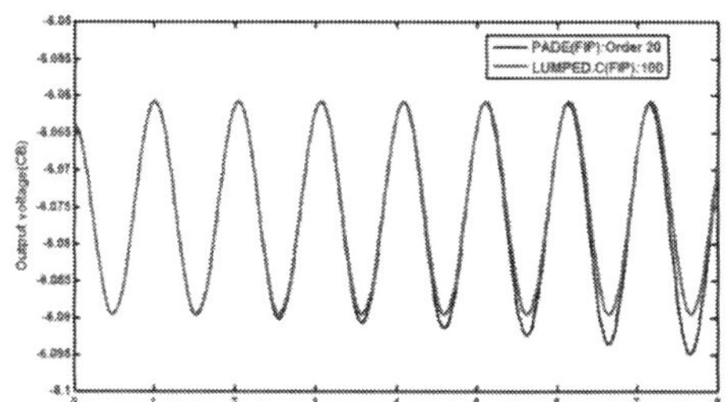

Fig. 8. Frequency response of the circuit shown in Fig.

3.2 Frequency–dependent parameters

Since the dielectric is homogeneous and lossless, only the line's per unit length (p.u.l) inductance and resistance are frequency-dependent. The frequency-dependent parameters of per unit length inductance and resistance are listed in Table 1.

Table 1. Per unit length and resistance parameters of the coupled transmission line

F(GHz)	1e-2	1e-3	1e-2	1e-1	1	3.5	5	10
R(Ω/m)	154	154	154	159	272	432	500	679
L(nH/m)	5.43e+3	5.31e+3	5.32e+3	5.31e+3	5.03e+3	4.90e+3	4.87e+3	4.83e+3

Table 2. Comparison between the lumped-circuit segmentation macro-model and the Pade Approximation

	Padé approximation macro-model	Lumped-circuit segmentation macro-model	Reduction
Number of variables used	109	293	63%

To validate our implementation of Pade approximation, we compared the results from Pade approximation and the traditional, lumped-circuit segmentation macro-model of transmission lines. The line was segmented into 100 segments, each of length $\Delta x = \frac{1}{100}$ cm, to model the frequency-dependent behaviour of the line resistance and inductance. Every segment contained a ladder of order 2.

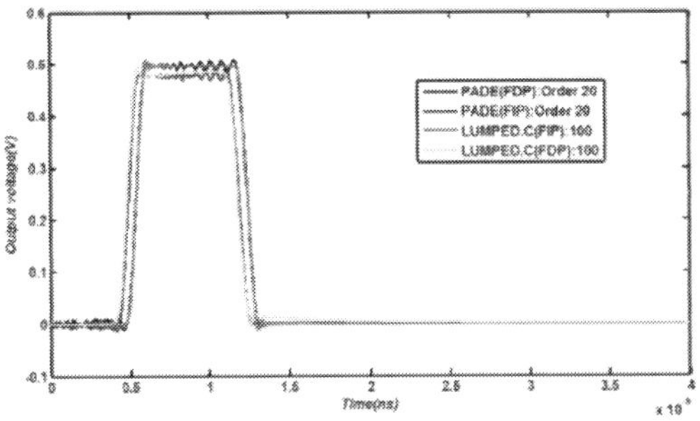

Fig. 9. Transient response of the circuit shown in Fig. 2

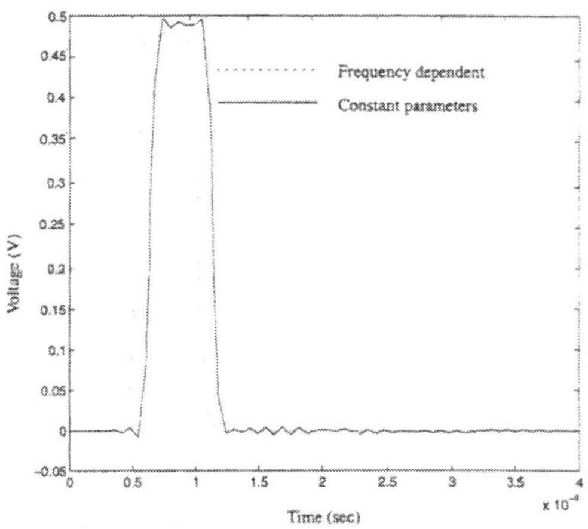

Fig. 10. Transient response of the circuit shown in Fig. 2 from [17]

Figure 9 shows excellent agreement between the Pade approximation and the lumped-circuit segmentation macro-model in the frequency-dependent case. Fig.9 compares the transient response of the transmission line for both the frequency-dependent and the frequency-independent cases. The comparison of the simulation results showed a difference in both amplitude and delay, which is usually related to the skin effect.

Comparing these with the analytical results published by Khazaka [17] shown in Figure 7 shows exact agreement between the results of the hybrid method and analytic method.

In this example, the lumped-segmentation model introduced 293 variables, whereas the proposed method introduced only 109 variables (62% savings), as seen in Table 2.

3.3 Nonlinear terminations

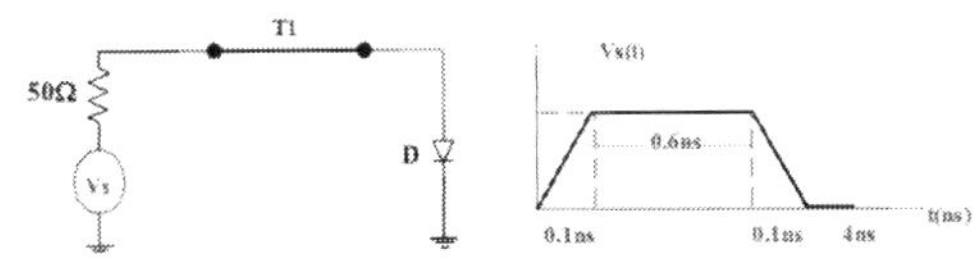

Fig. 11. Electrical representation of the line in Fig. 1 with nonlinear loads.

Figure 11 shows the same transmission structure in Fig.1 but terminated with nonlinear loads. The structure is excited in a similar way to the linear termination case. Figure 12 shows the transient responses obtained by using two different models; the Pade approximation and the traditional lumped-circuit segmentation macro-model.

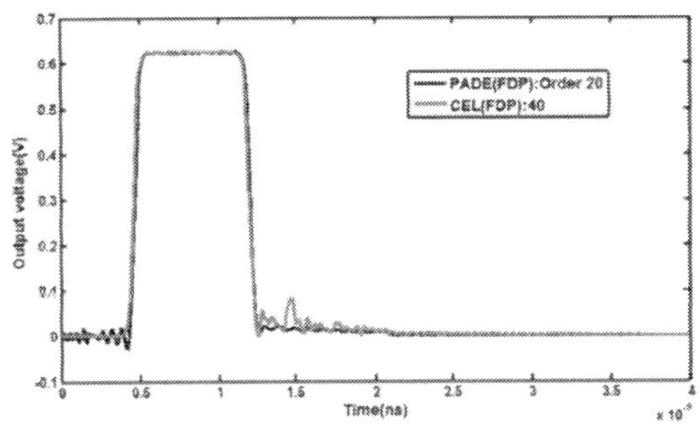

Fig. 12. Transient response of the circuit shown in Fig. 10

The results obtained by our method are in excellent agreement with those obtained by the lumped-circuit segmentation macro-model.

4 Conclusion

In this paper, a new method of modelling transmission lines in the form of a circuit model is presented. This method is based on the theory of transmission lines, as well as the Pade approximation of the exponential matrix. The method's validity was confirmed by comparing its results with those derived using other methods. It is easy to extend the proposed models to coupled transmission lines with linear and nonlinear terminations. We will discuss this in future papers.

5 References

1. A. Deutch et al., "Modeling and characterization of long on-chip interconnections for high performance microprocessors," IBM J. Res. Develop, 39, pp. 547-567, (1995).
2. A. Deutch et al., "When are transmission-line effects important for on-chip interconnections," IEEE Trans. Microwave 77zeor-yTech., Oct. 1997.
3. A. Semlyen & A. Dabuleanu, "Fast and accurate switching transient calculations on transmission lines with ground return using recursive convolutions," IEEE Transactions on Power Apparatus and Systems, PAS-94(2), pp. 561-571, 1975.
4. J. R. Marti, "Accurate modelling of frequency-dependent transmission lines in electromagnetic transient simulations," IEEE Transactions on Power Apparatus and Systems, PAS-101(1), pp. 147-157, 1982.
5. L. Marti, "Simulation of transients in underground cables with frequency dependent modal transformation matrices," IEEE Transactions on Power Delivery, 3(3), pp. 1099-1110, 1988.
6. Z.. Bouzidi, , A . El Idrissi,.; H .Rouijaa, M .Saih,. Transmission lines modeling approach based on the approximation of Pade. In Proceedings of the 2019 PhotonIcs & Electromagnetics Research Symposium, Rome, Italy, 17–20 June 2019; pp. 4009–4017.
7. C. W. Ho. C. W, "Theory and Computer-aided Analysis of Lossless Transmission Lines, " IBM J. Res. Dev., vol. 17, no. (3), pp. 249-255, May 1973.
8. F. H. Branin, "Transient analysis of lossless transmission lines," Proc. IEEE, 55, (11), pp. 2012-2013, 1967.
9. Y. Mejdoub, H. Rouijaa, && A. Ghammaz, "Transient analysis of lossy multi-conductor transmission lines model based by the characteristics method," Int. J. Eng. Technol., 3 (1), pp. 47–53, (2011).
10. A. Dounavis, X. Li, M. S. Nakhla, & R. Achar, "Passive Closed-Form Transmission-line Model for General-purpose Circuit Simulators," IEEE Trans. Microwave Theory Tech., 47 (12), Dec 1999.
11. Z. Bouzidi, A. El Idrissi, H. Rouijaa, & M. Saih, "Transmission lines modeling approach based on the approximation of Pade," In Proceedings of the 2019 Photonics & Electromagnetics Research Symposium, Rome, Italy, 17–20 June 2019, pp. 4009–4017.
12. C. Yen, Z. Fazarinc, & R. L. Wheeler, "Time-domain skin-effect model for transient analysis of lossy transmission lines," Proc. IEEE, 70, pp. 759-757. July 1982.
13. A. Deutsch, "Electrical characteristics of interconnections for high-performance systems," Proceedings of the IEEE, 86, pp. 315-355, Feb. 1998.
14. J. Vlach & K. Singhal, "Computer Methods for Circuit Analysis and Design". New York: Van Nostrand, 1983.
15. CW. Ho, A. Ruehli, P. Brennan, "The modified nodal approach to network analysis". IEEE Transaction on Circuit and System, vol. 22(No. 6): pp. 504-509, June 1975.
16. R. Khazaka, E. Chiprout, M. Nakhla and Q.J Zhang , "Analysis of high-speed interconnects with frequency dependent parameters" Proc. Int. Symposium on Electromagnetic Compatibility, pp. 203-208, Zurich Switzerland, March 1995.
17. L. INZOLI, H. ROUIJAA "Aseris: Emcap2000-Esacap software" Applications Handbook and Users Manual" DCR /B , janvier 2001

Plotting Lamb waves dispersion curves of an aluminum plate by the Semi-Analytical Finite Element (SAFE) method and comparison with analytical curves.

Ayoub Jabiri[1*], *Abdelali* Yacoubi[1], *Mhammed* El Allami[1-2]

[1] Labo MIET, Department of applied physics, FST, Settat, Morocco.

[2] CRMEF Settat, Morocco.

Abstract. It is well known that the propagation of Lamb waves in elastic solid medias depends on the frequency and on the material properties, but it does not depend on the propagation direction in isotropic medias and gives rise to two basic types of modes, symmetric and antisymmetric ones. This paper presents an application of the semi-analytical finite element (SAFE) method on an isotropic aluminum plate of thickness *e = 2mm*. The main goal of this paper is to make a quantitative evaluation of the SAFE method by carrying out the following tasks: a) calculation of the dispersion curves (*wavenumber, phase velocity, group velocity*, b) calculation of the relative error and validation of the proposed algorithm, c) comparison between the curves obtained by SAFE method on those obtained by Bisection method. It should be noted here that our study is restricted on the propagative modes.

Keywords. Semi-Analytical Finite Element (SAFE) method, Bisection method, Lamb waves, Dispersion curves, Isotropic plates.

1 Introduction

There are several numerical methods for calculating the Lamb waves dispersion curves. Among these methods, there is the semi-analytical finite element (SAFE) method which is useful for composite and isotropic plates. Lamb waves propagation is multimodal and when there are several modes present for a given excitation frequency, the signal interpretation would be considerably difficult, hence the need to calculate the Lamb waves dispersion curves to limit the choice of excitation frequency to a domain not presenting many modes. In this paper, the application of the SAFE method is made on a homogeneous isotropic aluminum plate of thickness *e=2mm* with free boundaries. Our goal is to make a quantitative evaluation of the semi-analytical finite element (SAFE) method by doing the following tasks: a) calculation of the dispersion curves (*wavenumber, phase velocity, group velocity*, b) calculation of the relative error and validation of the proposed algorithm, c) comparison between the curves obtained by SAFE method on those obtained by a Bisection method. Our study was limited to propagative modes, purely imaginary or complex solutions are not considered here.

2 Analytic formulation of SAFE method

In previous studies, the authors discussed dispersion curves and their multiple applications in non-destructive testing. They propose several techniques for plotting dispersion curves. These methods include the Newton-Raphson method [1], bisection method [2], spectral method [3], transfer matrix method [4], and others [5]. Furthermore, dispersion curves are plotted by iterative methods, even though they are very close to the analytical curves. For this reason, we propose an alternative method that can better estimate the dispersion curves.

The iterative methods (bisection and Newton Raphson) are applicable only on isotropic plate. However, these methods are simple in terms of coding compared with the spectral and SAFE methods. Both last methods provide results of high accuracy with short time of calculation compared with the iterative methods for the composite and isotropic plate. The transfer matrix method can be more useful for multi-layered anisotropic media.

*Corresponding author: a.jabiri@uhp.ac.ma

Bartoli et al. [6] modeling Lamb wave propagation in arbitrary section waveguides using the SAFE method. Ahmad et al. [7] applied the method for plotting the dispersion curves of an aluminum plate and a composite structure.

Recently, Xing et al. [8] suggested a defect localization method for rails and plotted rail dispersion curves by the SAFE method. The technique for calculating pipeline dispersion curves immersed in a fluid using SAFE method has been presented by Wenbo et al. [9].

The SAFE method combines the FE method with analytical expressions. Using this method, Lamb wave propagation curves for isotropic and composite plates can be calculated. However, when using the SAFE method, we don't need to discretize the entire plate, it is enough to do for the cross-section of the plate.

The advantage of this solution over other iterative methods in the literature appears in the fact that the SAFE method doesn't only allow us to calculate the dispersion curves, but also the nodal displacements, and using interpolation one can easily calculate the deformations and the stresses.

We consider an infinitely wide plate in the *y*-axis direction as shown in figure 1, the cross-section domain of the plate is subdivided into a system of multilayers which are modeled using 1D finite elements with three-node as shown in figure 1. In our study, the wave propagates along the *z*-axis with a wavenumber *k* and at frequency *ω*.

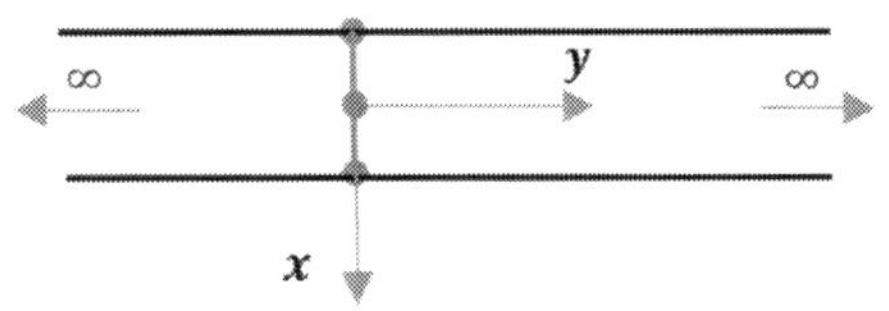

Fig. 1. Infinite (2D) plate with three nodes per element

The element displacement can be approximated by using the interpolation function $\mathbf{N}(x)$ and nodal unknown displacement $\mathbf{q}^{(\mathbf{e})}$ as follows

$$\mathbf{u}^{(\mathbf{e})}(x, z, t) = \mathbf{N}(x)\mathbf{q}^{(e)}e^{i(kz-\omega t)} \quad (1)$$

The element strain expression becomes

$$\varepsilon^{(e)} = \left[\mathbf{L}_x \frac{\partial}{\partial x} + \mathbf{L}_z \frac{\partial}{\partial z}\right] \mathbf{N}(x)\mathbf{q}^{(\mathbf{e})}e^{i(kz-\omega t)} = (\mathbf{B}_1 + ik\mathbf{B}_2)\,\mathbf{q}^{(\mathbf{e})}e^{i(kz-\omega t)} \quad (2)$$

where

$$\boldsymbol{B}_1 = \boldsymbol{L}_x\boldsymbol{N}_{,x}\ ,\ \boldsymbol{B}_2 = \boldsymbol{L}_z\boldsymbol{N}$$

and $\boldsymbol{N}_{,x}$ is the derivative of $\boldsymbol{N}$ concerning x.

with

$$\mathbf{L}_x = \begin{bmatrix}1 & 0\\ 0 & 0\\ 0 & 1\end{bmatrix} \qquad \mathbf{L}_z = \begin{bmatrix}0 & 0\\ 0 & 1\\ 1 & 0\end{bmatrix}$$

The equation of motion for the cross-section can be obtained by using Hamilton's principle:

$$[\boldsymbol{K}_1 + ik\boldsymbol{K}_2 + k^2\boldsymbol{K}_3 - \omega^2\boldsymbol{M}]\boldsymbol{U} = 0 \quad (3)$$

where

$$\boldsymbol{K}_1 = \cup_{e=1}^{n}\mathbf{k}_1^{(\mathbf{e})} \quad (4) \qquad \boldsymbol{K}_2 = \cup_{e=1}^{n}\mathbf{k}_2^{(\mathbf{e})} \quad (6)$$

$$\boldsymbol{K}_3 = \cup_{e=1}^{n}\mathbf{k}_3^{(\mathbf{e})} \quad (5) \qquad \boldsymbol{M} = \cup_{e=1}^{n}\mathbf{m}^{(\mathbf{e})} \quad (7)$$

and n denotes the number of elements with

$$\mathbf{k}_1^{(\mathbf{e})} = \int_x \mathbf{B}_1^{\mathrm{T}}\mathbf{C}_\mathbf{e}\mathbf{B}_1\,dx \quad (8)$$

$$\mathbf{k}_2^{(\mathbf{e})} = \int_x (\mathbf{B}_1^{\mathrm{T}}\mathbf{C}_\mathbf{e}\mathbf{B}_2 - \mathbf{B}_2^{\mathrm{T}}\mathbf{C}_\mathbf{e}\mathbf{B}_1)\,dx \quad (9)$$

$$\mathbf{k}_3^{(\mathbf{e})} = \int_x \mathbf{B}_2^{\mathrm{T}}\mathbf{C}_\mathbf{e}\mathbf{B}_2\,dx \quad (10)$$

$$\mathbf{m}^{(\mathbf{e})} = \int_x \mathbf{N}^{\mathrm{T}}\boldsymbol{\mu}_\mathbf{e}\mathbf{N}\,dx \quad (11)$$

where $\mathbf{C_e}$ and $\boldsymbol{\mu}_\mathbf{e}$ are the elastic stiffness tensor and the density for the generic element, respectively.

The integrals given above are approximated by using Gaussian quadrature integration.

The global matrices $\mathbf{K}_1$, $\mathbf{K}_3$, and $\mathbf{M}$ are symmetric. However, the global matrix $\mathbf{K}_2$ is skew-symmetric. To simplify the equation numerical resolution, $\mathbf{K}_2$ is transformed into a symmetric form by pre-multiplying and post-multiplying equation with transformation matrix $\mathbf{T}^{\mathrm{T}}$ and $\mathbf{T}$ respectively.

$$\begin{cases} \mathbf{K}_1 = \mathbf{T}^{\mathrm{T}}\mathbf{K}_1\mathbf{T} \\ \widehat{\mathbf{K}_2} = i\mathbf{T}^{\mathrm{T}}\mathbf{K}_2\mathbf{T} \\ \mathbf{K}_3 = \mathbf{T}^{\mathrm{T}}\mathbf{K}_3\mathbf{T} \\ \mathbf{M} = \mathbf{T}^{\mathrm{T}}\mathbf{M}\mathbf{T} \\ \widehat{\mathbf{U}} = \mathbf{T}^{\mathrm{T}}\mathbf{U} \end{cases} \quad (12)$$

$\mathbf{T}$ is the transformation matrix with each diagonal element equal to 1, except every second (corresponding to the displacement in the **z**-direction) being equal to *i*.

As an alternative approach, the equation (3) can be reformulated as follows

$$\left[\mathbf{K}_1 + k\widehat{\mathbf{K}_2} + k^2\mathbf{K}_3 - \omega^2\mathbf{M}\right]\widehat{\mathbf{U}} = 0 \quad (13)$$

where $\widehat{\mathbf{K}_2}$ and $\widehat{\mathbf{U}}$ are respectively the $\mathbf{K}_2$ symmetric matrix and the new eigenvector.

The equation of motion can be solved by using two approaches. The first one can be applied by fixing the wavenumber k and searching for the frequency ω.*The* second one is proceeded by fixing the frequency ω and searching for the positive wavenumber k. It should be noted that the last approach which is used in this paper.

In this approach, equation (13) is simplified as linear form [10]

$$(\mathbf{A}\text{-}k\mathbf{B})\mathbf{Q} = \mathbf{0} \quad (14)$$

where

$$\mathbf{A}=\begin{bmatrix} 0 & \boldsymbol{K}_1-\omega^2\boldsymbol{M} \\ \boldsymbol{K}_1-\omega^2\boldsymbol{M} & \widehat{\boldsymbol{K}_2} \end{bmatrix},\ \mathbf{B}=\begin{bmatrix} \boldsymbol{K}_1-\omega^2\boldsymbol{M} & 0 \\ 0 & -\boldsymbol{K}_3 \end{bmatrix}$$

and

$$\mathbf{Q}=\begin{bmatrix} \widehat{U} \\ k\widehat{U} \end{bmatrix}.$$

It should be noted that all the eigenvalues and eigenvectors are real. The eigenvalues represent the wavenumbers for all the propagating modes, while the corresponding eigenvectors represent the cross-sectional displacement.

From a coding point of view, during the research of the solutions which are in this case the wavenumbers k, it is necessary to create a matrix which will be filled by these solutions. Whereas among these solutions one can have complex or pure imaginary solutions or negative real values and since we are interested only by positive solutions one eliminates any undesirable solution by well-defined logical instructions.

The wavenumbers obtained by solving the eigenvalue problem can be used to calculate the phase velocity:

$$V_p = \omega \,/\, k \quad (15)$$

By simple manipulation, equation (14) leads to calculating the group velocity using the following relation

$$V_g = \partial\omega/\partial k = (\boldsymbol{Q}_R^T K' \boldsymbol{Q}_R)/(2\omega \boldsymbol{Q}_R^T \boldsymbol{M}\boldsymbol{Q}_R) \quad (16)$$

where $\boldsymbol{Q}_R$ indicates the right eigenvector of the system define by

$$\boldsymbol{Q}_R = \widehat{\mathbf{U}}$$

and

$$\mathbf{K}' = \widehat{\mathbf{K}}_2 + 2k\mathbf{K}_3\,.$$

3 Proposed Algorithm

An algorithm is proposed as shown in figure 2 to plot the dispersion curves:

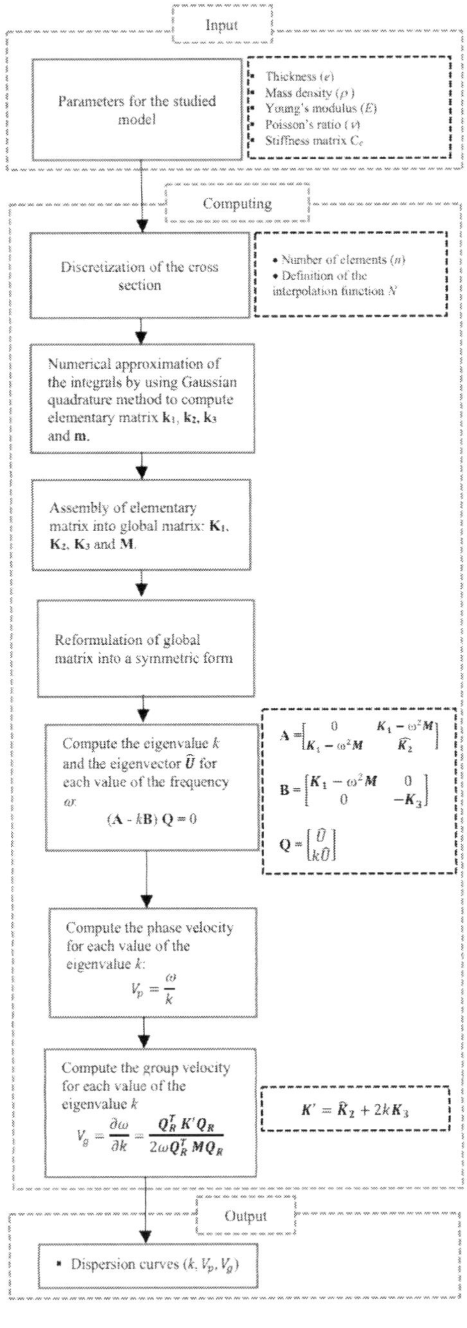

Fig. 2. Algorithm to plot the dispersion curves.

4 Dispersion curves

A Matlab code has been developed for plotting the dispersion curves. The wavenumber, phase velocity, and group velocity are plotted as a function of the frequency-thickness product (*fe*).

Let's take an aluminum plate with thickness *2mm*, Young's modulus E = 70GPa, Poisson's ratio ν = 0.32, and mass density ρ = 2700 kg/m^3.

The minimum number of elements required to ensure the appearance of all modes in the range of frequencies between 0 and 5000 kHz [11]

$$n_{min} = (\beta fe / V_T) \quad (17)$$

Here V_T is the transverse velocity of the propagative wave, and β depends on the order of the interpolation function.

In our case, we have used the quadrature interpolation function, thus $\beta = 4$ and $V_T = 3134\ m/s$. It follows that

$$n_{min} = 13.$$

In our case, we have taken $n = 18$ elements to compute the dispersion curves.

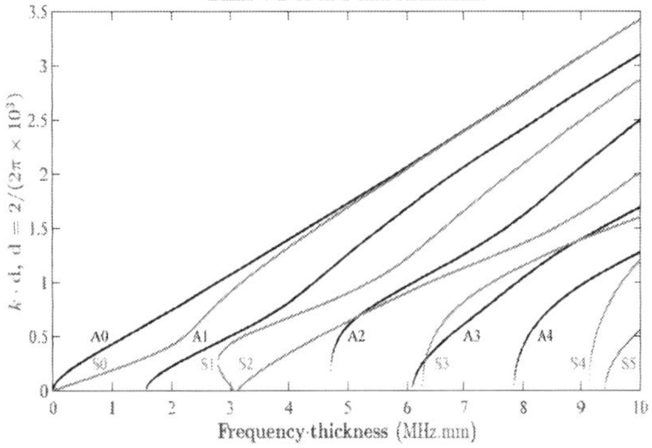

Fig. 3. Wavenumber curves for isotropic aluminum plate, *n* is taken to be equal to 18.

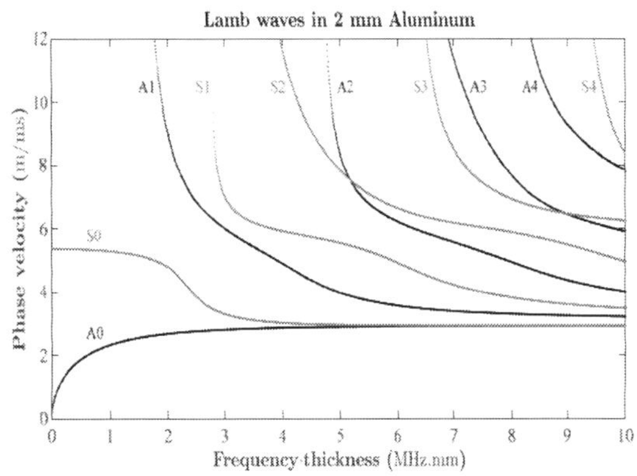

Fig. 4. Phase velocity curves for isotropic aluminum plate, *n* is taken to be equal to 18.

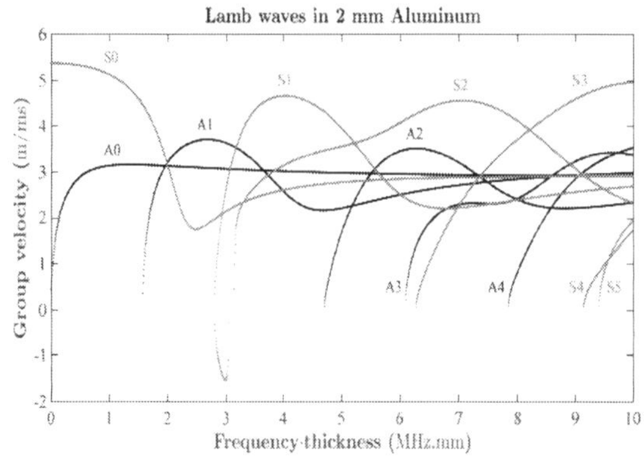

Fig. 5. Group velocity curves for isotropic aluminum plate, *n* is taken to be equal to 18.

The curves in figure 3 represent the wavenumber evolution of propagating modes versus the product frequency-thickness. For a given product (*fe*), the modes present in the plate are known due to the wavenumbers. The wavenumber of the first symmetric modes (S0, S1, S2, S3, and S4) and antisymmetric modes (A0, A1, A2, A3, and A4) are plotted in figure 3.

The curves in figure 4 represent the phase velocity evolution of propagating modes versus the product frequency-thickness. The knowledge of phase velocity is essential when modeling the propagation of Lamb waves. Indeed, to generate a mode of Lamb at a given frequency-thickness product, we must know exactly the phase velocity with which the mode propagates in the plate to avoid that it crosses with another mode other than the desired mode. The phase velocity of the first symmetric modes (S0, S1, S2, S3, and S4) and antisymmetric modes (A0, A1, A2, A3, and A4) are plotted in figure 4.

The curves in figure 5 represent the group velocity evolution of propagating modes versus the product frequency-thickness. The group velocity is the velocity with which a wave packet propagates along with the plate. The group velocity of the first symmetric modes (S0, S1, S2, S3, and S4) and antisymmetric modes (A0, A1, A2, A3, and A4) are plotted in figure 5.

5 Results and discussions

To plot the curves dispersion of homogeneous isotropic plate, we have developed an algorithm based on the SAFE method, then it has been transformed into code in Matlab software.

This method's precision of the solutions depends on the number of elements.

Table 1 illustrates that the average relative error for the two fundamental modes S0 and A0 by the SAFE method is a decreasing function of the number of elements.

Here we give the formula used to calculate the relative error and its average:

$$err = \left|\frac{k_{safe} - k_{ana}}{k_{ana}}\right| \times 100 \quad (18)$$

where k_{safe} is the wavenumber calculated by the SAFE method and k_{ana} is the analytical wavenumber. The average relative error is calculated by

$$err_a = \sum_{j=1}^{m} err_j / m \quad (19)$$

here m is the number of the considered solutions and is equal in our case to 40.

Table 1. The average relative error for modes S0 and A0 by the SAFE method

Number of elements (n)	*Average relative error (S0)*	*Average relative error (A0)*
1	6.82 %	4.01 %
4	1.42 %	1.36 %
7	0.47 %	0.45 %
10	0.17 %	0.17 %
13	0.074 %	0.073 %
14	0.057 %	0.056 %

To validate the proposed algorithm, we proceed by calculating the relative error between the values of the wavenumbers obtained by the SAFE method compared to those obtained by the bisection method.

It should be noted here that the relative error on the wavenumber for the minimum number of elements (*n = 13*) increases as an exponential function as shown in figure 6. The error increases as the frequency increases.

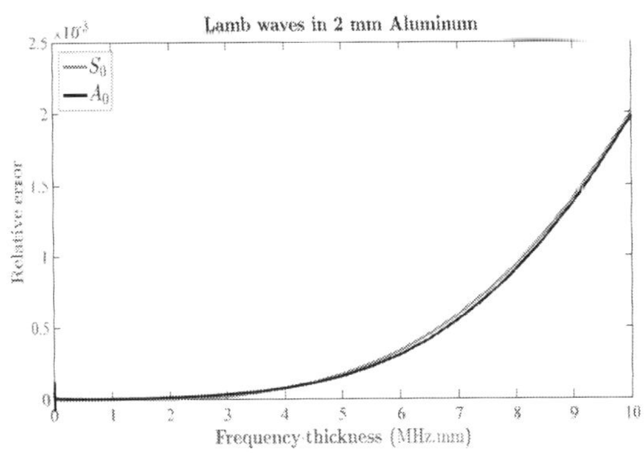

Fig. 6. Relative error on the wavenumber for *n = 13*

On the other hand, the dispersion curves obtained are compared with those calculated by the bisection method. The results are shown in figure 7, and one can see a good convergence of the SAFE method to the analytical curves (bisection method).

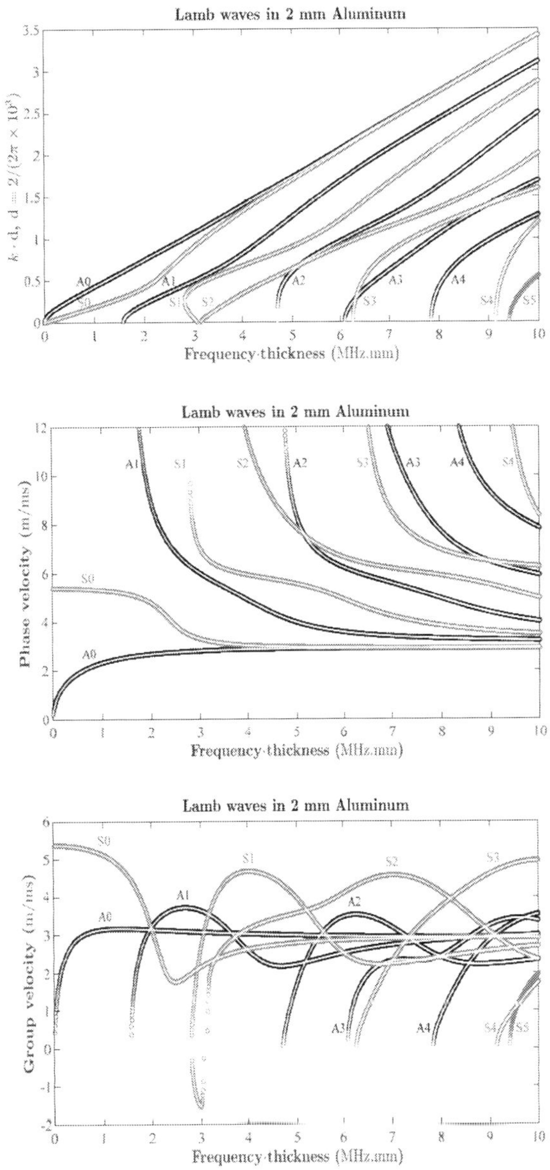

Fig. 7. Dispersion curves obtained by the SAFE method superposed on the analytical curves (yellow) for n = 18.

6 Conclusion

In this paper, the SAFE method is applied to calculate propagating modes with higher precision and with a large number of elements within plate thickness with a great frequency range. The results show a very good agreement between the results obtained by the code developed using the SAFE method and those obtained by using the bisection method. In addition, the bisection method is simple in terms of coding compared to the

SAFE method. However, the SAFE method can be more useful for composite or multilayer materials.

It should also be mentioned that the dispersion curves obtained by SAFE method are crucial for any theoretical, numerical or experimental study and for detecting damage in industrial structures.

The SAFE method doesn't only allow us to calculate the dispersion curves, but also the nodal displacements, and using interpolation one can easily calculate the deformations and the stresses. This purpose can be the subject of a future paper.

As perspective, it is planned to apply the SAFE method in industrial case.

References

1. E. B. Ndiaye, H. Duflo, "Non-destructive testing of sandwich composites: adhesion defects evaluation; experimental and finite element method simulation comparison", in Acoustics 2012 Nantes, Nantes, France, 2012, p. 2659-2664.
2. S. Nissabouri, M. El Allami, M. Bakhcha, "Lamb waves propagation plotting the dispersion curves", in Conference: ICCWCS16 Volume: proceedings issn, Morocco, 2016, p. 153-156.
3. B. Sinha, H. P. Valero, F. Karpfinger, A. Bakulin, B. Gurevich, "Spectral-method algorithm for modeling dispersion of acoustic modes in elastic cylindrical structures", Geophysics 75 (2010), p. H19-H27.
4. A. Nayfeh, "The general problem of elastic wave propagation in multilayered anisotropic media", J. Acoust. Soc. Am. (1991), p. 1521-1531.
5. P. Hora, "Determination of Lamb wave dispersion curves using Fourier transform", Applied and Computational Mechanics 6 (2012) 5–16.
6. I. Bartoli, M. Alessandro, L. Francesco, V. Erasmo, "Modeling wave propagation in damped waveguides of arbitrary cross-section", J. Sound Vibrat. 295 (2006), p. 685-707.
7. Z. A. B. Ahmad, J. M. Vivar-Perez, U. Gabbert, "Semi-analytical finite element method for modeling of lamb wave propagation", CEAS Aeronaut. J. 4 (2013), p. 21-33.
8. B. Xing, Y. Zujun, X. Xining, Z. Liqiang, S. Hongmei, "Research on a rail defect location method based on a single mode extraction algorithm", Appl. Sci. 9 (2019), p. 1-16
9. D. Wenbo, K. Ray, "Guided wave propagation in buried and immersed fluid-filled pipes: Application of the semi analytic finite element method", Comput. Struct. 212 (2019), p. 236-247
10. E. Viola, A. Marzani, and I. Bartoli, "Semi analytical Formulation for Guided Wave Propagation" in Mechanical Vibration: Where do we Stand, vol. 488, I. Elishakoff, Ed. Vienna: Springer Vienna, 2007, pp. 105–121.
11. J. M. Galán and R. Abascal, "Numerical simulation of Lamb wave scattering in semi-infinite plates: Numerical simulation of Lamb wave scattering" Int. J. Numer.Methods Eng., vol. 53, no. 5, pp. 1145–1173.

Silicon Die Transient Thermal Peak Prediction Approach

Aziz Oukaira[1], *Amrou Zyad* Benelhaouare[1*], *Dariush* Amirkhani[1], *Jamal* Zbitou[2] and *Ahmed* Lakhssassi[1]

[1]University of Quebec in Outaouais, Department of Engineering and Computer Science, Gatineau, 18X 3X7, Canada
[2]LABTIC, ENSA of Tangier, Abdelmalek Essaadi University, Tangier, Morocco

Abstract. It is well known that Field Programmable Gate Arrays (FPGA) are good platforms for implementing embedded systems because of their configurable nature. However, the temperature of FPGAs is becoming a serious concern. Improvements in manufacturing technology led to increased logic density in integrated circuits as well as higher clock frequencies. As logic density increases, so do power density, which in turn increases the temperature, FPGAs follow the same path. A prediction of the thermal state of the Altera Cyclone V System-on-Chip (SoC) is presented in this work. The prediction study employs a numerical technique called Finite Element Method (FEM), which is a discretization method to approximate the real solution of the Partial Differential Equation (PDE) for heat transfer around the board's critical sources. The DE1 5CSEMA5F31C6N board was simulated using the COMSOL Multiphysics® tool for predicting thermal peaks during 13 hours of normal operation. Using the NISA tool, we obtained very similar results to those previously obtained with a margin of error of 2 %. As a result, a Verilog code implementation that describes the same approach used by the last two simulation tools is uploaded to the FPGA to verify the results of these simulations. This paper provides a more accurate vision of the level of operating stability of our FPGA board, which are currently the most important source for prototyping and designing the world's largest systems.

1 Introduction

With the increase in temperature, Integrated circuits (ICs) leakage power increases exponentially; furthermore, device lifetime decreases exponentially; ICs also operate at a slower speed at high temperatures; therefore, embedded systems based on ICs perform poorly. Additionally, 50 % of all electronic failures are caused by an increase in the internal temperature of the chip [1]. Such as Altera boards (Fig. 1) is devastated if their temperature becomes high [2-3]. All these factors have forced embedded system designers to apply temperature management methods.

Fig. 1. DE1-SoC development board (top view).

Thermal prediction, monitoring, and management are being investigated by researchers across many disciplines. The approaches of these researchers differ from each other in several aspects, including the experimental model used to study the detection and prediction either by modeling and numerical simulation FEM or other methods. Thermal analysis by the finite element method is one of the most efficient numerical simulation methods practiced today. It consists of a simple approximation of the geometry and variables describing the physical phenomenon such as displacement, velocity, pressure, and temperature [4-6]. In the previous works [7], the method of evaluating the thermal peak distributions and temperature distribution of all monitored areas using Gradient Direction Sensor (GDS) is presented along with a method of monitoring the thermal and thermomechanical stress of a Very-Large-Scale Integration (VLSI) chip.

This paper is organized as follows: Section II describes the thermal model of the board and its simulation under COMSOL Multiphysics®. The validation by the NISA tool of the prediction results obtained by the COMSOL tool is detailed in section III. The implementation of a Verilog code that translates the same algorithm used on these last two numerical tools, which is in turn simulated on the NCLaunch Cadence tool, is reported in section IV. Section V concludes the paper by summarizing our main contributions and findings.

2 Simulation of the thermal evolution of the Altera FPGA with COMSOL tool

Figures COMSOL Multiphysics® is a simulation platform that provides fully coupled Multiphysics and single-physics modeling capabilities. However, engineers and scientists use the COMSOL tool to simulate designs, and processes in all areas of engineering, manufacturing, and scientific research [8-11]. In our case, the use of Multiphysics simulation should allow us to gain a deeper understanding of the thermal evolution in a time interval of the Altera Cyclone V FPGA board design. It also represents a design tool thanks to its attitude to managing complex 2D and 3D geometry, and different physical modules existing in COMSOL, among which we find fluid mechanics, electricity, electromagnetism, chemistry, heat transfer, etc. It is possible to combine several physical phenomena in the same numerical simulation, which is the strength of this tool. The global heat transfer equation used in our thermal modeling can be described simply as follows:

$$\frac{\delta T}{\delta x}\left(K_x * \frac{\delta T}{\delta x}\right)+\frac{\delta T}{\delta y}\left(K_y * \frac{\delta T}{\delta y}\right)+\frac{\delta T}{\delta z}\left(K_z * \frac{\delta T}{\delta z}\right) + P(t) = \rho * C_p * \frac{\delta T}{\delta t} \qquad (1)$$

Where Kx, Ky, and Kz are the x, y, and z thermal conductivities, P(t) is the rate at which heat is generated per unit volume, t is time, ρ is the mass density of the material, and Cp is the specific heat. Fig. 2 shows our model of the Altera FPGA board with its four critical zones that generate heat

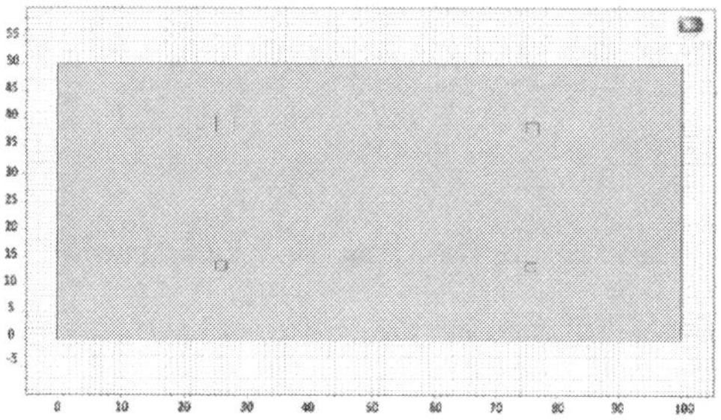

Fig. 2. Modelling the Altera board with the four critical heat sources.

The FEM used must impose simplifying assumptions on the material properties (Table I) and boundary conditions. The finite element method is based on the discretization of space and time. The main advantage of this method is its very high generality, it can deal with complex geometries by considering boundary conditions and temperature-dependent material properties. To solve thermal equations, the boundary conditions must be defined. In our case, we set a constant temperature of 25 °C around the structure of the Altera FPGA board, representing the ambient temperature of the environment where it is placed.

Table 1 : Physical properties of various significant materials.

Material Type	Width (mm)	Depth (mm)	Length (mm)
FR4	6.25	0.3	11.25
Al_2O_3	5.9	0.462	11.6
Gold	5.4	0.038	10.2
Si	5.4	0.13	10.2
SiO_2	9.66	4.86	13.23

Fig. 2 illustrates the modeling of our FPGA board and the four sources of heat that very often generate high temperatures that can damage our circuit and negatively influence the normal operation of the board. This representation is made with different materials that compose them, which are well indicated in Table 1. However, each material has a different thermal conductivity; therefore, this part is essential before we can simulate the thermal model under the COMSOL tool. Fig. 3 shows the thermal evolution of the first simulation of the Altera 5CSEMA5F31C6 FPGA board.

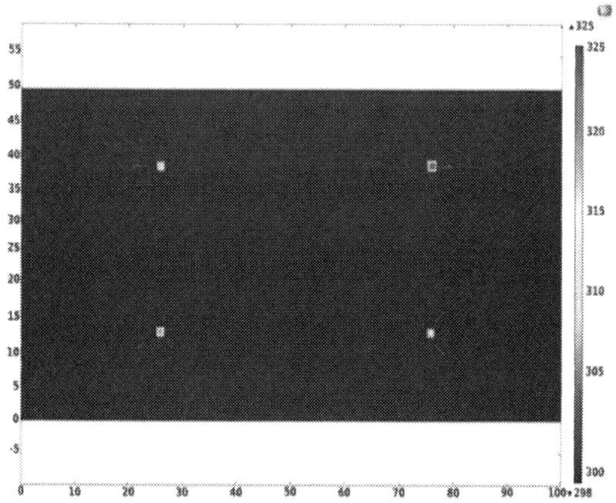

Fig. 3. Thermal distribution of the four critical heat sources on the FPGA board.

This first simulation of our Altera FPGA board under COMSOL gives a good idea of the behavior and thermal diffusion of the heat sources in an FPGA board, and it also shows an unimaginable temperature increase to 325 K (51.85 °C), one of the heat sources in the upper right. It is a thermal peak that was reached after the normal operation of the FPGA board lasted for 13 hours. Fig. 4 shows the thermal evolution of the Altera board versus time.

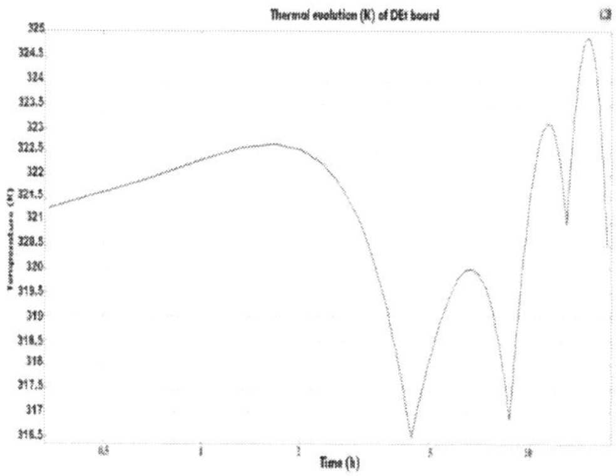

Fig. 4. Thermal evolution of the Altera board versus time.

The graph shown in Fig. 4 is used to clarify the prediction of these thermal peaks and their evolution as a function of time. We can clearly see that after 4 hours of operation of the board, which represented the phase of the transient regime of the FPGA board, we enter the permanent regime or the stationary regime. As indicated, a thermal peak occurs just after 7 hours of ignition of the card, which reaches 320 K (46.85 °C). Then, after around 11 hours of work, another thermal peak is generated, which reaches 323 K (49.85 °C). Next, leaving the card to work well powered, and as soon as it reaches 13 hours of work, another thermal peak is predicted by the COMSOL tool, which can reach 325 K (51.85 °C).

3 Validation of numerical results by the NISA tool

Numerically Integrated System Analysis (NISA) is a finite element-based program that is composed of a series of modules that, depending on the case, have different uses. NISA uses the NISAU/HEAT TRANSFER module, which is a program for the finite element analysis of linear and nonlinear heat transfer in steady and transient conditions [12-16]. In this part, the finite element program NISA is used to predict the thermal behavior of the Altera FPGA board to validate the results obtained by the COMSOL tool, as previously shown.

The approach used with the COMSOL tool is also used with the NISA tool. Fig. 5 shows the results of the thermal evolution of a heat source in the FPGA board. As shown in Fig. 5, the temperature of the heat source on the top right reaches 52.95 °C, which is equivalent to 326.10 K. This confirms and validates the results obtained with the COMSOL tool (Fig. 3).

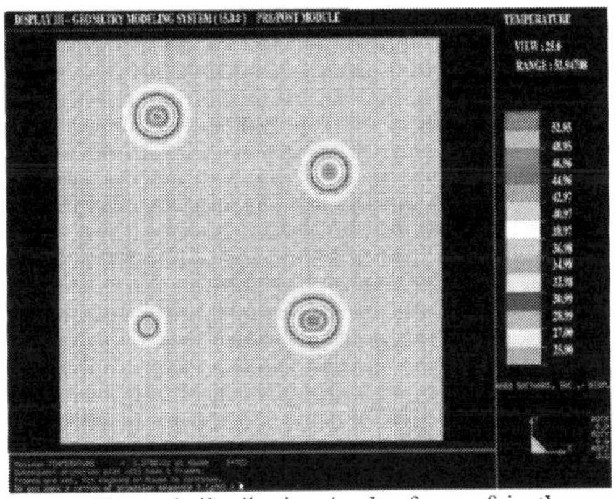

Fig. 5. Thermal distribution in the form of isotherms around a heat source.

This second simulation of our Altera FPGA board with NISA gives a good idea of the behavior and thermal diffusion of the heat sources in the FPGA board.

4 Simulation and FPGA implementation

Recently, the design of specialized applications on FPGAs has been done quickly and efficiently, as they currently play an essential role in the field of rapid prototyping. For this reason, a Verilog code is developed and implemented under the NCLaunch Cadence tool, which is a graphical interface that will allow us to configure and launch the analysis environment to verify and validate the predictions found by COMSOL and NISA. The following part is dedicated to the implementation of the code in the FPGA board under the Altera cyclone V family 5CSEMA5F31C6 platform.

4.1 Logical simulation of the Verilog code

This part has as its main goal the simulation and implementation of the theoretical and numerical results obtained by COMSOL and NISA. For this reason, we have developed a Verilog code.

The examination and validation of the simulation results are realized by the NCLaunch Cadence tool, this last one allows for simulation of the real-time behavior of the system. The implemented Verilog code has been validated against (1) and the study is based on the FEM. Fig. 6 displays the logic simulations showing the equivalent of the results obtained analytically before.

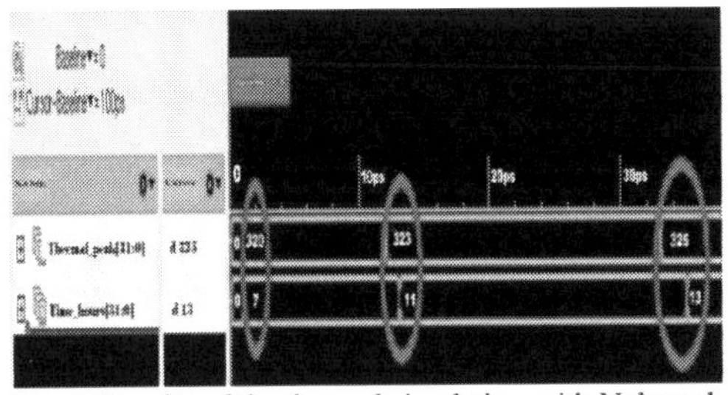

Fig. 6. Results of the thermal simulation with Nclaunch Cadence.

The simulation and synthesis of the heat transfer equation, which is described by the Verilog code, are carried out by the tools of the LIMA laboratory such as the test bench, which verifies the capacity of our algorithm to function according to the initial specifications. The implementation in the FPGA was carried out after the Synthesis Placement and Routing (P&R) steps. Fig. 6 summarizes the results of the three thermal peaks. The most critical one is the one obtained after 13 hours of continuous operation of the board.

4.2 Implementation of the Verilog code on the FPGA board

Once the compilation is complete, after assigning the pins, our program is ready to be uploaded to the DE1 Cyclone V family 5CSEMA5F31C6 board.

Now our Verilog program is available on the Altera DE1 board after setting the clock to 50 Mhz. Fig. 7 shows the value of the critical thermal peak displayed after 13 hours of board operation that is logically simulated and implemented on the FPGA board without floating points. After using the synthesis, planning, placement, and routing tools, the specialized post-synthesis verification tools integrated into the Quartus Prime tool were used to program the FPGA board.

The simulation by the NCLaunch cadence provides us with an insight into the implementation compared to the logic simulation.

Fig. 7. The value of the thermal peak available on FPGA.

This FPGA implementation clearly shows that the temperature values are almost the same compared to the two simulation results via the two tools COMSOL and NISA. The maximum error is about 2 %, which indicates a good result for the validation of our proposal.

5 Conclusion

In this paper, we have developed a simple 2D model for a rapid heat flow analysis of the Altera Cyclone V board. We have also presented COMSOL and NISA simulations for the prediction of thermal peaks occurring on the board under standard working conditions. A maximum error of 2 % was observed between measured and simulated temperatures. This step is important because it affects the overall performance of the board, and it is for the first time that a practical implementation on this type of board of this simple analysis is carried out to have a clear idea of the thermal diffusion around the four sources that produce the critical heat and that can damage the normal operation of the board if exceeded.

References

1. Sun, Yalong, et al. "A review on fabrication and pool boiling enhancement of three-dimensional complex structures." *Renewable and Sustainable Energy Reviews* 162 (2022): 112437.
2. Ettahri, Ouafaa, et al. "A real-time thermal monitoring system intended for embedded sensors interfaces." *Sensors* 20.19 (2020): 5657.
3. Yan, Zhaorui, et al. "Review on thermal analysis in laser-based additive manufacturing." *Optics & Laser Technology* 106 (2018): 427-441.
4. Oukaira, Aziz. *Conception et implémentation d'un système de contrôle thermique appliqué au scanner LabPET II à Haute Performance*. Diss. Université du Québec en Outaouais, 2020.
5. Chen, Kun-Chih, et al. "Temperature tracking and management with number-limited thermal sensors for thermal-aware noc systems." *IEEE Sensors Journal* 20.21 (2020): 13018-13028.
6. Oukaira, A., et al. "Simulation and FPGA implementation of thermal convection equation for complex system design." *IREA) International Journal on Engineering Applications* 2.6 (2016): 307-312.
7. Saydé, Michel, et al. "Silicon-die thermal monitoring using embedded sensor cells unit." *2011 IEEE 9th International New Circuits and Systems conference*. IEEE, 2011.
8. Oukaira, Aziz, et al. "Foster-based Transient Thermal Analysis of SiP for Thermomechanical Studies." *2021 19th IEEE International New Circuits and Systems Conference (NEWCAS)*. IEEE, 2021.
9. Shariffuddin, Sk, et al. "Review on Arbiter Physical Unclonable Function and its Implementation in FPGA for IoT Security Applications." *2022 6th International Conference on Devices, Circuits and Systems (ICDCS)*. IEEE, 2022.
10. Oukaira, Aziz, et al. "Simulation and FPGA Implementation of a Ring Oscillator Sensor for Complex System Design." *Adv. Sci. Technol. Eng. Syst. J* 3 (2018): 317-321.
11. Oukaira, Aziz, et al. "Thermal cooling system development for LabPET II scanners by forced convection flow." *2017 15th IEEE International New Circuits and Systems Conference (NEWCAS)*. IEEE, 2017.
12. López-Buedo, Sergio et al. "Dynamically inserting, operating, and eliminating thermal sensors of FPGA-based systems." *IEEE Transactions on Components and Packaging Technologies* 25 (2002): 561-566.
13. Oukaira, Aziz, et al. "Thermal management and monitoring based on embedded ring oscillator network sensors for complex system design." *International Journal of Computer Engineering and Information Technology* 9.7 (2017): 127.
14. Nisa, Claudia, Celeste Varum, and Anabela Botelho. "Promoting sustainable hotel guest behavior: A systematic review and meta-analysis." *Cornell Hospitality Quarterly* 58.4 (2017): 354-363.
15. Bougataya, Mohammed, et al. "Mixed fluid-heat transfer approach for VLSI steady-state thermal analysis." *IEEE CCECE2002. Canadian Conference on Electrical and Computer Engineering. Conference Proceedings (Cat. No. 02CH37373)*. Vol. 1. IEEE, 2002.
16. Taliyan, S. S., et al. "Finite element based thermal analysis of sealed electronic rack and validation." *2010 2nd International Conference on Reliability, Safety and Hazard-Risk-Based Technologies and Physics-of-Failure Methods (ICRESH)*. IEEE, 2010.

Atmospheric pressure plasma jet based on the dielectric barrier discharge

Najlae Seddaoui[1*], *Moustapha* Ouali[1], and *Youssef* Lagmich[1]

[1]Laboratory of Sciences and Advanced Technologies Polydisciplinary Faculty of Larache, Abdelmalek Essaadi University, Morocco

Abstract. The dielectric barrier discharge has attracted the interest of many researchers in several fields since it produces a non-thermal plasma at atmospheric pressure. Its field of application is totally vast and includes medicine, biotechnology, chemistry, and various other fields of science and technology because of the physical and chemical properties that characterize the plasma jet based on the dielectric barrier discharge. These are characterized by different possible configurations and can be powered by radio frequency, alternating current, and even pulsed direct current depending on the application. In this work, we have presented the atmospheric pressure plasma jet as a new micro discharge that can be used in several fields. We have presented the different possible configurations of the plasma jets based on the dielectric barrier discharge that vary according to the targeted application field. In addition, we have presented some fields of study and specialties in the different application areas such as medicine, biotechnology and food preservation, agriculture and combustion domains. Finally, we have presented the results of the simulation with COMSOL Multiphysics of a dielectric barrier discharge at atmospheric pressure for different frequencies and different discharge gaps between the electrodes, since this is the discharge used in the plasma jet.

1 Introduction

Atmospheric pressure plasma jets allow the non-thermal plasmas to be effectively formed and are easy to use in ambient air conditions since they are considered as new micro discharges. Numerous theoretical and experimental studies based on the chemical and physical properties of plasma jets at atmospheric pressure are carried out in various fields such as medicine, biotechnology, chemistry, nanotechnology, surface modification, and many other fields of science and technology [1][2][3][4]. Various types of gases can be chosen to generate cold atmospheric plasma like helium, nitrogen, argon, air, and heliox a mix of helium and oxygen [5]. Most of the researches reported in several journals has used a plasma jet propagating in a helium stream surrounded by free air at ambient temperature for its potential capacity to deliver high flows of reactive species to a target located a few centimeters away which is interesting in the different fields of application. In this paper, two simulations of a dielectric barrier discharge have been realized with a 2D model at atmospheric pressure using helium with an applied sinusoidal voltage of 1kV, in order to study the influence of the applied voltage frequency which varies between 200 Hz and 10kHz, and the influence of the discharge diameter between the electrodes which varies between 0.8 mm and 2 mm, on the gas breakdown voltage and on the discharge current intensity.

2 DBD plasma jet

The dielectric barrier discharge is an electrical discharge created between two electrodes separated by a dielectric material. In general, the discharge is created by an alternating current applied to the terminals of the electrodes. If a very high potential is applied, a breakdown voltage can be reached. When this voltage is attained between two simple electrodes, an electric arc is created between them. This arc is due to a concentration of charges localized in one place on the surface of an electrode. To avoid this arc, dielectrics are placed on one or both electrodes or even between the electrodes, the role of this dielectric is to distribute the charges accumulated on their surface. Moreover, by accumulating charges on their surfaces, the dielectric helps to create an electric field opposite to the electric field applied to the terminals of the electrodes. This last property stops the discharge before the transition to the arc at the voltage predicted by Paschen's law. Instead of having a very energetic and localized discharge, we will have a stable and distributed discharge on the total surface of the electrodes hidden by the dielectrics.[6][7].

Several types with different configurations of atmospheric pressure plasma jet have been referred to in different research works, most of them operating with a pure noble gas or mixed with a little portion of reactive gases, such as air. Plasma jets operating with noble gases can be classified into four categories [8][9], i. e.

* Corresponding author: najlaeseddaoui@gmail.com

 https://doi.org/10.1051/itmconf/20224802008

2.1 Dielectric-free electrode jets (DFEJ)

Dielectric-free electrode jet (DFEJ) is powered by a 13.56 MHz radiofrequency energy source. It is composed by an internal electrode, which is coupled to the energy source, and an external grounded electrode as depicted in Figure 1.

A mixture of He and reactive gases is introduced into the annular space between the two electrodes at 51L/min. Cooling water is required to prevent jet superheating, and the plasma jet gas temperature ranges from 50 to 300 °C, depending on the radio frequency power. The researchers found for this type of dielectric free electrode jet that the power delivered to the plasma is much higher which leads to a strong increase of the gas temperature and subsequently, the plasma will be very reactive which is not acceptable for biomedical applications. The electric field inside the discharge space and along the direction of propagation of the plasma plume is relatively weak since this jet is powered by a radio frequency. Thus, the propagation of this plasma plume is probably driven by the gas flow than by electricity. [5][10][11]

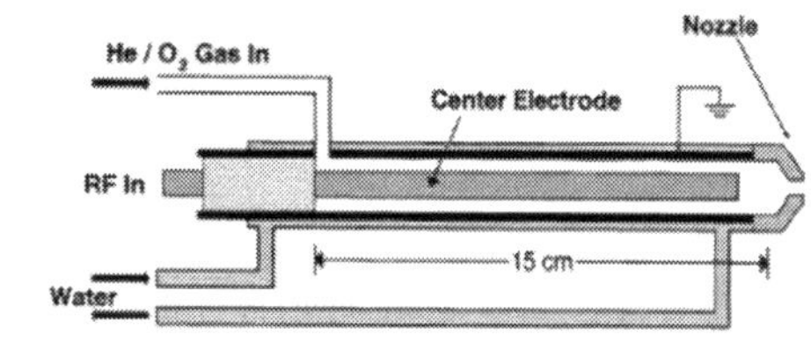

fig 1: Schematic of the dielectric-free electrode atmospheric-pressure plasma jet[11]

2.2 Dielectric barrier discharge jets (DBDJ)

Dielectric barrier discharge jets (DBDJ), characterized by several jet configurations with one or two dielectric barriers. Figure 2(a,b) represents the first configuration of dielectric barrier plasma jets which is characterized by a dielectric tube surrounded on the outer side by two metallic ring electrodes. When a high voltage of kHz is applied across the electrodes and a working gas (He or Ar) passes through the dielectric tube, a cold plasma jet is created in the surrounding air. This type was first described[12] and reviewed [5],[13]. The authors found that the plasma jet consumes a low power of a few watts. The temperature of the plasma gas is equivalent to the ambient temperature, and that the plasma jet, which appears to the naked eye to be homogeneous, is, in reality, a bullet-like volume of plasma with a high propagation Velocity higher than 10 km s^{-1}. By elimination of a ring electrode as shown in figure 2(e) the discharge inside the dielectric tube becomes weaker [14]. To improve the electric field along the plasma jet which is very favorable for generating longer plasma jet and for more active plasma chemistry, [15][16] replaced a high voltage ring electrode with a pin electrode centered inside a dielectric tube, as shown in Figure 2(c,d). If the second ring electrode is eliminated as shown in figure 2(f), the discharge inside the dielectric tube becomes weaker[15].

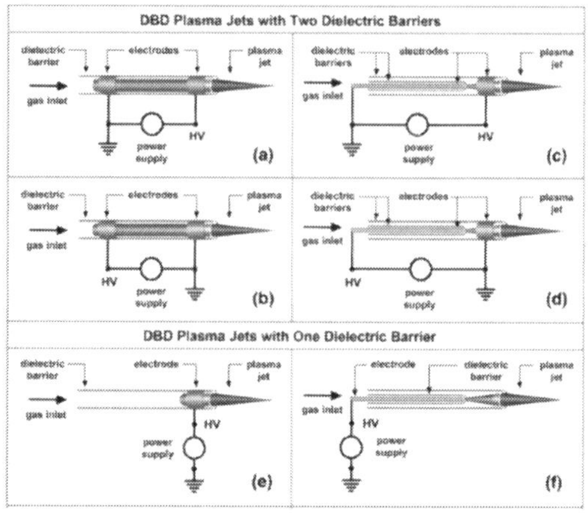

fig 2 : Different configurations of DBD plasma jet [13]

2.3 DBD-like plasma jets

DBD-like plasma jets in figure 3, are called DBD-like because, when the plasma jet is free of contact with a conductive object, the discharge looks approximately like a dielectric barrier discharge. Whereas if the plasma jet is in contact with a conductive object, the discharge passes between the conductive object and the high-voltage electrode, in this stage, it is not a dielectric barrier discharge anymore. Although other DBD systems do not present any risk of having an arc between the object and the dielectric whatever the distance between them, on the other hand for plasma jets of DBD-like there is a risk of having an electric arc depending on the material used is conductive or not and even depending on the field of application. If this arc is avoided, the use of this type presents many advantages [5] [17]. In addition, DBD plasma jets can be powered by radio frequency power, kHz AC power, and even pulsed DC power. The second configuration is to replace the high-voltage electrode with a hollow electrode as shown in figure 3(b), they found that the plasma jet is longer by injecting a noble gas in the first inlet and a reactive gas in the second inlet [18][19].

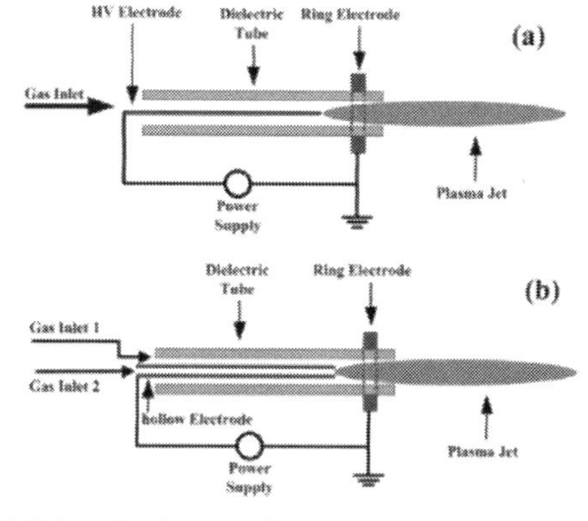

fig 3: Schematic of a DBD-like plasma jet.[5]

2.4 Single electrode jets (SE)

Single electrode jets (SE), Single electrode jets, or plasma needle jets, are basically the same as DBD-like jets, the only difference being that the ring electrode outside the electric tube is eliminated in this model, so the dielectric tube plays a unique role in directing the gas flow [20][21]. This type of jet can also be powered by direct current, kHz alternating current, radio frequency or pulsed direct current [22].

3 Research areas and applications

The applications of atmospheric pressure plasma based on the dielectric barrier discharge in several fields, depending on the technology used, have attracted the interest of many researchers. Among these fields of application, we find the decomposition of volatile organic compounds (VOCs) and the most well-known combustion gases such as butane, toluene, ethanol (90° alcohol), acetone and benzene, which are found in the industrial field during various industrial processes[23][24]. The conversion of greenhouse gases into liquid fuel [25]. The sterilization of water [26]. The treatment of gases and wastewater by the destruction of dangerous organic pollutants [24], the modification of surfaces and the treatment of soils and wastes. Plasma can also be used in medicine in order to treat several diseases, disinfection and sterilization [27]. In the agricultural field and food industry, plasma is used for germination of seeds, growth of plants, storage of products, sterilization of packaging, disinfection, pasteurization and food preservation[28][29].

3.1 Plasma medicine

Plasma medicine is an exciting field where life science and physics intersect. The ability to produce a cold plasma at atmospheric pressure, which is characterized by a multitude of charged particles, reactive gaseous species, electric current, and ultraviolet rays, has been the basis for the fast growth of plasma-related application areas in biomedicine [30]. Plasma can be used in oncology or cancerology; it has been demonstrated that plasma causes apoptosis or programmed physiological cell death of tumor cells as well as reducing skin tumors [31]. To this end, several dermatology studies have been conducted to investigate the efficacy and safety of cold plasma in dermatooncology. To date, plasma has been used for the treatment of eczema and has not caused any negative side effects, according to experiments conducted on more than 40 patients who were treated daily with plasma for two minutes [32][33]. Atmospheric pressure plasma is also used for disinfection, numerous studies have shown the lethality of plasma on fungi and bacteria[34], such as group A streptococcus, Escherichia coli, Pseudomonas Aeruginosa, and (MRSA) methicillin-resistant staphylococcus aureus, as well as the elimination of fungal strains clinically relevant by atmospheric pressure plasma[35][36][37] according to experiments realized on 19 patients with onychomycosis of the nails. Another use of atmospheric pressure plasma on human skin concerns the hydrolipidic film composed by several bacteria, produced by the sebaceous and perspiratory glands. This hydrolipidic film protects the skin against dryness with a pH value between 5.4 and 5.9 for healthy skin [38]. An increase in the pH of the hydrolipidic film leads to a change in the microbial load which causes diseases, such as atopic dermatitis and ichthyosis[39]. Since pH values above 6 are favored by pathogenic bacteria[40].

The interaction of the hydrolipidic film of diseased skin with chemical compounds in the plasma causes a decrease in pH values after five seconds according to experiments on patients with ichthyosis [41], the decrease in pH leads to inhibition of pathogens, and consequently, healing of the skin. Atmospheric pressure plasma helps to accelerate the healing of wounds[42][43] the treatment of scars and acne by decreasing nearly 80% of the sebum production [44]. Generally, the field of plasma medicine is very interesting and exciting, and numerous experiments provide the use of cold plasma for the treatment of various dermatological and oncological diseases.

3.2 Food preservation

The preservation treatments applied to food aim to preserve their edibility and their gustatory and nutritional properties by preventing the development of bacteria, fungi and microorganisms that they contain and that can in some cases lead to food poisoning. The methods used for the preservation of food are based on: heat (pasteurization, sterilization, appertization ...), cold (deep freezing, freezing, and refrigeration ...), there are also vacuum packaging or modified atmosphere, dehydration and drying, fermentation, salting, pickling, smoking, ionization, etc... Spores and spore-forming organisms, when exposed over a long period of time to these mentioned treatments, become resistant. Which prompted researchers [45][46], to develop new preservation technologies among them we find the use of plasma at atmospheric pressure based on dielectric barrier discharge. These studies were based on the influence of different parameters such as the type of electrode and the method of direct or indirect exposure as well as the distance from the sample, the power used, the dose and duration of the treatment and also the interaction of the reactive species of the plasma with bacteria and spores[47][48].

It has been proven that plasma preservation at atmospheric pressure has several advantages such as the simplicity of using the equipment in an ambient temperature and the rapidity of treatment, it improves the stability of food with better nutritional value and the reduction of microbial load [45][47][49].

3.3 Agricultural sector

Atmospheric pressure plasma can be used in the agricultural field. Several devices have been developed according to their application either at the pre-planting, pre-harvest or post-harvest level [50][51]. Atmospheric pressure plasma technologies can be used to disinfect crops or food after harvest [51], decontaminate seeds during storage, improve growth and germination of seeds, reduce soil invasion by pathogens by adding reactive species which leads to a decrease in pH, treatment and sterilization of water used for washing products[52], volatile organic compounds in storage installations and food preparation equipment, and finally the destruction of dangerous food residues.

3.4 Combustion sector

Dielectric barrier discharge is one of the most useful alternative methods for atmospheric pressure plasma combustion. With the aim of satisfying the urgent needs of having a clean and stable combustion, while preserving ozone, this combustion method is attracting attention in the last decades from several researchers[53][54]. Several parameters allow the dielectric barrier discharge to manipulate and control combustions, first, its simple structure and flexibility at variable pressures, second, by using different voltage waveforms and with adjustable frequencies and amplitudes the DBD causes significant ionic winds[55], furthermore, the DBD allows the modification of the kinetic pathway through the generation of radicals and active species [56][57]. Even at the configuration of the DBD used, [58], showed that the configuration plays an important role in the control and stability of the flame.

4 Modelization of plasma at atmospheric pressure

In order to understand the atmospheric pressure plasma based on the dielectric barrier discharge, a two-dimensional model has been used at atmospheric pressure (760 Torr) using Helium as a noble gas, with sinusoidal voltage, a discharge gap between two electrodes varying between 0.1 mm and 2 mm, a relative dielectric permeability of 10. This dielectric permit to avoid the concentration of localized charges in a single place on the surface of the electrodes and subsequently it allows to avoid the creation of an arc between the electrodes.

fig 4: simulation model

In order to calculate the different physical characteristics such as electron density, energy density of electrons. the local energy approximation is used which is based on the solution of a system that contains three fluid equations called moment of Boltzman equation.

$$\frac{\partial(n_e)}{\partial t} + \nabla\,\Gamma_e = R_e \quad (1)$$

$$\Gamma_e = -n_e(\mu_e.\boldsymbol{E}) - \boldsymbol{D}_e.\nabla n_e \quad (2)$$

With n_e: electronic density (1/m3). μ_e: electron mobility (m2/ (V.s)). R_e: electron rate (1/ (m3. S)). $\boldsymbol{D}_e$: electron diffusivity. E: electric field (V/m). $[-(\mu_e.\boldsymbol{E})n_e]$: indicates the electron migration due to an electric field. $(-\boldsymbol{D}_e.\nabla n_e)$: the diffusion of electrons from high to low electron density regions.

$$\frac{\partial(n_\varepsilon)}{\partial t} + \nabla\,\Gamma_\varepsilon + \boldsymbol{E}.\Gamma_e = R_\varepsilon \quad (3)$$

$$\Gamma_\varepsilon = -n_\varepsilon(\mu_\varepsilon.\boldsymbol{E}) - \boldsymbol{D}_\varepsilon.\nabla n_\varepsilon \quad (4)$$

With n_ε: electron energy density (V/m^3). μ_ε: energetic mobility of electrons (m^2/(V.s)). R_ε: Loss/gain of energy due to inelastic collisions. $\boldsymbol{D}_\varepsilon$: energetic diffusivity of electrons. $\boldsymbol{E}.\Gamma_e$: the electrons heating due to external E, it can be a heat source or a heat sink, depending on if the electrons drift or not in the same direction as the electric field.

The energy diffusivity, the energy mobility and the electron diffusivity are calculated from the electron mobility using the Maxwellian electron energy distribution function.

$$\boldsymbol{D}_e = \mu_e T_e \quad (5)$$

$$\mu_\varepsilon = \left(\frac{5}{3}\right)\mu_e \quad (6)$$

$$\boldsymbol{D}_\varepsilon = \mu_\varepsilon T_e \quad (7)$$

T_e: temperature: is a function of the mean electron energy, ε

The accumulation of charges on the dielectric surface leads to the following boundary condition:

$$n.(\boldsymbol{D}_1 - \boldsymbol{D}_2) = \rho \quad (8)$$

$$n.(\boldsymbol{E}_1\,\boldsymbol{\varepsilon}_1 - \boldsymbol{E}_2\,\boldsymbol{\varepsilon}_2) = \rho \quad (9)$$

$\boldsymbol{\varepsilon}_1$ and $\boldsymbol{\varepsilon}_2$: the relative permittivities of the dielectric and the gas, respectively.

The chemical reactions taken into consideration are respectively of the types: Elastic, Excitation and Ionization.

e+He => e+He (10)

e+He => e+Hes (11)

e+Hes => e+he$^+$ (12)

5 Results

The high-pressure discharge with a higher species density and a fairly important space charge generates more interactions between the particles and a low mean free path. hence the considerable electric field plays an important role in the breakdown of the gas. The mechanism involved in this case is called "Streamer mechanism".

The electrical characteristics calculated when two cycles of voltage are applied at atmospheric pressure (760 torr) with a discharge distance equal to 0.5 mm are shown in Figure 5 and 6 at different frequencies (a) 200Hz, (b) 1kHz and (c) 10kHz.

https://doi.org/10.1051/itmconf/20224802008

We notice from the simulation results obtained in fig 5 that the increase of the frequency of the applied voltage causes an increase of the intensity of the discharge current which varies from 15.10^{-6} (A) for a 200Hz frequency to 9.10^{-5} (A) for a 1kHz and then the current intensity reaches 0.003 (A) at 10kHz.

These results can be explained by the fact that when we increase the frequency the period between two maximums is reduced and then the applied voltage reaches the breakdown voltage quickly which reduces the discharge rate which is illustrated on the figures by the number of peaks which is decreased, and finally, we pass from the filamentary regime to a homogeneous regime.

From fig 6 we find that the variation of positive ion density varies in the same way as that of the discharge current in fig 5 with the same peaks, which is totally natural, explained by the fact that the addition of the dielectric stops the accumulation of charges on the surface of the electrodes to avoid the creation of an electric arc. Then it creates an electric field in the opposite direction to the applied electric field. So, the ions propagate in the direction of the applied electric field and the electrons propagate in the opposite direction which requires that the electrons and the ions must be created in an identical way to preserve the charge balance, which explains why the peaks are identical.

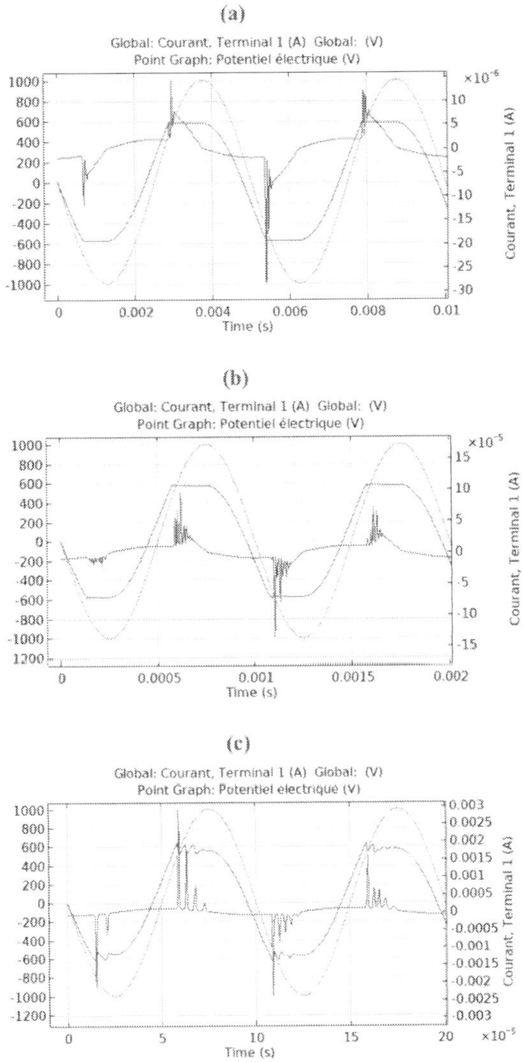

fig 5 : the variation of the breakdown voltage (in red) and of the electric current (in blue) as a function of the applied voltage (in green) with d = 0.5 mm and P = 760 Torr at different frequencies (a) 200Hz, (b) 1kHz and (c) 10kHz.

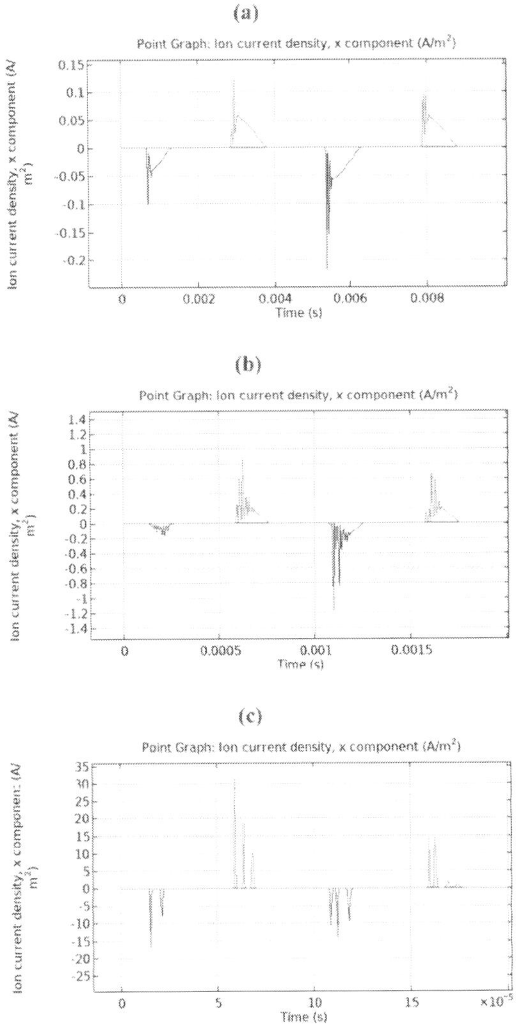

fig 6 : the variation of positive helium ion density, (in blue) next to the dielectric and (in green) next to the cathode with d = 0.5 mm and P = 760 Torr at different frequencies (a) 200Hz, (b) 1kHz and (c) 10kHz.

On fig 7 and 8 another study is presented to better understand the influence of the distance of discharge which separates the electrodes on the intensity amplitude of the discharge current, while keeping the same frequency on the applied voltage. The results of simulation on fig 7 show that with the enlargement of the distance of discharge, the intensity amplitude of current decreases from 25.10^{-5} for a distance of 0.8 mm to 6.10^{-5} for a distance of 2mm passing by 15.10^{-5} for 1mm of distance. This diminution of the discharge distance provokes a reduction of the number of peaks.

These results can be explained by the fact that with the same applied voltage and the same frequency when the discharge distance is very small the particles inside are very close which generates more interaction between them which causes an increase in the number of peaks and an increase in the intensity of the discharge current. and inversely, more the electrode moves away more the particles move away too, which makes the interactions between the particles weaker and consequently decreases the number of peaks.

The simulation results on figure 8 concerning the variation of ions density compared by the results presented on figure 7 confirms what we concluded previously that ions propagate in the direction of the applied electric field and electrons propagate in the opposite direction which requires that electrons and ions must be created in an identical way to preserve the charge balance.

These obtained simulations results are in agreement with several experimental studies performed which presented the variation of discharge current intensity and applied voltage for different frequencies: 500 Hz [21], 18 kHz[31], 27 kHz [30], 10 kHz [59].

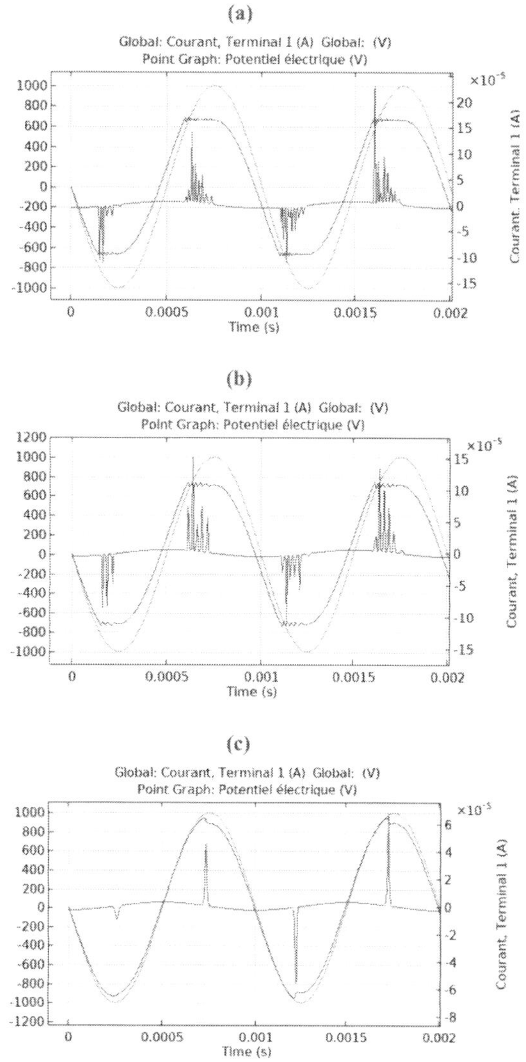

fig 7 : the variation of the breakdown voltage (in red) and of the electric current (in blue) as a function of the applied voltage (in green) with f = 1kHz and P = 760 Torr at different discharge distances (a)0.8mm, (b) 1mm and (c)2mm.

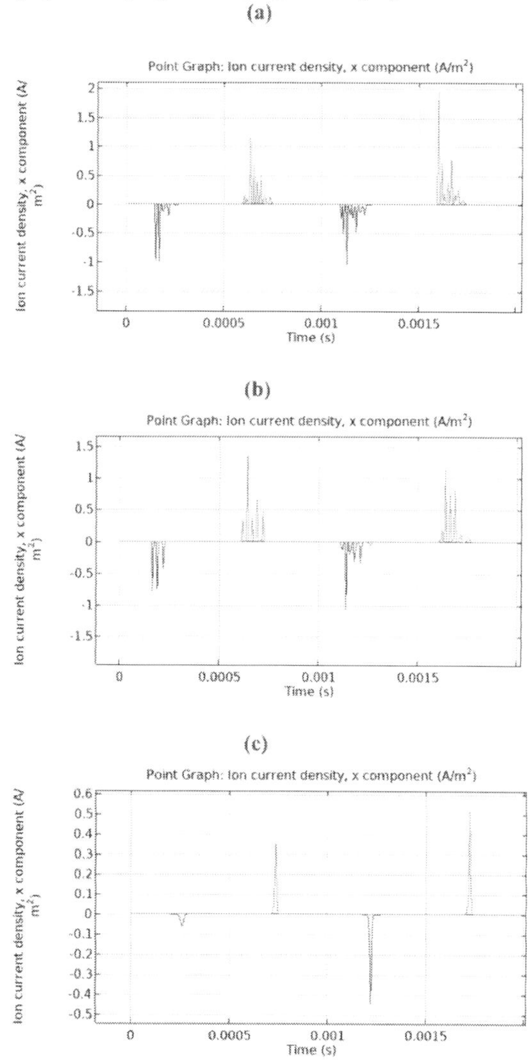

fig 8 : the variation of positive helium ion density, (in blue) next to the dielectric and (in green) next to the cathode with f = 1kHz and P = 760 Torr at different discharge distances (a)0.8mm, (b) 1mm and (c)2mm.

6 Conclusion

The atmospheric pressure plasma based on the dielectric barrier discharge can be used in different fields and with different configurations depending on the application desired due to its physical and chemical properties. In this work two numerical studies were carried out using a two-dimensional model with helium inside with a sinusoidal voltage applied of 1kV.

The first study was dedicated to determine the influence of the applied frequency on the gas breakdown voltage and on the intensity of the discharge current, keeping the same distance of discharge between the electrodes. It was concluded that the increase of the frequency provokes an increase of the current intensity amplitude, but the number of peaks decreases and consequently the rate of discharge decreases as well.

The second study was allocated to determine the influence of the discharge distance between the electrodes on the intensity amplitude of the discharge current while keeping the same frequency on the applied voltage. It was concluded that the widening of the discharge distance decreases the intensity of the discharge current and minimizes the number of peaks. Therefore, from these two studies we conclude that we can go from a filamentary or non-homogeneous regime to a homogeneous one, either by widening the discharge distance or by increasing the frequency of the applied voltage.

References

[1] D. Gidon, H. S. Abbas, A. D. Bonzanini, D. B. Graves, J. Mohammadpour Velni, and A. Mesbah, "Data-driven LPV model predictive control of a cold atmospheric plasma jet for biomaterials processing," *Control Eng. Pract.*, vol. 109, no. December 2020, p. 104725, 2021, doi: 10.1016/j.conengprac.2021.104725.

[2] A. Dubuc *et al.*, "Use of cold-atmospheric plasma in oncology: a concise systematic review," *Ther. Adv. Med. Oncol.*, vol. 10, pp. 1–12, 2018, doi: 10.1177/1758835918786475.

[3] M. Laroussi, X. Lu, and M. Keidar, "Perspective: The physics, diagnostics, and applications of atmospheric pressure low temperature plasma sources used in plasma medicine," *J. Appl. Phys.*, vol. 122, no. 2, 2017, doi: 10.1063/1.4993710.

[4] M. Kehrer, J. Duchoslav, A. Hinterreiter, A. Mehic, T. Stehrer, and D. Stifter, "Surface functionalization of polypropylene using a cold atmospheric pressure plasma jet with gas water mixtures," *Surf. Coatings Technol.*, vol. 384, no. December 2019, p. 125170, 2020, doi: 10.1016/j.surfcoat.2019.125170.

[5] X. Lu, M. Laroussi, and V. Puech, "On atmospheric-pressure non-equilibrium plasma jets and plasma bullets," *Plasma Sources Sci. Technol.*, vol. 21, no. 3, 2012, doi: 10.1088/0963-0252/21/3/034005.

[6] X. Xu, "Dielectric barrier discharge - Properties and applications," *Thin Solid Films*, vol. 390, no. 1–2, pp. 237–242, 2001, doi: 10.1016/S0040-6090(01)00956-7.

[7] S. Brandt, F. D. Klute, A. Schütz, and J. Franzke, "Dielectric barrier discharges applied for soft ionization and their mechanism," *Anal. Chim. Acta*, vol. 951, pp. 16–31, 2017, doi: 10.1016/j.aca.2016.10.037.

[8] J. Asenjo *et al.*, "Atmospheric-Pressure Non-Thermal Plasma Jet for biomedical and industrial applications," *J. Phys. Conf. Ser.*, vol. 591, no. 1, 2015, doi: 10.1088/1742-6596/591/1/012049.

[9] A. I. Muhammad, Q. Xiang, X. Liao, D. Liu, and T. Ding, "Understanding the Impact of Nonthermal Plasma on Food Constituents and Microstructure—A Review," *Food Bioprocess Technol.*, vol. 11, no. 3, pp. 463–486, 2018, doi: 10.1007/s11947-017-2042-9.

[10] J. Y. Jeong *et al.*, "Etching materials with an atmospheric-pressure plasma jet," *Plasma Sources Sci. Technol.*, vol. 7, no. 3, pp. 282–285, 1998, doi: 10.1088/0963-0252/7/3/005.

[11] S. E. Babayan *et al.*, "Deposition of silicon dioxide films with a non-equilibrium atmospheric-pressure plasma jet," *Plasma Sources Sci. Technol.*, vol. 10, no. 4, pp. 573–578, 2001, doi: 10.1088/0963-0252/10/4/305.

[12] M. Teschke, J. Kedzierski, E. G. Finantu-Dinu, D. Korzec, and J. Engemann, "High-speed photographs of a dielectric barrier atmospheric pressure plasma jet," *IEEE Trans. Plasma Sci.*, vol. 33, no. 2 I, pp. 310–311, 2005, doi: 10.1109/TPS.2005.845377.

[13] V. Horvatic, C. Vadla, and J. Franzke, "Discussion of fundamental processes in dielectric barrier discharges used for soft ionization," *Spectrochim. Acta - Part B At. Spectrosc.*, vol. 100, pp. 52–61, 2014, doi: 10.1016/j.sab.2014.08.010.

[14] Q. Li, J. T. Li, W. C. Zhu, X. M. Zhu, and Y. K. Pu, "Effects of gas flow rate on the length of atmospheric pressure nonequilibrium plasma jets," *Appl. Phys. Lett.*, vol. 95, no. 14, pp. 95–98, 2009, doi: 10.1063/1.3243460.

[15] X. Lu, Z. Jiang, Q. Xiong, Z. Tang, and Y. Pan, "A single electrode room-temperature plasma jet device for biomedical applications," *Appl. Phys. Lett.*, vol. 92, no. 15, 2008, doi: 10.1063/1.2912524.

[16] J. L. Walsh and M. G. Kong, "Contrasting characteristics of linear-field and cross-field atmospheric plasma jets," *Appl. Phys. Lett.*, vol. 93, no. 11, pp. 1–4, 2008, doi: 10.1063/1.2982497.

[17] S. Dikmen, N. Şahin, Z. Dikmen, and M. Tanışlı, "Characterization of Ag-exchanged clinoptilolite treated with a plasma jet at atmospheric pressure," *Clay Miner.*, vol. 55, no. 3, pp. 238–247, 2020, doi: 10.1180/clm.2020.33.

[18] A. Shashurin, M. N. Shneider, A. Dogariu, R. B. Miles, and M. Keidar, "Temporal behavior of cold atmospheric plasma jet," *Appl. Phys. Lett.*, vol. 94, no. 23, pp. 1–4, 2009, doi: 10.1063/1.3153143.

[19] V. Léveillé and S. Coulombe, "Design and preliminary characterization of a miniature pulsed RF APGD torch with downstream injection of the source of reactive species," *Plasma Sources Sci. Technol.*, vol. 14, no. 3, pp. 467–476, 2005, doi: 10.1088/0963-0252/14/3/008.

[20] R. Zaplotnik, M. Bišćan, Z. Kregar, U. Cvelbar, M. Mozetič, and S. Milošević, "Influence of a sample surface on single electrode atmospheric plasma jet parameters," *Spectrochim. Acta - Part B At. Spectrosc.*, vol. 103–104, pp. 124–130, 2015, doi: 10.1016/j.sab.2014.12.004.

[21] C. Jiang *et al.*, "Single-electrode He microplasma jets driven by nanosecond voltage pulses," *J. Appl. Phys.*, vol. 119, no. 8, 2016, doi: 10.1063/1.4942624.

[22] D. Hassanpour and S. J. Pestehe, "The effects of grounded electrode geometry on RF-driven cold

atmospheric pressure plasma micro-jet," *J. Theor. Appl. Phys.*, vol. 14, no. 4, pp. 387–398, 2020, doi: 10.1007/s40094-020-00395-0.

[23] M. Gavrilescu, K. Demnerová, J. Aamand, S. Agathos, and F. Fava, "Emerging pollutants in the environment: Present and future challenges in biomonitoring, ecological risks and bioremediation," *N. Biotechnol.*, vol. 32, no. 1, pp. 147–156, 2015, doi: 10.1016/j.nbt.2014.01.001.

[24] B. Jiang *et al.*, "Review on electrical discharge plasma technology for wastewater remediation," *Chem. Eng. J.*, vol. 236, pp. 348–368, 2014, doi: 10.1016/j.cej.2013.09.090.

[25] F. Carmen, H. Szilagyi, and A. Hegri, "Environment and Pollution Management of Pollution Volatile Organic Compounds in Cluj-Napoca," *Walter De Gruyter*, vol. 10, pp. 207–217, 2016, [Online]. Available: http://archive.sciendo.com/PESD/pesd.2016.10.issue-2/pesd-2016-0038/pesd-2016-0038.pdf.

[26] S. Dorevitch *et al.*, "Solar powered microplasma-generated ozone: Assessment of a novel point-of-use drinking water treatment method," *Int. J. Environ. Res. Public Health*, vol. 17, no. 6, pp. 1–13, 2020, doi: 10.3390/ijerph17061858.

[27] N. Mohd Nasir, B. K. Lee, S. S. Yap, K. L. Thong, and S. L. Yap, "Cold plasma inactivation of chronic wound bacteria," *Arch. Biochem. Biophys.*, vol. 605, pp. 76–85, 2016, doi: 10.1016/j.abb.2016.03.033.

[28] K. Kučerová, M. Henselová, Ľ. Slováková, and K. Hensel, "Effects of plasma activated water on wheat: Germination, growth parameters, photosynthetic pigments, soluble protein content, and antioxidant enzymes activity," *Plasma Process. Polym.*, vol. 16, no. 3, pp. 1–14, 2019, doi: 10.1002/ppap.201800131.

[29] A. Starek *et al.*, "Evaluation of selected microbial and physicochemical parameters of fresh tomato juice after cold atmospheric pressure plasma treatment during refrigerated storage," *Sci. Rep.*, vol. 9, no. 1, pp. 1–11, 2019, doi: 10.1038/s41598-019-44946-1.

[30] A. G. Yahaya, T. Okuyama, J. Kristof, M. G. Blajan, and K. Shimizu, "Direct and Indirect Bactericidal Effects of Cold Atmospheric-Pressure Microplasma and Plasma Jet," *Molecules*, vol. 26, pp. 13–18, 2021.

[31] A. J. Kenari *et al.*, "Therapeutic effect of cold atmospheric plasma and its combination with radiation as a novel approach on inhibiting cervical cancer cell growth (HeLa cells)," *Bioorg. Chem.*, vol. 111, no. March, p. 104892, 2021, doi: 10.1016/j.bioorg.2021.104892.

[32] J. Heinlin *et al.*, "A randomized two-sided placebo-controlled study on the efficacy and safety of atmospheric non-thermal argon plasma for pruritus," *J. Eur. Acad. Dermatology Venereol.*, vol. 27, no. 3, pp. 324–331, 2013, doi: 10.1111/j.1468-3083.2011.04395.x.

[33] G. Isbary, T. Shimizu, J. L. Zimmermann, H. M. Thomas, G. E. Morfill, and W. Stolz, "Cold atmospheric plasma for local infection control and subsequent pain reduction in a patient with chronic post-operative ear infection," *New Microbes New Infect.*, vol. 1, no. 3, pp. 41–43, 2013, doi: 10.1002/2052-2975.19.

[34] G. Daeschlein *et al.*, "In vitro susceptibility of important skin and wound pathogens against low temperature atmospheric pressure plasma jet (APPJ) and dielectric barrier discharge plasma (DBD)," *Plasma Process. Polym.*, vol. 9, no. 4, pp. 380–389, 2012, doi: 10.1002/ppap.201100160.

[35] G. Daeschlein *et al.*, "In vitro killing of clinical fungal strains by low-temperature atmospheric-pressure plasma jet," *IEEE Trans. Plasma Sci.*, vol. 39, no. 2, pp. 815–821, 2011, doi: 10.1109/TPS.2010.2063441.

[36] Z. Xiong, J. Roe, T. C. Grammer, and D. B. Graves, "Plasma Treatment of Onychomycosis," *Plasma Process. Polym.*, vol. 13, no. 6, pp. 588–597, 2016, doi: 10.1002/ppap.201600010.

[37] S. R. Lipner, G. Friedman, and R. K. Scher, "Pilot study to evaluate a plasma device for the treatment of onychomycosis," *Clin. Exp. Dermatol.*, vol. 42, no. 3, pp. 295–298, 2017, doi: 10.1111/ced.12973.

[38] M. S. H. C. Korting, "The pH of the Skin Surface and Its Impact on the Barrier Function," pp. 296–302, 2006, doi: 10.1159/000094670.

[39] H. Öhman and A. Vahlquist, "The pH gradient over the stratum corneum differs in X-linked recessive and autosomal dominant ichthyosis; a clue to the molecular origin of the 'acid skin mantle'?," *J. Invest. Dermatol.*, vol. 111, pp. 674–677, 1998.

[40] C. M. Stewart, M. B. Cole, J. D. Legan, L. Slade, M. H. Vandeven, and D. W. Schaffner, "Staphylococcus aureus growth boundaries: Moving towards mechanistic predictive models based on solute-specific effects," *Appl. Environ. Microbiol.*, vol. 68, no. 4, pp. 1864–1871, 2002, doi: 10.1128/AEM.68.4.1864-1871.2002.

[41] A. Helmke, D. Hoffmeister, N. Mertens, S. Emmert, J. Schuette, and W. Vioel, "The acidification of lipid film surfaces by non-thermal DBD at atmospheric pressure in air," *New J. Phys.*, vol. 11, 2009, doi: 10.1088/1367-2630/11/11/115025.

[42] S. Vandersee *et al.*, "Laser scanning microscopy as a means to assess the augmentation of tissue repair by exposition of wounds to tissue tolerable plasma," *Laser Phys. Lett.*, vol. 11, no. 11, 2014, doi: 10.1088/1612-2011/11/11/115701.

[43] H. R. Metelmann *et al.*, "Scar formation of laser skin lesions after cold atmospheric pressure plasma (CAP) treatment: A clinical long term observation," *Clin. Plasma Med.*, vol. 1, no. 1, pp. 30–35, 2013, doi: 10.1016/j.cpme.2012.12.001.

[44] C. Chutsirimongkol, D. Boonyawan, N. Polnikorn, W. Techawatthanawisan, and T. Kundilokchai, "Non-thermal plasma for acne treatment and aesthetic skin improvement," *Plasma Med.*, vol. 4, no. 1–4, pp. 79–88, 2014, doi: 10.1615/PlasmaMed.2014011952.

[45] U. Shah *et al.*, "Effects of cold plasma treatments on spot-inoculated Escherichia coli O157:H7 and quality of baby kale (Brassica oleracea) leaves," *Innov. Food Sci. Emerg. Technol.*, vol. 57, no. November 2018, p. 102104, 2019, doi: 10.1016/j.ifset.2018.12.010.

[46] M. Gavahian and P. J. Cullen, "Cold Plasma as an Emerging Technique for Mycotoxin-Free Food: Efficacy, Mechanisms, and Trends," *Food Rev. Int.*, vol. 36, no. 2, pp. 193–214, 2020, doi: 10.1080/87559129.2019.1630638.

[47] C. Smet *et al.*, "Combined effect of cold atmospheric plasma, intrinsic and extrinsic factors on the microbial behavior in/on (food) model systems during storage," *Innov. Food Sci. Emerg. Technol.*, vol. 53, no. June 2017, pp. 3–17, 2019, doi: 10.1016/j.ifset.2018.05.016.

[48] M. Umair, S. Jabbar, M. M. Nasiru, A. M. Senan, H. Zhuang, and J. Zhang, "Sequential Application of

High-Voltage Electric Field Cold Plasma Treatment and Acid Blanching Improves the Quality of Fresh Carrot Juice (Daucus carota L.)," *J. Agric. Food Chem.*, vol. 68, no. 51, pp. 15311–15318, 2020, doi: 10.1021/acs.jafc.0c03470.

[49] M. Umair *et al.*, "Recent Advances in Plasma Technology: Influence of Atmospheric Cold Plasma on Spore Inactivation," *Food Rev. Int.*, vol. 00, no. 00, pp. 1–23, 2021, doi: 10.1080/87559129.2021.1888972.

[50] R. Laurita, D. Barbieri, M. Gherardi, V. Colombo, and P. Lukes, "Chemical analysis of reactive species and antimicrobial activity of water treated by nanosecond pulsed DBD air plasma," *Clin. Plasma Med.*, vol. 3, no. 2, pp. 53–61, 2015, doi: 10.1016/j.cpme.2015.10.001.

[51] V. Scholtz, J. Pazlarova, H. Souskova, J. Khun, and J. Julak, "Nonthermal plasma - A tool for decontamination and disinfection," *Biotechnol. Adv.*, vol. 33, no. 6, pp. 1108–1119, 2015, doi: 10.1016/j.biotechadv.2015.01.002.

[52] J. A. Buendia, E. Perez-Lopez, and A. Venkattraman, "System-level model and experiments for irrigation water alkalinity reduction and enrichment using an atmospheric pressure dielectric barrier discharge," *Water Res.*, vol. 144, pp. 728–739, 2018, doi: 10.1016/j.watres.2018.07.073.

[53] R. Brandenburg, "Corrigendum: Dielectric barrier discharges: progress on plasma sources and on the understanding of regimes and single filaments (Plasma Sources Science and Technology (2017) 26 (053001) DOI: 10.1088/1361-6595/aa6426)," *Plasma Sources Sci. Technol.*, vol. 27, no. 7, 2018, doi: 10.1088/1361-6595/aaced9.

[54] I. V. Adamovich and W. R. Lempert, "Challenges in understanding and predictive model development of plasma-assisted combustion," *Plasma Phys. Control. Fusion*, vol. 57, no. 1, p. 14001, 2015, doi: 10.1088/0741-3335/57/1/014001.

[55] M. Simeni Simeni, Y. Tang, Y. C. Hung, Z. Eckert, K. Frederickson, and I. V. Adamovich, "Electric field in Ns pulse and AC electric discharges in a hydrogen diffusion flame," *Combust. Flame*, vol. 197, pp. 254–264, 2018, doi: 10.1016/j.combustflame.2018.08.004.

[56] T. Ono, T. Segawa, N. Saito, E. Takahashi, and M. Nishioka, "Effect of long-lived species generated by non-thermal plasmas on the auto-ignition delay of liquid Hydrocarbon fuel-air pre-mixtures," *Combust. Sci. Technol.*, vol. 189, no. 9, pp. 1624–1638, 2017, doi: 10.1080/00102202.2017.1318856.

[57] S. Nagaraja, V. Yang, Z. Yin, and I. Adamovich, "Ignition of hydrogen-air mixtures using pulsed nanosecond dielectric barrier plasma discharges in plane-to-plane geometry," *Combust. Flame*, vol. 161, no. 4, pp. 1026–1037, 2014, doi: 10.1016/j.combustflame.2013.10.007.

[58] M. G. De Giorgi, A. Ficarella, A. Sciolti, E. Pescini, S. Campilongo, and G. Di Lecce, "Improvement of lean flame stability of inverse methane/air diffusion flame by using coaxial dielectric plasma discharge actuators," *Energy*, vol. 126, pp. 689–706, 2017, doi: 10.1016/j.energy.2017.03.048.

[59] P. Svarnas, A. Spiliopoulou, P. Koutsoukos, K. Gazeli, and E. Anastassiou, "Acinetobacter baumannii Deactivation by Means of DBD-Based Helium Plasma Jet," *Plasma*, vol. 2, no. 2, pp. 77–90, 2019, doi: 10.3390/plasma2020008.

Semi-Analytical Finite Element (SAFE) method for plotting Lamb waves dispersion curves of an aluminum plate and comparison with Disperse software.

Abdelali Yacoubi [1*], *Ayoub* Jabiri [1], *Mhammed* El Allami[1-2]

[1] Labo MIET, Department of applied physics, FST, Settat, Morocco.

[2] CRMEF Settat, Morocco.

Abstract. Investigation into ultrasonic Non-destructive evaluation (NDE) methods for assessing complex designs depends firmly on displaying apparatuses that calculate dispersion curves. These expectations are vital for empowering the best assessment methodologies to be distinguished and their aversions to being assessed. This study, the main purpose of this work is to do a quantitative evaluation of the SAFE technique by calculating dispersion curves (wavenumber, phase velocity, and group velocity) and comparing them to those generated by Disperse Software.

Keywords. Semi-Analytical Finite Element (SAFE), Disperse software, Lamb waves, Dispersion curves

1 Introduction

Lately, there is a developing interest in the use of elastic-guided waves in non-destructive testing (NDT). These waves show up in structures wherever one aspect is smaller compared to the others, concerning model wires, pipes, bars, plates, or shells [1], [2].

The wavelength, dispersivity, phase, and group velocities of guided waves are all displayed on dispersion curves at a specific frequency [3]. When performing inspection measures, it's helpful to create a visual model of the structure in question. This often entails determining the theoretical wave motion modality and dispersivity for plates and other waveguides.

These relationships, however, do not hold for increasingly complicated geometries and material models. Various matrix algorithms were developed to describe guided waves in layered media [4]. These techniques have been applied to a wide range of applications from thin film coatings to global seismology. These techniques were used to develop the software Disperse [5], a widely used research tool for plate modeling.

The authors have previously studied dispersion curves and their various uses in NDT. They proposed several iterative methods for plotting dispersion curves.

However, even though the dispersion curves generated by iterative approaches are very close to analytical curves, they still show errors in calculating the cutoff frequencies. As a result, we suggest an alternative strategy for a better dispersion curve estimate of the solution [6].

Many researchers utilized the SAFE method to examine Lamb wave propagation in isotropic and composite structures. The method was used by Ahmad et al. [7] to plot the dispersion curves of an aluminum plate and composite construction. The SAFE approach was used by Bartoli et al. [8] to model the wave propagation of Lamb waves in waveguides of any section.

Wenbo et al. [9] recently published a methodology for estimating pipeline dispersion curves immersed in a fluid that incorporates the SAFE method approach and the perfect match layer technique. The SAFE method was used by Hayashi et al [10]. to plot rail dispersion curves for railroads.

Computer models and visualization have also revealed sophisticated directed wave movements. Particular calculation techniques were developed for these numerical analyses to express guided wave propagation

* Corresponding author: a.yacoubi@uhp.ac.ma

over long distances, such as the hybrid method and the finite element method (FEM) or boundary element method (BEM), and the semi-analytical finite element (SAFE) approach, which treats wave motions in the propagation direction theoretically and subdivides the cross-section [11].

The SAFE approach is used to simulate guided-wave propagation of homogeneous isotropic aluminum plate with free borders in this study. Analytically, the displacements along the wave propagation direction can be characterized as harmonic exponential functions. Only the cross-section needs to be discretized using this procedure[12] [8], by constructing a large number of large complex stiffness matrices [13]. The SAFE codes generate the solutions from an eigenvalue problem, they do not require the root finding procedures utilized in iterative approaches. Our goal is to do an evaluation of the semi-analytical finite element (SAFE) and compare the solutions to those obtained by the Disperse Software.

2 SAFE Method

Lamb wave propagation curves for isotropic and composite plates can be determined using this method. When solving the dispersive relation using SAFE, The methodology includes both FE and analytical formulations. Only the cross-section of the plate is discretized with Finite Element (FE). The simulated displacements are using analytical harmonic functions in the wave propagation direction, i.e. normal to the cross-section plane or line, hence the name semi-analytical [14]. This offers computational savings, especially for short wavelengths when a very thin mesh is required [15].

In the SAFE method, at every frequency, a discrete number of directed modes is acquired. The waveguide cross-section is placed on the x-y plane for a Cartesian reference framework (See Fig.1).

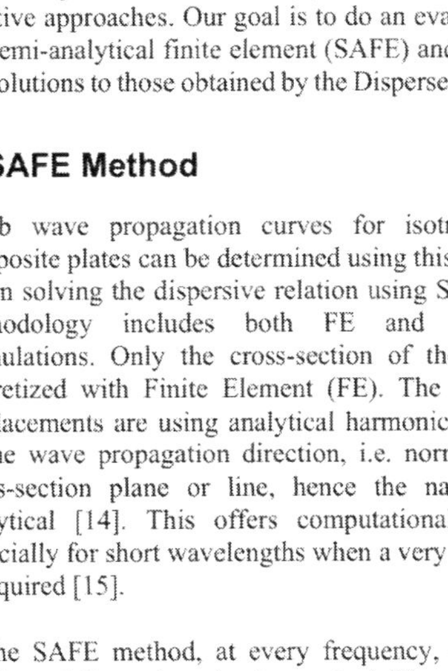

Fig. 1. Cross-section of a plate waveguide in Cartesian coordinates for SAFE

The harmonic motion is represented by introducing a harmonic displacement field along the propagation direction, z, and interpolating the displacement inside the element in the cross-sectional plane x-y [8].

The displacement field's final expression is:

$$\mathbf{U}(x,y,z,t)=\begin{bmatrix} u(x) \\ w(x) \end{bmatrix} e^{i(kz-\omega t)} \quad (1)$$

where $i^2 = -1$, ω is the angular frequency and k is the wavenumber.

It can be observed that by assuming an exponential harmonic term for the displacement field in (z, t) for an infinite plate in y, the x dependence can be described using the interpolation functions with discrete technique[14].

The plate cross-section domain is partitioned into a multilayer system[16]. Interpolation functions matrix $\mathbf{N}(x)$ and nodal unknown displacements can be used to approximate the element displacement. For a three-node one-dimensional element, the displacement field can be expressed as:

$$\mathbf{U}^{(e)}(x,z,t) = \mathbf{N}(x)\mathbf{q}^{(e)} e^{i(kz-\omega t)} \quad (2)$$

The element strain can be expressed as:

$$\varepsilon^{(e)} = \left[\mathbf{L}_x \frac{\partial}{\partial x} + \mathbf{L}_z \frac{\partial}{\partial z}\right] \mathbf{N}(x)\mathbf{q}^{(e)} e^{i(kz-\omega t)}$$

$$= (\mathbf{B}_1 + ik\mathbf{B}_2)\, \mathbf{q}^{(e)} e^{i(kz-\omega t)} \quad (3)$$

in which $\mathbf{B}_1 = \mathbf{L}_x \mathbf{N}_{,x}$ (4) $\mathbf{B}_2 = \mathbf{L}_z \mathbf{N}$ (5)

$\mathbf{N}_{,x}$ is the derivative of the shape functions in x.

The quadratic system of eigenvalue equations is obtained by substituting these functions into the equation of motion and building the elementary matrices by using the Hamilton principle:

$$[\mathbf{K}_1 + ik\mathbf{K}_2 + k^2\mathbf{K}_3 - \omega^2\mathbf{M}]\mathbf{U} = 0 \quad (6)$$

where $\mathbf{U}$ is the vector of amplitudes of the global displacement field; the matrices $\mathbf{K}_1$, $\mathbf{K}_2$, $\mathbf{K}_3$, and $\mathbf{K}_4$ are defined by

$$\mathbf{K}_1 = \bigcup_{e=1}^{nel} \mathbf{k}_1^{(e)} \quad (7) \qquad \mathbf{K}_2 = \bigcup_{e=1}^{nel} \mathbf{k}_2^{(e)} \quad (8)$$

$$\mathbf{K}_3 = \bigcup_{e=1}^{nel} \mathbf{k}_3^{(e)} \quad (9) \qquad \mathbf{M} = \bigcup_{e=1}^{nel} \mathbf{m}^{(e)} \quad (10)$$

with

$$\mathbf{k}_1^{(e)} = \int_x^0 \mathbf{B}_1^{\mathbf{T}} \mathbf{C}_e \mathbf{B}_1 \, dx \quad (11)$$

$$\mathbf{k}_2^{(e)} = \int_x^0 (\mathbf{B}_1^{\mathbf{T}} \mathbf{C}_e \mathbf{B}_2 - \mathbf{B}_2^{\mathbf{T}} \mathbf{C}_e \mathbf{B}_1) \, dx \quad (12)$$

$$\mathbf{k}_3^{(e)} = \int_x^0 \mathbf{B}_2^{\mathbf{T}} \mathbf{C}_e \mathbf{B}_2 \, dx \quad (13)$$

$$\mathbf{m}^{(e)} = \int_x^0 \mathbf{N}^{\mathbf{T}} \mu_e \mathbf{N} \, dx \quad (14)$$

$\mathbf{C_e}$ and μ_e are the elastic stiffness tensor and the density for the generic element, respectively.

$\mathbf{K}_2$ is changed into a symmetric form by pre-multiplying and post-multiplying equations using transformation matrices $\mathbf{T^T}$ and $\mathbf{T}$, respectively, to simplify computational savings' numerical resolution.

$$\mathbf{T^T K_1 T = K_1} \quad (15) \qquad \mathbf{T^T K_3 T = K_3} \quad (16)$$

$$T^T K_2 T = -i\widehat{K_2} \quad (17) \qquad \mathbf{T^T M T = M} \quad (18)$$

$$\mathbf{T^T U} = \widehat{\mathbf{U}} \quad (19)$$

The equation (6) can also be rewritten as:

$$[\mathbf{K_1} + k\widehat{\mathbf{K_2}} + k^2\mathbf{K_3} - \omega^2\mathbf{M}]\widehat{\mathbf{U}} = 0 \quad (20)$$

The $\mathbf{K}_2$ symmetric matrix and the new eigenvector are represented by $\widehat{\mathbf{K_2}}$ and $\widehat{\mathbf{U}}$, respectively.

The integrations over the elements were computed using the Gauss quadrature rule in this study [17]. Fixing the angular frequency and solving the problem

The equation of motion can be solved by fixing the frequency ω and solving for the positive wavenumber k.

The answer can be written as an eigensolution, with the eigenvalues and eigenvectors having to satisfy certain conditions:

$$\left(\begin{bmatrix} 0 & \mathbf{K_1} - \omega^2\mathbf{M} \\ \mathbf{K_1} - \omega^2\mathbf{M} & \widehat{\mathbf{K_2}} \end{bmatrix} - k\begin{bmatrix} \mathbf{K_1} - \omega^2\mathbf{M} & 0 \\ 0 & -K_3 \end{bmatrix}\right)\begin{bmatrix} \widehat{\mathbf{U}} \\ k\widehat{\mathbf{U}} \end{bmatrix} = \begin{bmatrix} 0 \\ 0 \end{bmatrix}$$

(21)

The global matrices $\mathbf{K}_1$, $\mathbf{K}_3$, and $\mathbf{M}$ are symmetric. However, the global matrix $\mathbf{K}_2$ is skew-symmetric.

The phase velocity can be calculated using the wavenumbers obtained by solving the eigenvalue problem:

$$V_p = {}^{\omega}/_{k} \quad (22)$$

This leads to the calculation of group velocity using the relation after some algebra manipulation[14].

$$V_g = \frac{\partial\omega}{\partial k} = \frac{\mathbf{Q_R^T K' Q_R}}{2\omega\mathbf{Q_R^T M Q_R}} \quad (23)$$

where the right eigenvector of the system is denoted by $\mathbf{Q_R}$.

and $$\mathbf{K'} = \widehat{\mathbf{K}}_2 + 2k\mathbf{K_3} \quad (24)$$

Proposing a real positive and solving for the wavenumber k, as well as proposing a real or complex wavenumber and solving for the angular frequency, can lead to an eigensolution. Typically, one offers an angular frequency and solves for the wavenumber $\boldsymbol{k = k(\omega)}$, which allows the analysis of both propagating and non-propagating waves **[18]**.

3 Numerical results

In this part, a numerical example demonstrating the validity and correctness of the proposed SAFE for the analysis of guided waves in elastic-free plate problems is shown. To begin, these results are compared to numerical data acquired using the usual SAFE approach and analytical results generated using the software Disperse.

The SAFE formulation has been validated for a homogeneous aluminum plate of 2mm thickness with the following isotropic elastic characteristics: density $\rho = 2700\ kg/m3$, longitudinal velocity$V_L =$ 6091m/s, and shear velocity $V_T = 3134\ m/s$.

By solving the eigenvalue problem established in the previous section, the SAFE formulation may produce dispersion curves for Lamb waves in a simple manner. There are two methods to go through frequency sweep or wavenumber sweep.

The modeling method is separated into many sections, as shown in Fig. 2. The first step is to define the structure's conceptual model, as well as select a material model for the structure. The second step, create a FE model of the conceptual model and solve the eigenvalue problem that results. Finally, the collected solution is post-processed to save useful data. Physically meaningful roots are chosen, whereas false roots are rejected.

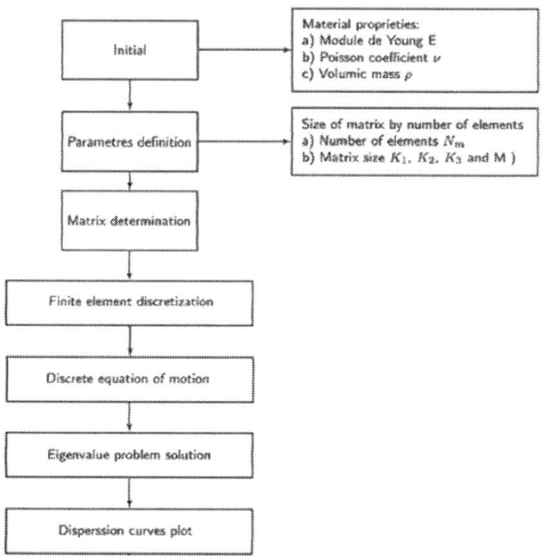

Fig. 2. Flowchart for the modeling process

3.1 Impact of number of elements

To more understand the impact of the number of elements. This last option is being considered as a variant in this research to see how it affects the results [19].

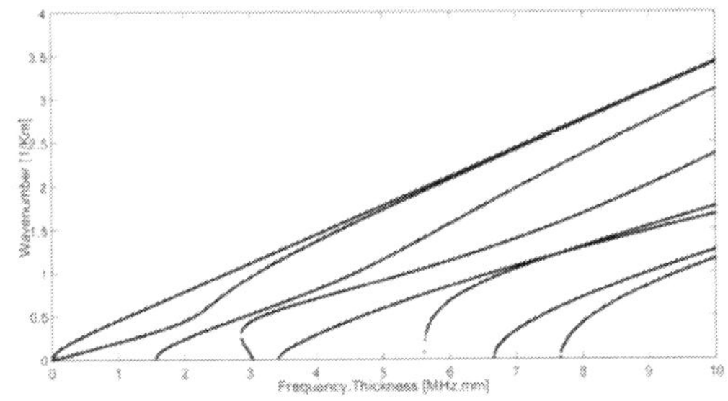

Fig. 3. The wavenumber dispersion curves by the SAFE method for $N=2$

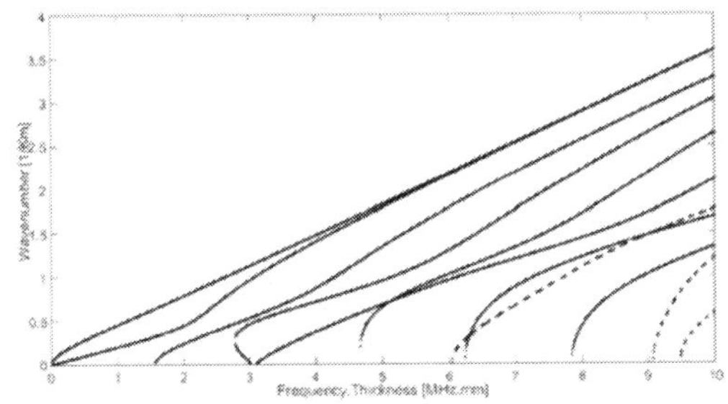

Fig. 4. The wavenumber dispersion curves by the SAFE method for $N=8$

The wavenumber curves generated by the SAFE approach with different numbers of elements $N=2$, and $N=8$ are shown in Fig. 3 and Fig 4. Which illustrate the apparition of new modes.

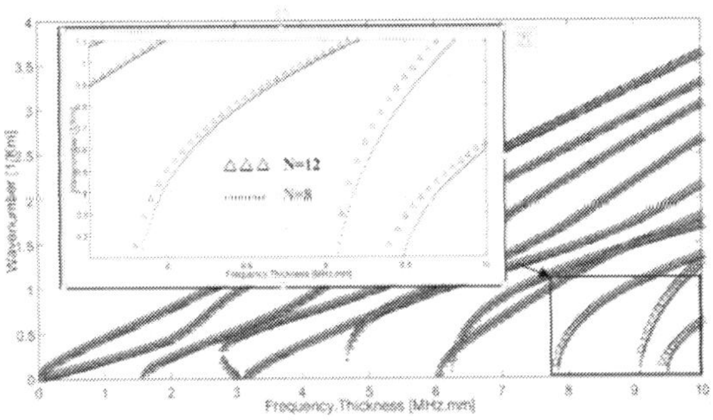

Fig. 5. Superposed wavenumber curves for different numbers of elements by the SAFE method

Comparing dispersion curves of a different numbers of elements Fig.5 justifies that the application of a lower number of elements in the input, it may not be capable to detect some roots, or, not achieving the desired precision, especially for the high product of thickness and frequency.

3.2 Minimum number of elements N_{min}

Some calculations may reduce precision as a result of speed optimizations performed by the compiler, the main factor is the number of subdivisions in the SAFE method related to the number of elements.

The following meshing criterion is provided based on the examination of the element stiffness matrix error **[20]**.

$$\frac{\lambda_T}{l} = \beta \tag{25}$$

where l denotes the element's length, $\lambda_T = \frac{V_T}{f}$ denotes the wavelength of transverse waves propagating at velocity V_T, and $\beta = 4$ signifies the degree of the polynomial interpolation.

replacing l with e/N, where N is the number of elements the formulation in inequality is

$$\frac{NV_T}{fe} > \beta \tag{26}$$

As a result, the required minimum number of elements can be calculated. to ensure the SAFE method's accuracy for a given product $F = fe$.

$$N_{min} = \left|\frac{\beta}{V_T} F\right| \tag{27}$$

By replacing $\beta = 4$ and $V_T = 3134\ m/s$ in our example, and F = 10 MHz.mm, the minimal number of elements is: $N_{min} = 13$.

The application of this number of elements in a calculation does not ensure the desired precision.

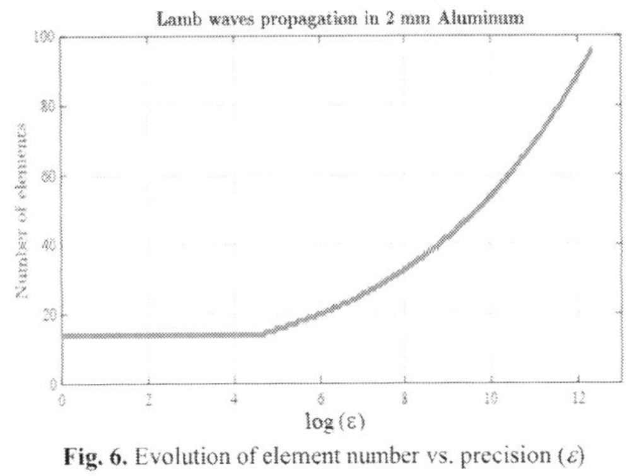

Fig. 6. Evolution of element number vs. precision (ε)

The precision of roots from the SAFE technique is shown against the number of elements in Fig. 6. The curve describes that to obtain a high precision an important number of elements is required.

Only the real roots of the dispersion relations that correspond to Lamb's propagating modes are taken into account in this study. The non-propagating (pure imaginary wavenumber) or attenuated (complex wavenumber) modes are not taken into account [17].

https://doi.org/10.1051/itmconf/20224802009

To compute the dispersion curves in our example, the number of elements used is $N = 18$ elements.

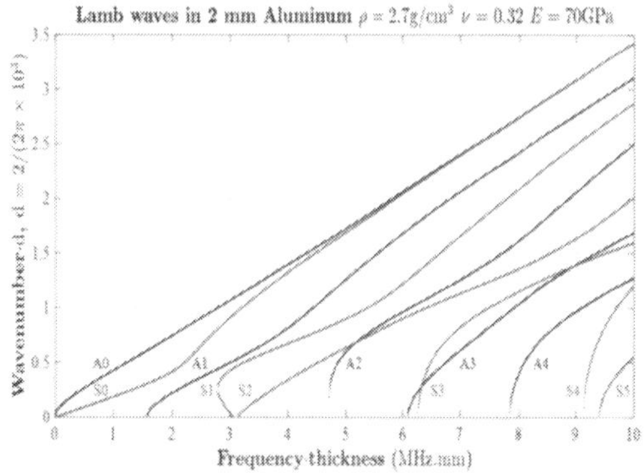

Fig. 7. Wavenumber curves for isotropic aluminium plate

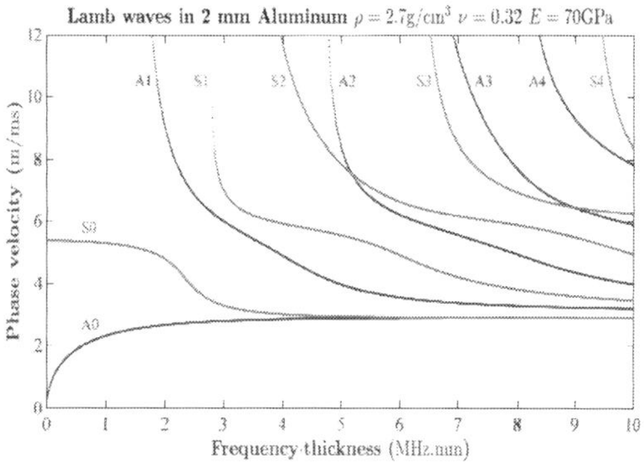

Fig. 8. Phase velocity curves for isotropic aluminium plate

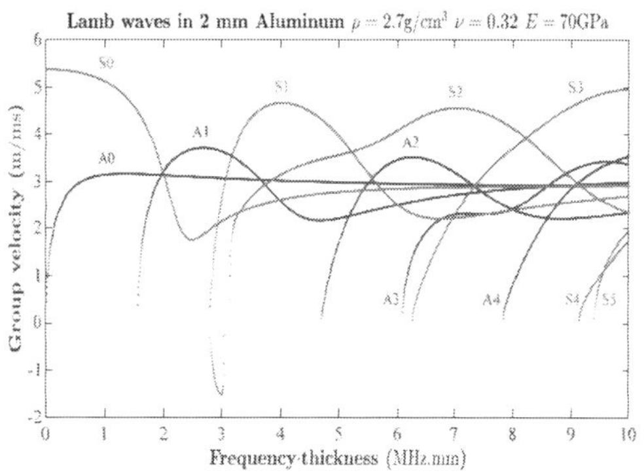

Fig. 9. Group velocity curves for the isotropic aluminium plate

Fig.7 shows the evolution of propagating wavenumber vs. the product frequency-thickness. The wavenumbers reveal the modes present in the plate for a specific product (f.e).

Fig.7 shows the wavenumbers of the first symmetric (S0, S1, S2, S3, and S4) and antisymmetric (A0, A1, A2, A3, and A4) modes.

Fig. 8 shows the phase velocity evolution of propagating modes vs. the frequency-thickness product. When modeling the propagation of Lamb waves, the understanding phase velocity is crucial. Indeed, to generate a Lamb mode at a certain frequency-thickness product, we must know the exact phase velocity with which the mode propagates in the plate so that it does not cross with another mode that is not the intended mode. Fig.8 shows the phase velocity of the first symmetric (S0, S1, S2, S3, and S4) and antisymmetric (A0, A1, A2, A3, and A4) modes.

Fig. 9 describes the evolution of propagating modes' group velocity vs. frequency-thickness product. The group velocity is the speed at which the wave packet containing the group pulse propagates along with the plate. Along with the plate, it also represents the wave's propagating energy speed. Fig.9 shows the group velocity of the initial symmetric modes (S0, S1, S2, S3, and S4), as well as the antisymmetric modes (A0, A1, A2, A3, and A4).

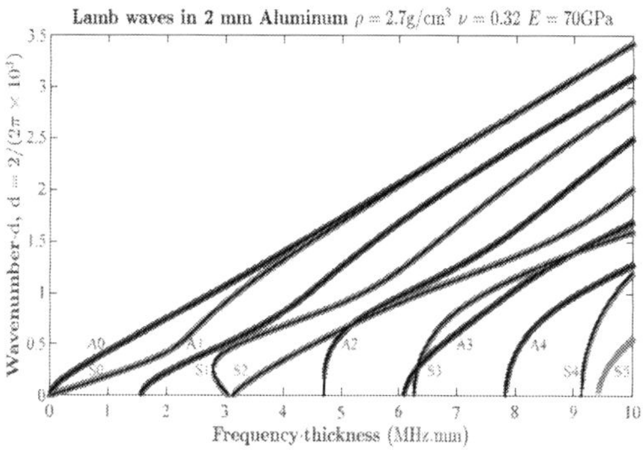

Fig. 10. Wavenumber curves superposed with Disperse

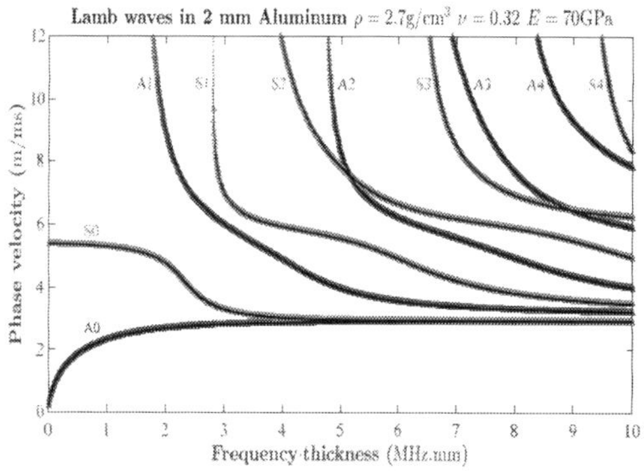

Fig. 11. Phase velocity curves superposed with Disperse

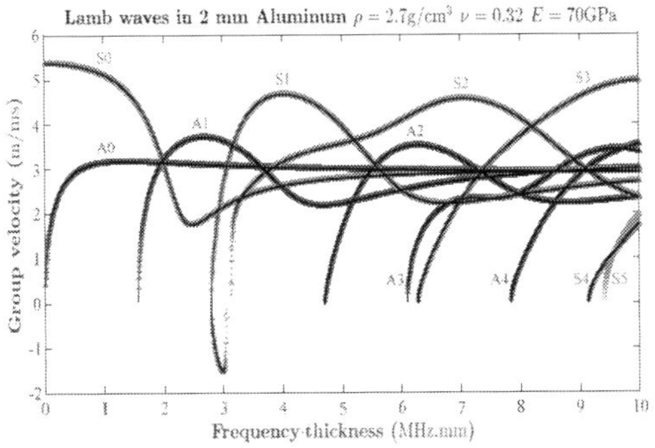

Fig 12. Group velocity curves superposed with disperse

The dispersion relations are accurately captured by the SAFE approach. The phase velocity of real-valued roots from the SAFE method is plotted against the Disperse dispersion curves for the five lowest modes, i.e. as shown in Fig.11. Along the dispersion curves, SAFE group velocity solutions can be discovered. For the five lowest modes, i.e. those illustrated in Fig.12, the wavenumber of real-valued roots from the SAFE technique is shown against the Disperse wave number in Fig. 10. As a result, due to the computing benefits, using a SAFE approach of cross-sections is recommended.

4 Conclusion

In this study, the SAFE dispersion results match those of the Disperse software for a continuous homogeneous isotropic linear elastic free plate. The results revealed that the method's accuracy is dependent on the number of meshing elements used. The number of elements per plate thickness must be tuned to achieve acceptable precision.

The approach proposed in this study can be used to investigate progressively the nodal displacement, deformation, and strain of a plate or complicated structures like composites or functionally graded material in the future.

References

[1] J. D. Achenbach, *Wave propagation in elastic solids*. Amsterdam, New York: North-Holland Pub. Co.; American Elsevier Pub. Co, 1973.

[2] K. Worden, "Rayleigh and Lamb Waves - Basic Principles," *Strain*, vol. 37, no. 4, pp. 167–172, Nov. 2001, doi: 10.1111/j.1475-1305.2001.tb01254.x.

[3] J. L. Rose, *Ultrasonic Guided Waves in Solid Media*. New York: Cambridge University Press, 2014. doi: 10.1017/CBO9781107273610.

[4] M. J. S. Lowe, "Matrix techniques for modeling ultrasonic waves in multilayered media," *IEEE Trans. Ultrason. Ferroelectr. Freq. Control*, vol. 42, no. 4, pp. 525–542, Jul. 1995, doi: 10.1109/58.393096.

[5] B. Pavlakovic, M. Lowe, D. Alleyne, and P. Cawley, "Disperse: A General Purpose Program for Creating Dispersion Curves," in *Review of Progress in Quantitative Nondestructive Evaluation*, D. O. Thompson and D. E. Chimenti, Eds. Boston, MA: Springer US, 1997, pp. 185–192. doi: 10.1007/978-1-4615-5947-4_24.

[6] S. Nissabouri, M. El Allami, and E. H. Boutyour, "Quantitative evaluation of semi-analytical finite element method for modeling Lamb waves in orthotropic plates," *Comptes Rendus Mécanique*, vol. 348, no. 5, pp. 335–350, Nov. 2020, doi: 10.5802/crmeca.13.

[7] Z. A. B. Ahmad, J. M. Vivar-Perez, and U. Gabbert, "Semi-analytical finite element method for modeling of lamb wave propagation," *CEAS Aeronaut. J.*, vol. 4, no. 1, pp. 21–33, Apr. 2013, doi: 10.1007/s13272-012-0056-6.

[8] I. Bartoli, A. Marzani, and H. Matt, "Modeling Guided Wave Propagation for Structural Health Monitoring Applications," p. 9, 2006.

[9] W. Duan and R. Kirby, "Guided wave propagation in buried and immersed fluid-filled pipes: Application of the semi analytic finite element method," *Comput. Struct.*, vol. 212, pp. 236–247, Feb. 2019, doi: 10.1016/j.compstruc.2018.10.020.

[10] T. Hayashi, W.-J. Song, and J. L. Rose, "Guided wave dispersion curves for a bar with an arbitrary cross-section, a rod and rail example," *Ultrasonics*, vol. 41, no. 3, pp. 175–183, May 2003, doi: 10.1016/S0041-624X(03)00097-0.

[11] L. Gavrić, "Computation of propagative waves in free rail using a finite element technique," *J. Sound Vib.*, vol. 185, no. 3, pp. 531–543, Aug. 1995, doi: 10.1006/jsvi.1995.0398.

[12] A. Marzani, I. Bartoli, E. Viola, and F. L. di Scalea, "Finite element based technique for modeling propagative stress waves in axisymmetric viscoelastic waveguides," p. 5.

[13] I. I. Setshedi, P. W. Loveday, C. S. Long, and D. N. Wilke, "Estimation of rail properties using semi-analytical finite element models and guided wave ultrasound measurements," *Ultrasonics*, vol. 96, pp. 240–252, Jul. 2019, doi: 10.1016/j.ultras.2018.12.015.

[14] E. Viola, A. Marzani, and I. Bartoli, "Semi-analytical Formulation for Guided Wave Propagation," in *Mechanical Vibration: Where do we Stand?*, vol. 488, I. Elishakoff, Ed. Vienna: Springer Vienna, 2007, pp. 105–121. doi: 10.1007/978-3-211-70963-4_6.

[15] A. Marzani, E. Viola, P. Bocchini, I. Bartoli, S. Coccia, and S. Salamone, "A SOFTWARE FOR THE COMPUTATION OF ACOUSTIC WAVES IN CYLINDRICAL, PLATE AND ARBITRARY CROSS-SECTION WAVEGUIDES," p. 10.

[16] Asrar Bin Ahmad, "Numerical Simulations of Lamb Waves in Plates Using a Semi-Analytical Finite Element Method," 2011.

[17] Z. A. B. Ahmad, J. M. Vivar-Perez, and U. Gabbert, "Semi-analytical finite element method for modeling of lamb wave propagation," *CEAS Aeronaut. J.*, vol. 4, no. 1, pp. 21–33, Apr. 2013, doi: 10.1007/s13272-012-0056-6.

[18] T. Hayashi and D. Inoue, "Calculation of leaky Lamb waves with a semi-analytical finite element method," *Ultrasonics*, vol. 54, no. 6, pp. 1460–

1469, Aug. 2014, doi: 10.1016/j.ultras.2014.04.021.
[19] M. V. Predoi, "Guided waves dispersion equations for orthotropic multilayered pipes solved using standard finite elements code," *Ultrasonics*, vol. 54, no. 7, pp. 1825–1831, Sep. 2014, doi: 10.1016/j.ultras.2014.01.019.
[20] J. M. Galán and R. Abascal, "Numerical simulation of Lamb wave scattering in semi-infinite plates: NUMERICAL SIMULATION OF LAMB WAVE SCATTERING," *Int. J. Numer. Methods Eng.*, vol. 53, no. 5, pp. 1145–1173, Feb. 2002, doi: 10.1002/nme.331.

A framework for recommending tourist attractions using deep learning and association rule mining-based methods

Hanjo Jeong*

Mokpo National University, Department of Software Convergence Engineering, 1666 Yeong san-ro, Cheonggye-myeon, Muan-gun, Jeollanam-do, 58554, Korea

Abstract. Many of tourism recommendation researches are based on the user rating and review data on the tourism platforms, and these approaches might be only suitable for a discrete recommendation for the tourist attractions. It is because each rating and review data on the platforms is created for a tourist place, not for multiple places on a travel itinerary. A travel blog data often contains information about the multiple places on a travel itinerary, but it is difficult to analyse the data compared to the rating and review data since it is like a text document having longer text than the review. In this paper, we introduce a framework consisting of a deep learning-based tourist-attraction extraction method from the blog text and an association rule mining-based recommendation method to recommend a list of tourist attractions that might be favourable to visit together in a travel itinerary.

1 Introduction

There have been many researches on the tourism recommendation problems. Many of them use user rating data and text reviews on the tourism platforms such as TripAdvisor. Such approaches are based on collaborative or hybrid filtering [1-3], sentimental analysis [4-6], and semantic clustering or classification [7-8] algorithms. However, the approaches based on the ratings and review data might be only suitable for a discrete recommendation for the tourist attractions. It is because they recommend the discrete list of the tourist attractions in the order of their reputations and/or the expected user preference on them, i.e., they do not consider the associations among the recommending tourist attractions.

Many of tourists do not visit only a place at a time, they want to visit more places and it might be better if the related places are recommended together. Such demand can be fulfilled with tourist route recommendation approaches. The most of the researches on the tourist route recommendation are focused on the location prediction based on the smart device and IoT data such as smart phones, RFID, and WI-FI tracking data [9-11]. In general, these approaches recommend the tourist route based only on the findings of the most co-visited tourist attractions, and they do not consider the user reviews on the tourist attractions.

In this paper, we introduce a framework for recommending a set of the tourist attractions, which might be better to visit together. The framework includes a deep learning-based automatic extraction of tourist attractions and an association rule mining (ARM) -based recommendation of the related tourist attractions. The tourist attractions for the recommendation include tourist spots, famous restaurant, and accommodations.

Fig. 1 shows the overall process of the recommendation framework. We collect the travel blog data from Naver open API (openapi.naver.com). We use the blog data instead of the ratings and the review data from TripAdvisor. It is because the rating and the review data is only for one tourist attraction thereby impossible to extract the associations among the tourist attractions. A travel blog generally introduces the tourist spots, restaurants, and accommodations visited in a travel itinerary, so places frequently mentioned in common are highly likely to be good places to visit together in a travel itinerary. Thus, a deep learning-based method is introduced to extract the tourist places from the collected blog data and an ARM-based tourist-attraction recommendation method is also introduced to find the favourable associations of the extracted tourist attractions.

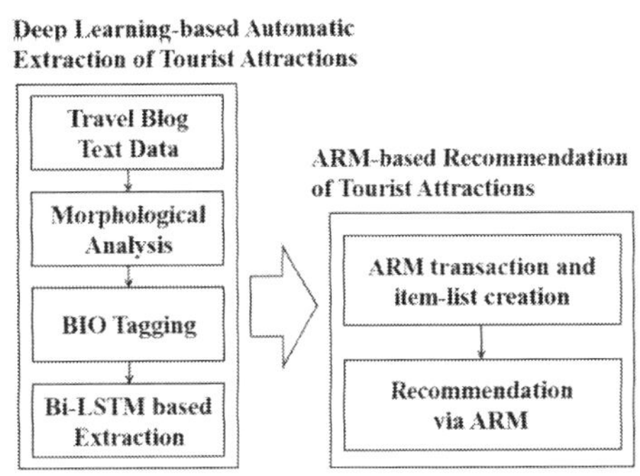

Fig. 1. Overall process of the recommendation framework

* Corresponding author: hanjojeong@mnu.ac.kr

 https://doi.org/10.1051/itmconf/20224803001

2 Deep learning-based tourist attraction extraction

This section describes a deep learning-based method of automatic extraction of the names of tourist attractions, restaurants and accommodations from the travel blog text data.

2.1 Data collection

A total of 22,000 travel blog data were collected through search terms with "여행_travel" and "관광_tourism" including the names of 22 major cities in Jeonnam province through Naver Open API.

For example, the search terms are generated as "전남_Jeonnam 강진_Gangjin 여행_travel 관광_tourism", "전남_Jeonnam 고흥_Goheung 여행_travel 관광_tourism", "전남_Jeonnam 곡성_Gokseong 여행_Travel 관광_Tourism", etc.

2.2 Data preprocessing and training data creation

Fig. 2 shows the examples of the travel blog data collected from Naver open API and only the description column is used to extract tourist attractions, restaurant, and accommodations.

Title	Link	Description
<b>전남 강진 여행	https://blog.na\	개별 ? 소규모 ? 맞춤형' <b>관광</b>
<b>전남 강진 여행	https://blog.na\	남도답사 1번지 <b>관광</b>택시 발
힐링<b>여행</b>	https://blog.na\	생활하며 <b>강진</b>의 문화를 이해
<b>전남</b> 가볼	https://blog.na\	한 농가에서 농촌 생활<b>관광</b>을
<b>전남</b><b>'	https://blog.na\	<b>강진</b>의 맛 <b>전남</b><b>
<b>전남 강진</b>	https://blog.na\	<b>전남 강진</b> 병영<b>여행</b>
[지역N문화 테마<b	https://blog.na\	<b>전남 강진 여행</b>코스 <b>강진
<b>전남 강진여행·	https://blog.na\	<b>전남 강진</b> 가볼만한곳으로 <b
<b>전남 강진</b>	https://blog.na\	명의 <b>여행</b>객이 방문하는 서남

Fig. 2. Examples of travel blog data collected from Naver open API.

Html tags are removed on the description and the description sentences are split into morphemes via a Korean morphological analysis. Lastly, we tagged the morphemes with the BIO tags shown in Table 1 in order to use them for a deep learning-based training.

Table 1. BIO tags denoting the word/s for tourist attractions, restaurants, and accommodations

Tag	Description
B-TA	Beginning word of tourist attractions (TA) including tourist spot, historical place, etc.
I-TA	Intermediate word of TA
B-RS	Beginning word of restaurant (RS)
I-RS	Intermediate word of RS
B-HT	Begging word of Hotel (HT)
I-HT	Intermediate word of HT
O	Other words besides of TA, RS, HT

Followings are an example training data consisting of the paired list of the morphemes of a sentence and their corresponding BIO tags as defined Table 1:

['원시/Noun', '체험/Noun', '의/Josa', '섬/Noun', '시호도/Noun', '와/Josa', '가족/Noun', '의/Josa', '섬/Noun', '우도/Noun', '를/Josa', '거닐다/Verb', '과역/Noun', '터미날/Noun', '앞/Noun', '에/Josa', '있는/Adjective', '해주/Noun', '식당/Noun', '에서/Josa', '갈치/Noun', '반/Noun', '으로/Josa', '점심식사/Noun', '를/Josa', '마쳤다/Verb', './Punctuation']

['O', 'O', 'O', 'O', 'B-TA', 'O', 'O', 'O', 'O', 'B-TA', 'O', 'O', 'O', 'O', 'O', 'O', 'O', 'B-RS', 'I-RS', 'O', 'O', 'O', 'O', 'O', 'O', 'O']

2.3 Deep learning-based tourist attraction extraction

2.3.1 Long short-term memory (LSTM)

Long short-term memory (LSTM) algorithm [12] is a recurrent neural networks (RNNs)-based algorithm, which alleviates the vanishing gradient problem in the RNNs. LSTM uses the memory cell consisting of input gate, forget gate, output gate as the hidden node and the forget gate is lastly introduced to reset the memory cell with the more relevant memory contents [13]. The LSTM's update equations are as follows:

$$f_t = \sigma(W_f x_t + U_f h_{t-1} + b_f) \quad (1)$$

$$i_t = \sigma(W_i x_t + U_i h_{t-1} + b_i) \quad (2)$$

$$o_t = \sigma(W_o x_t + U_o h_{t-1} + b_o) \quad (3)$$

$$g_t = tanh(W_g x_t + U_g h_{t-1} + b_g) \quad (4)$$

$$c_t = f_t \odot c_{t-1} + i_t \odot g_t \quad (5)$$

$$h_t = o_t \odot \tanh(c_t) \quad (6)$$

where f, i, o represents the forget gate, input gate, and output gate respectively, and operator $\odot$ represents the Hadamard product [14], which is the element-wise multiplication between two matrices.

The hidden state, $h_{t-1}$of the LSTM's memory cell memorizes all of the information until the $t-1$ time, and it combines the current input information and transfers the merged information to next state. The forget gate controls which previous information is reset or not, and the combined information with current input information, which calculated by $i_t \odot g_t$ as shown in Equation (5), will be transferred to the next cell's hidden state h_t by combining the value of output state calculated by $o_t \odot \tanh(c_t)$ as shown in Equation (6).

2.3.2 Bidirectional LSTM CRF (Bi-LSTM-CRF) based tourist attraction extraction

While the LSTM alleviates the vanishing gradient problem, it only takes account of the previous information because is recurrent only in one way. Bidirectional LSTM is introduced to use both previous and next information by recurring in both backward and forward [15], i.e., it can apply the effect of the later words to determine the previous words. Bidirectional LSTM with a CRF layer is also introduced to utilize the sentence level information for the sequence tagging [16].

We use the Bi-LSTM-CRF network to have the advantages of the sentence level information since the travel blogs consist of multiple sentences. Fig. 3 represents how we perform the named-entity recognition of the tourist attractions, restaurants and accommodations using the Bi-LSTM-CRF. Each morpheme consisting of a word and a part of speech pair is defined as a term in vocabulary to form the input vector as in Fig. 3. The output vector consists of the seven BIO tags defined in Table 1 including the 'o' tag.

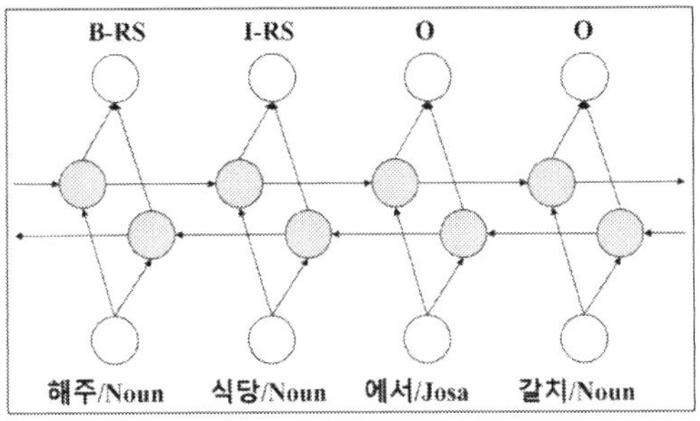

Fig. 3. Overall process of tourist attractions recommendation.

3 ARM-based tourist attraction recommendation

This section describes an association rule mining (ARM)-based recommendation method for recommending the tourist attractions, restaurants, and accommodations with the extracted information using the deep learning-based method introduced in Section 2.

3.1 ARM transaction and item-list creation

Each blog is created as a transaction and the extracted tourist attractions from the blog via the deep learning method in Section 2 are created as the containing items of the transaction.

The blogs containing none of the extracted tourist attractions are excluded from the transaction dataset, and also the blogs containing less than two items are also excluded because it does not indicate the correlation of co-occurrence among the items.

Additionally, the named entity having exact syllables with general words such as "Gwangju, restaurant, Jeong, moon, production, sunset, park, map, painter, space, etc." are ignored when extracting the item set. These general nouns are excluded from the itemset because these general terms can be incorrectly extracted, i.e., it is unclear whether the extracted word represents the name of the attraction visited by the blogger or the general word just used in a sentence.

Table 2. Examples of extracted transaction dataset

TR	List
1	{'Naru restaurant', 'Nam river', 'BunhongNaru restaurant', 'Saeojae historic-place', 'Haeyoun restaurant'}
2	{'Nam river', 'Jindo island'}
3	{'Gaudo island', 'Dasanchodang historic-place', 'baekryon temple', 'Udo island'}
4	{'Nam river', 'Haesang hotel'}
5	{'Gaudo island', 'Nam river', 'Daschodang historic-place', 'baekryon temple', 'Saeojae historic-place', 'Udo island'}

3.2 Recommendation via ARM

Each travel-blog post is defined as a transaction and the list of tourist attractions extracted from a blog is defined as an itemset of a transaction. The association rules for tourist attraction recommendation is extracted in the form of X => Y while X and Y denotes a set of the attraction items. Thus, the three major measures of ARM can be defined as follows:

- The support (X) represents the ratio of the number of transactions containing X itemset to the total number of entire transactions:

$$Support(X) = \frac{|\{t \in T;\ X \subseteq t\}|}{|T|} \quad (7)$$

- The confidence (X => Y) represents the conditional probability of the occurrence of Y itemset given the occurrence of X itemset:

$$Confidence(X => Y) = \frac{Support(X \cup Y)}{Support(X)} \quad (8)$$

- The Lift (X => Y) represents independence of X and Y and it also prevents incorrect rules, which can be generated by common items:

$$Lift(X => Y) = \frac{Support(X \cup Y)}{Support(X) \times Support(Y)} \quad (9)$$

4 Experiments and Results

ARM is performed using the apriori [17-18] method in Python mlxtend package. The support value is set to a small number, 0.001 even it is important to measure how much the rule itself is reliable. It is because the number of the tourist attraction items are relatively large in the size of blog data and there are a few items that can have a large number of the support value.

Confidence and lift values are also set to a small number, 0.01 because our proposed system performs recommendation incrementally based on user-selected tourist attractions. Thus, we find all of the rules having the higher values of the minimum threshold values and sort the rules based on lift value. Lastly, the top N rules are used to recommend the tourist attractions, which is mostly visited by the bloggers based on the user-selected tourist attractions.

As shown in Table 3, the lift value is also used to sort the rules because it enables to generate better recommendations by penalizing the famous tourist attractions, i.e., the famous tourist attractions tend to get higher possibilities of higher confidence values since they can appear at many of the blog posts. Using the lift values can alleviate such biases and make it lower values for the famous items.

Table 4 represents the Top 10 association rules with items {'Gatbawi rock', 'Yudal mountain'} at left hand sides of rules, i.e., if a user wants to be recommended places with the given two places, the ARM-based method can recommend the most associated places according to the user-provided places as shown in Table 4.

Table 3. Top 10 association rules among the entire rules

Rule#	LHS	RHS	support	confidence	lift
60	{'Kwangju temple'}	{'Kwangju park'}	0.001222	1.000000	818.500000
446	{'Jayounae restaurant'}	{'Seomji farm'}	0.001222	0.615385	503.692308
1132	{'Jangsung lakefront street'}	{'Backryondong cypress forest'}	0.002443	1.000000	385.176471
1228	{'Sinan monument', 'Jeungdo island'}	{'Taiping salt garden'}	0.001680	0.846154	369.374359
182	{'Bada garden'}	{'Dajungwon garden'}	0.002902	1.000000	344.631579
1024	{'Jangsung lakefront street', 'Backryondong cypress forest'}	{'Nambaek restaurant'}	0.002443	1.000000	284.695652
670	{'Chuam farm'}	{'Chuamgol inn'}	0.001069	0.304348	284.695652
1174	{'Sanbada restaurant', 'Bulgabsa temple'}	{'Seonunsa temple'}	0.001069	1.000000	272.833333
1060	{'Daegijeomdo island', 'Jeungdo island'}	{'Sogijeomdo island'}	0.002749	0.900000	267.872727
780	{'Gosado island', 'Daegijeomdo island'}	{'Sogijeomdo island'}	0.002291	0.882353	262.620321

Table 4. Top 10 association rules with items {'Gatbawi rock', 'Yudal mountain'} at LHS

Rule#	LHS	RHS	support	confidence	lift
14348	{'Gatbawi rock', 'Yudal mountain'}	{'Nojeokbong peak'}	0.000153	0.0625	68.072917
14412	{'Gatbawi rock', 'Yudal mountain'}	{'Street of culture'}	0.000153	0.0625	45.381944
14434	{'Gatbawi rock', 'Yudal mountain'}	{'Samhakdo island'}	0.000459	0.1875	40.843750
14538	{'Gatbawi rock', 'Yudal mountain'}	{'Chonsa bridge'}	0.000306	0.1250	30.254630
14324	{'Gatbawi rock', 'Yudal mountain'}	{'Kimdaejung nobel-peace-prize memorial'}	0.000153	0.0625	24.025735
14520	{'Gatbawi rock', 'Yudal mountain'}	{'Ullimsanbang historical place'}	0.000306	0.1250	6.188447
14351	{'Gatbawi rock', 'Yudal mountain'}	{'Dadohae national park'}	0.000153	0.0625	5.236378
14531	{'Gatbawi rock', 'Yudal mountain'}	{'Jindo island'}	0.000459	0.1875	1.708944

14541	{'Gatbawi rock', 'Yudal mountain'}	{'Dokchon restaurant'}	0.000153	0.0625	0.557975
14542	{'Gatbawi rock', 'Yudal mountain'}	{'Hangukwan restaurant'}	0.000153	0.0625	0.557975

5 Conclusion

In this paper, we presented a framework consisting of a deep learning-based tourist attraction extraction and an ARM-based recommendation method to recommend a list of tourist attractions that might be favourable to visit together in a travel itinerary. Firstly, we used an RNN-based deep learning framework, Bi-LSTM-CRF network, to train and extract tourist attractions based on the travel-blog text data.

In addition, an ARM-based recommendation method is also presented in this paper to recommend the favourable tourist attractions incrementally, i.e., if a user select one place in a travel itinerary and the list of the places preferable to co-visit with the given place can be recommended based on the ranks of the rules. If a user adds one more place in the travel itinerary, and the new list of favourable places are recommended by finding the rules containing all of the currently selected places at the LHS without re-generating the entire rules.

For future work, a more elaborated deep learning-based tourist-attraction extraction method should be addressed to improve the precision of the automatic extraction from the text data.

This work was supported by the National Research Foundation of Korea (NRF) grant funded by the Korea government (MSIT) (No. 2021-0058).

References

1. Q. Shambour, M. Hourani, S. Fraihat, *An item-based multi-criteria collaborative filtering algorithm for personalized recommender systems*, International Journal of Advanced Computer Science and Applications, **7(8)**, 274-279 (2016).
2. M. E. B. H. Kbaier, H. Masri, S. Krichen, *A personalized hybrid tourism recommender system*, In 2017 IEEE/ACS 14th International Conference on Computer Systems and Applications (AICCSA), 244-250 (2017).
3. M. Nilashi, A. Ahani, M. D. Esfahani, E. Yadegaridehkordi, S. Samad, O. Ibrahim, E. Akbari, *Preference learning for eco-friendly hotels recommendation: A multi-criteria collaborative filtering approach*, Journal of Cleaner Production, **215**, 767-783 (2019).
4. A. Valdivia, M. V. Luzón, F. Herrera, *Sentiment analysis in tripadvisor*, IEEE Intelligent Systems, **32(4)**, 72-77 (2017).
5. V. Taecharungroj, B. Mathayomchan, *Analysing TripAdvisor reviews of tourist attractions in Phuket, Thailand*, Tourism Management, **75**, 550-568 (2019).
6. N. Z. Dina, *Tourist sentiment analysis on TripAdvisor using text mining: A case study using hotels in Ubud, Bali*, African Journal of Hospitality, Tourism and Leisure, **9(2)**, 1-10 (2020).
7. Z. Abbasi-Moud, H. Vahdat-Nejad, J. Sadri, *Tourism recommendation system based on semantic clustering and sentiment analysis*, Expert Systems with Applications, **167**, 114324 (2021).
8. Z. Abbasi-Moud, H. Vahdat-Nejad, J. Sadri, *Tourism recommendation system based on semantic clustering and sentiment analysis*, Expert Systems with Applications, **167**, 114324 (2021).
9. Y. Su, X. Li, W. Tang, J. Xiang, Y. He, *Next check-in location prediction via footprints and friendship on location-based social networks*, In 2018 19th IEEE International Conference on Mobile Data Management (MDM), 251-256 (2018).
10. T. Arreeras, M. Arimura, T. Asada, S. Arreeras, *Association rule mining tourist-attractive destinations for the sustainable development of a large tourism area in Hokkaido using Wi-Fi tracking data*, Sustainability, **11(14)**, 3967 (2019).
11. C. Bin, T. Gu, Y. Sun, L. Chang, L. Sun, *A travel route recommendation system based on smart phones and IoT environment*, Wireless Communications and Mobile Computing (2019).
12. S. Hochreiter, J. Schmidhuber, *Long short-term memory*, Neural computation, **9(8)**, 1735-1780 (1997).
13. F. A. Gers, J. Schmidhuber, F. Cummins, *Learning to forget: Continual prediction with LSTM*, Neural computation, **12(10)**, 2451-2471 (2000).
14. R. A. Horn, *The hadamard product*. In Proc. Symp. Appl. Math, **40**, 87-169 (1990).
15. A. Graves, J. Schmidhuber, *Framewise phoneme classification with bidirectional LSTM and other neural network architectures*, Neural networks, **18(5-6)**, 602-610 (2005).
16. Z. Huang, W. Xu, K. Yu, *Bidirectional LSTM-CRF models for sequence tagging*, arXiv preprint arXiv:**1508**.01991 (2015).
17. R. Agrawal, R. Srikant, *Fast algorithms for mining association rules*, in Proceedings of 20th international conference very large data bases, VLDB, **1215**, 487-499 (1994).
18. M. Al-Maolegi, B. Arkok, *An improved Apriori algorithm for association rules*, arXiv preprint arXiv:**1403**.3948 (2014).

A comparative study of Machine learning Algorithms on the UNSW-NB 15 Dataset

Rachid Tahri[1*] *Abdessamad* Jarrar[2], *Abdellatif* Lasbahani[3], and *Youssef* Balouki[1]

[1]Faculty of Sciences and Technics, Hassan First, Settat, Morocco
[2]Faculty of Sciences, Mohammed First University, Oujda, Morocco[1]
[3]National School of Applied Sciences, Sultan Moulay Slimane University, Bni Mellale, Morocco

Abstract.

The world has experienced a radical change due to the internet. Internet became a crucial element in our daily life, therefore, the security of our DATA could be threatened at any time. This safety is handled using systems to detect network intrusion called Intrusion Detection Systems (IDS). Machine learning techniques are being implemented to improve these systems. In order to enhance the performance of IDS, different classification algorithms are applied to detect various types of attacks. Choosing a good one for building IDS is not an easy task. The best method is to test the performance of the different classification algorithms. Nevertheless, most researchers have focused on the confusion matrix as measurements of classification performance. Therefore, many papers use this matrix to present a detailed comparison with the dataset, data preprocessing, feature selection technique, algorithms classification and performance evaluation. The goal of this paper is to present a comparison of application of different Machine Learning algorithms used to build and improve intrusion detection systems in terms of confusion matrix, accuracy, recall, precision, FAR, specificity and sensitivity using the UNSW-NB15 Dataset. Furthermore, we introduce some lesson learnt to shoot more researchers in their future works.

1 Introduction

Many researches especially concerning NIDS (Network Intrusion Detection Systems), have been conducted under specific conditions in terms of systems, algorithms, and data sets used, limiting the extrapolation of results to a larger target of systems. These researches are based on statistical learning methods allowing to design algorithms never implemented in a real system. As a result, new attacks are detected and many problems due to the constant delay of the security tools on the threats are frequently faced (data theft, espionage...).these informatics attacks are grouped together as a dataset.

Indeed, datasets represent instances that consist of several features and are related to the intrusion detection system. So, it is essential to realize the type of data containing different types of attacks and features. The most popular data set that is being used for the intrusion detection system is UNSW-NB 15.After classification; the dataset will be treated using machine learning.

Machine Learning (ML) techniques widely used in computer security data sets have recently become a trend in security technology. It contributes to analyses and handling the massive amount of data and extracts the essential features that are used in various techniques for feature selection. IDS is a commonly used machine learning classifier to distinguish between various attacks as a class. Many supervised classification algorithms are applied to IDS, such as Decision Trees, Naïve Bayes, K-Nearest Neighbor, Random Forest, Support Vector Machine, and Logistic Regression. Evaluation of classification algorithms depends on various statistical metrics, especially confusion matrix results, to classify and predict different types of threats.

2 Classification method

Classification is one of the tasks of machine learning. It is a supervised learning model. It is used for intrusion detection systems based on binary or multiple classes [1]. In supervised learning, data is always labeled.

Each record in a data set is assigned to a particular class. An IDS based on a classification model classifies all network traffic into normal or anomalous classification algorithms. The obstacle to building the model is the massive amount of data. Classification algorithms, facing many problems when building a model, need a data-preprocessing step, especially when the dimension of the data is high.

The choice of the best classification algorithm depends on the performance evaluation in terms of confusion matrix and accuracy.

* Corresponding author: rachid.tahritr@gmail.com

The process of classifying the data in the dataset includes the two stages of training and testing. During the training and learning phase, a classifier is learned as a target, while during the second stage, the testing phase, the constructed model is used to predict class labels.

It is essential to analyze the time required for each classifier for both training and testing stages. Before applying the classifiers, data preprocessing helps the classification model to reduce time and complexity by eliminating irrelevant data.

Before applying the classifiers, data preprocessing helps the classification model reduce time and complexity by eliminating irrelevant data to improve the efficiency of the classification algorithms.

The cross-validation process of the dataset is divided into two equal groups for the classification of the network traffic dataset.

One group for testing, and the rest will be used for the training model. Few algorithms are able to distinguish different attacks from normal attacks with sufficient results.

The most popular classifiers are Decision Tree (DT), Random Forest, Support Machine Learning (SVM), K-Nearest Neighbor (KNN), Naïve, Decision Tree and Logistic Regression Supervised Algorithms

3 Supervised Algorithms

Machine learning (ML) and data mining have proven many effective applications in gene expression analysis. ML is used in every area of computational work where algorithms are designed and performance is increased. The main kinds of machine learning are supervised learning (SL) and unsupervised learning (USL). The classification pattern is used in discrete value problems, while the regression pattern is used to make decisions in persistent value problems.

From the above mentioned machine learning models supervised learning used in this paper. Supervised learning is learning that involves an expert well-versed in the environment and the expected response from the learning system [2]. The modelling of some maps between the input and output vectors is a problem of learning under the supervision of some real phenomena. Supervised learning techniques include classification, regression, and band techniques where the target variable is categorical in classification and continues to decline

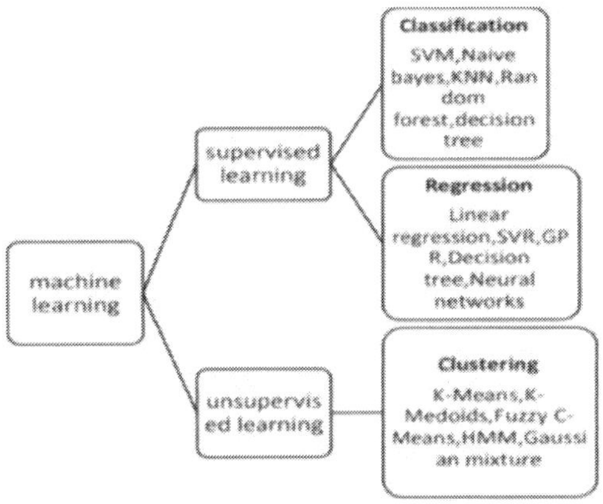

Fig. 1. Machine Learning Algorithms

4 Evaluation Performance

The performance of the intrusion detection models was achieved by evaluating the measurements of the values of the coincidence matrix also known as the confusion matrix. The confusion matrix shows the distribution of instances that are either correctly classified or misclassified by the models, the confusion matrix is an N X N matrix, where N is any integer greater than 1, The diagonal elements represent the number of points for which the predicted label is equal to the true label, while the off-diagonal elements are those that are mislabelled by the classifier. The higher the diagonal values of the confusion matrix, indicating a large number of correct predictions

It produced four results, which are:

- True positive (TP) means correct detection of the intrusion.
- False positive (FP) means normal traffic is considered as a cyber-attack.
- True negative (TN) refers to normal traffic correctly labelled as normal.
- False negative (FN) means that the intrusion disclosure fails.

Authors should use the forms shown in Table 1 in the final reference list.

Table 1. Metrics for classification algorithms.

Element Formula	Description Evaluation Metrics
Accuracy=(TP+TN)/(TP+TN+FP+FN)	Total correct classified over the total number of records
Precision = TP / (TP+FP)	True positive that are correctly predicted from the total predicted patterns in a positive class
Book Recall = TP / (TP+FN)	Positive patterns that are correctly attack classified.
Specificity= TN / (TN+FP)	Negative patterns that are correctly classified.
Sensitivity or True Positive Rate (TPR) = TP / (TP+FN)	Sensitivity: correctly classified over the total amount of abnormal network. True Positive Rate (TPR): Attacks correctly classified as predicted attacks It called (detection rate)

False Positive Rate (FPR) =FP / (FP+TN)	False positive (FP): Incorrectly classified normal as predicted attacks
Page True Negative Rate (TNR) =TN/(TN+FP)	True negative (TN): Normal correctly classified as normal (false alarm)
Year False Negative Rate (FNR)=FN/(FN+TP)	False negative (FN): Incorrectly classified attacks as a normal

5 Review of Classification Algorithms for IDS

In the last decade, many works have been presented to improve IDSs to detect and prevent various malicious attacks of accessing computer information. This section discusses some of the machine learning techniques and algorithms in the classification process used for intrusion detection, including data preprocessing, feature selection techniques, classification algorithms, and metric evaluation algorithms

- In 2018, Belouch et al. [4] paper evaluated the performance of four classification algorithms, namely SVM, Naïve Bayes, Decision Tree, and Random Forest. The approach applies Apache Spark tools to classify intrusion detection on network traffic. The public dataset for network intrusion detection UNSW-NB15 is applied with 42 features to build the model. The experimental results demonstrate a random forest classifier to be the best among other classifiers with the accuracy of 97.49% sensitivity 93.53%, and specificity of 97.75%
- In 2020, Zina et al. [5] , proposed a novel method for classification and feature selection applying Regression Trees (CART) combining with Random Forest. This system is called the Hybrid Anomaly-based Intrusion Detection System (HAIDS). The hybrid approach is used to improve the efficiency of the model rather than in a single algorithm. Moreover, the process of removing irrelevant features is applied to overcome the case of high dimensionality. The proposed model was applied to the UNSW-NB15 dataset and selected the highest-ranked thirteen features. The hybrid method achieved the highest performance and accuracy in terms of false alert rate with 11.86% and accuracy rate 87.74%.
- In 2020, Jie Gua and Shan Luc. [6] proposed a novel method, an embedding system for intrusion detection system based on Support Vector Machine (SVM) with Naive Bayes feature embedding.
  The naïve Bayes is used for feature transformation to convert data state. The SVM algorithm is implemented as a classifier. The embedding model was applied to multiple data sets to detect different types of attacks such as UNSW-NB15, NSL-KDD, CICIDS2017, and Kyoto 2006+ using different features for each data dataset. The proposed method, embedding system result compared
  with a single SVM algorithm, concluded that detection's highest accuracy gets with embedding Naive Bays with SVM. The experiment demonstrated NSL-KDD as the best data set with the highest accuracy of 99.36%. DR 99.25, FAR0.54
- In 2020, Pokharel and P. et al. [7] presented IDS depended on a hybrid classification algorithm and profile improvement to detect anomalous user behavior. The hybrid approaches contain Naïve Bayes and Support Vector Machine (SVM) algorithms for classification. Moreover, it provides data preprocessing. The excellent effect on model accuracy such as data normalization scaled features between (0,1) and selecting the right features on the real-time data set. In this hybrid approach, classifiers get a total accuracy of 0.931 and a precision of 0.958. Also, it provides the accuracy for Classifier Enhancement (CE) 0.953 and precision 0.958.

6 Comparison and Discussion

The implementation of classification algorithms for IDS to classify different types of attack are presented in the table [1] below.to improve intrusion detection systems, Machine learning techniques have been applied to the field of network security. Previous sections reviewed some researches about classification algorithms used to build the IDS model and evaluated the performance by different metrics in terms of accuracy, recall, precision, f-score, specificity, sensitivity, error rate, and dependable tool confusion matrix.

Indeed, the dimension reduction and feature selection had a good effect on the classification model performance because it reduces training and testing time via removing the irrelevant features, making the classification process more accurate and less complicated. So a combination of multi-classification algorithms and called hybrid classification could be the optimal solution to classify attacks type. The different data types of attacks may deal with different types of classification algorithms. Most studies now have focused on the hybrid classification algorithm rather than a single classification because it provides very satisfying results in different performances measurement.

The best results for most reviewed studies showed that the Random Forest algorithm achieved the best accuracy of classification because it combines many decision trees, then decide the type of attack, leading to the decrease of the risk of overfitting. The random forest can deal with various big types of features that do not require data scaling. Moreover, the Practical Swarm Optimization (PSO) gets the best result for feature selection. In this paper, the comparison is performed in terms of data set; data pre-processing techniques, a number of features selected, feature selection techniques, classification algorithms, and evaluation metrics. This study aims to show different classification algorithms' performance by using different measurements to select a suitable classifier best model in order to gain speed and accuracy.

Table 2. Comparison of evaluation of different classification algorithms performance.

Ref	Data set	Data preprocessing Techniques	Number of features selected	Feature Selection Techniques	Classification Algorithm	Evaluation Metrics
[4] 2018	UNSW NB15	Apache Spark processing tools	42 features out of 49	-	SVM, Naïve Bayes, Decision Tree and Random Forest	best results Random Forest accuracy 97.49, Sensitivity 93.53, specificity 97.75
[5] 2020	UNSW NB15	categorical features remove redundant and irrelevant features	top rank 13	Random Forest	Classification and Regression Trees (CART)	accuracy 87.74
[6] 2020	UNSWNB , CICIDS2017 NSL-KDD Kyoto 2006+	Data normalization Data transformation	Different number for each data set	Naïve Bayes embedding feature	Embedding SVM Naive Bayes	the highest score on NSL KDD data set with accuracy 99.36%., DR 99.25%, FAR 0.54%

7 Conclusion

To conclude, IDS improvement performance depends on different machine learning techniques. Classification algorithms have a significant role in helping IDS to distinguish different types of attacks. The present document aims to test different classifier algorithms and find the evaluation performance by using different metrics, by applying various metric measurements to evaluate classifiers' performance, noticed that the random forest algorithm achieved sufficient results and the highest accuracy to classify different types of attacks. Obtaining high performance of the model, most researchers used the hybrid classification algorithm for building intrusion detection systems rather than individual classification. The effectiveness of dimension reduction to reduce big data sets' complexity leads to select optimal features to obtain better performance in classification in terms of accuracy and speed.

References

1. B. Charbuty, and A. Adnan. "*Classification based on decision tree algorithm for machine learning*." Journal of Applied Science and Technology Trends 2.01, pp. 20-28, (2021).
2. Krishnan, Deepa. "*Analysis of Accuracy of Supervised Machine Learning Algorithms in Detecting Denial of Service Attacks*." Advances in Signal and Data Processing. Springer, Singapore, pp. 519-529, (2021).
3. A. Agarwal, et al. "*Classification model for accuracy and intrusion detection using machine learning approach*." PeerJ Computer Science **7** (2021): e437
4. M. Belouch, S. El Hadaj, & M. Idhammad, *Performance evaluation of intrusion detection based on machine learning using Apache Spark*. Procedia Computer Science, **127**, pp. 1-6, (2018).
5. Z. Chkirbene, S. Eltanbouly, M. Bashendy. N. AlNaimi, & A. Erbad. *Hybrid machine learning for network anomaly intrusion detection*. IEEE International Conference on Informatics, IoT, and Enabling Technologies (ICIoT), pp. 163-170. IEEE, (2020).
6. J. Gu, & S. Lu. *An effective intrusion detection approach using SVM with naïve Bayes feature embedding*. Computers & Security, **103**, 102158, (2021).

SCMA Codebook Design Based on a 16 Star-QAM with MED Maximization

Afilal *Meriem* [1], Hatim *Anas* [1], Latif *Adnane* [1] *and* Arioua *Mounir* [2].

[1]TIM Team, ENSA, Cadi Ayyad University, Marrakech Morocco.
[2]ENSA, Abedelmalek Essaidi University, Tetouan, Morocco.

Abstract. The SCMA codebook design is based on star-QAM signalling constellations with operating several segmentations to generate the different user's codebook while considering the MED maximization as the criterion design. In addition, we constructed the mapping matrix F using Progressive Edge Growth (PEG) algorithm which facilitate iterative decoding performance of the message passing algorithm (MPA). At the end, a comparison with similar studies is conducted on the MEDs of the designed codebooks.

1 Introduction

Non-orthogonal multiple access (NOMA) has been considered as an enabling scheme for massive connectivity and ultra-low-latency in future communication networks [1]. NOMA serves multiple users simultaneously using different power levels or/and codebooks, taken advantage of the non-orthogonal spreading technique, Sparse code Multiple Access (SCMA) is proposed as a multi-dimensional constellation codebook.

In fact, two major research lines are conducted on the SCMA system: the codebook design and the multiuser detection. With the fact that the performance of the SCMA technique remains in the codebook design, we construct these codebooks under multi- stages: First, construct a mother constellation (MC), after assign the MC to different users using specific operators (permutation, rotation …). Finally, applying a mapping matrix which decide on the occupation of users to the resource elements.

To more improve the system performance such as spectral efficiency, bit error rate, and peak-to-average power ratio (PAPR), the codebook design could be optimized considering various key performances, mainly the Minimum Euclidean Distance (MED).

For instance, authors in [2]–[4] focused on optimizing the MED of the MC to generate codebooks with maximized MED, the large MED results in minimizing the complexity of the decoder by facilitating the decision making of the MPA.

Various work considered the maximization of the MC MED thanks to its better Bit to Error Ratio (BER) performance. Therefore, further studies as [5] and [6], are extended to a newer definition of MED: MED of superimposed codewords. Instead of optimizing the Euclidean Distance between the point-to-point within the same dimension of each codebook, the studies are conducted on the user's codewords overlapping on the same subcarrier to establish a larger MED.

While most recent studies as [7], [8], investigated maximization of the two different MEDs, simultaneously. As well as, in this paper we optimize the codebook design using the MED maximization criterion under the point-to-point of same codebook along with superimposed codewords.

In this paper, from a 16 star-QAM we design our codebook with segmentation while always maximizing MED either at the MC parameters construction, the sub-constellation generation, or at the sub-constellations mapping to form the final codebooks. The contributions of this study are:

Design the mapping matrix F using PEG algorithm which impose a respectable minimum distance bound that enhances decoding performances.

Construct an MC with large MED, after that, generates codebooks with large MED between the interfering users.

The rest of this paper is organized as follows. Section II represents the SCMA system model. Section III, the construction of the mapping matrix F using PEG algorithm is discussed. Section IV shows the steps of generating the SCMA codebooks from the constellation type choice of the MC to the final codebooks mapping, while maximizing MEDs. Section V, presents the results of the paper in comparison with other studies. Finally, Section VII presents the conclusion.

2 System model

The SCMA encoder maps a $\log_2(M)$ data/input bits to a K -dimensional complex codeword. The codeword is selected from a pre-defined codebook of size M, and it is a sparse vector with N<K non-zero entries. Finally, the SCMA codewords are transmitted over shared orthogonal resources. The received signal on subcarrier i can be expressed as:

$$y_i = \sum_{j=1}^{J} h_{ij}.x_{ij} + n \qquad (1)$$

Here with $x_{ij}(w_j)$ denotes the i-th value within the codeword w_j of user j, h_j denotes the channel coefficient of user j and n is the ambient noise.

3 Constructing the matrix F using PEG

The relationship between users and subcarriers is determined based on a sparse mapping matrix F. Here, we propose to design the matrix F with PEG algorithm. The high flexibility of this algorithm makes it a good candidate to generate codes of any block length [7], hence the SCMA mapping matrix F. Further experimentations with this algorithm will help forge connections of a higher user's number and subcarriers. A bipartite graph with K check nodes (rows) and J symbol nodes (columns) can be created using F. Such a graph is also called a Tanner graph, Fig. 1. The symbol degree sequence $D_s = \{d_{s1}, d_{s2}, \dots, d_{sJ}\}$ denotes the number of check nodes connected to each symbol node.

Algorithm 1 Progressive Edge Growth to create F.

1. **Inputs:** Number of check nodes, Number of symbol nodes and symbol degree sequence.
2. **Outputs:** F.
3. **Initialization:** Set the symbol nodes edges to 0.
   //*start with the first symbol node.
4. **If** (Current symbol node has no edges).
5. Select check node with the lowest symbol degree.
6. Establish an edge between the two nodes.
   //*For each subsequent edge to be placed, a subtree (subgraph) is expanded from the variable node in question up to a depth l).
7. **Else**
8. **If** (The subgraph expanded from the current symbol node has not established edges with all the check nodes).
9. Select lowest check node from the set of uncovered check nodes.
10. **Else** (The current subgraph node has established edges with all the check nodes).
11. Choose check node farthest from symbol node.

The edge-selection procedure is made edge by edge, and progresses through columns from left to right, according to the Algorithm. 1.

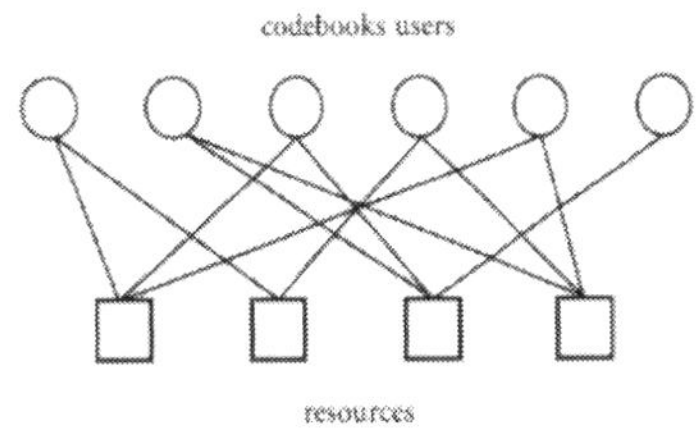

Fig. 1. Factor graph representation of 4 check nodes connected to 6 symbol nodes and $D_s = \{2, 2, 2, 2, 2, 2\}$.

4 SCMA codebook design

4.1 Mother Constellation (MC): Constellation type choice

To construct the multidimensional constellation there are several constellation choices such as: QAM, QPSK…, and for the SCMA choice it has more freedom degrees over previous techniques [3]. Here with, we propose to design the mother constellation using a 16 Star-QAM constellation of 2 radiuses, then we extend the design to four radiuses where the first quadrant points can be connected in a straight line to the third quadrant points, and the remaining constellation points are distributed on the abscissa axis [4], as shown in Fig. 2.

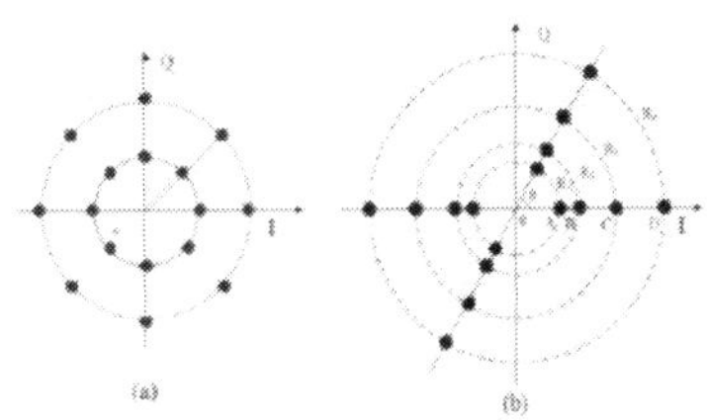

Fig. 2. (a)Constellation of round 16 QAM. (b) Proposed constellation.

Where: |OA| = R1, |OB| = R2, |OC| = R3, |OD| = R4.
The MC parameters can be written as:

$$MC = \begin{pmatrix} x_1 & x_2 & \alpha x_1 & \beta x_2 \\ -x_1 & -x_2 & -\alpha x_1 & -\beta x_2 \\ x_1 e^{i\theta} & x_2 e^{i\theta} & \alpha x_1 e^{i\theta} & \beta x_2 e^{i\theta} \\ -x_1 e^{i\theta} & -x_2 e^{i\theta} & -\alpha x_1 e^{i\theta} & -\beta x_2 e^{i\theta} \end{pmatrix} \qquad (2)$$

The length of the sequences is equal to M, with $x_2 = \beta x_1$.. Also, we define:
The ratio between two successive points is: $\frac{R2}{R1} = \frac{R4}{R3} = \beta$, The ratio between non-successive points is: $\frac{R3}{R1} = \frac{R4}{R2} = \alpha$. And $\theta \in [0, 2\pi]$.

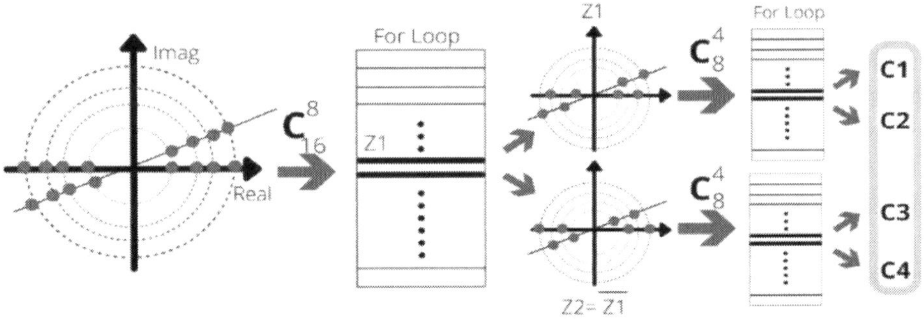

Fig. 3. Sub-constellations generation from the MC.

4.2 Mother constellation parameters

We propose to study the case of six users (J=6) associated with four subcarriers (K=4), (N = 2) entries non-null and M=4 representing the constellation points for occupation, each time single user transmits m bits with $M = 2^m$.

with (J, M, N, K) = (6, 4, 2, 4), to generate the MC defined by equation (2), we need to determine the three parameters (α, β and θ), which help calculate the four radiuses (R1, R2, R3 and R4). We set $x_1 = R_1$ and $x_2 = R_2$.

For a 16 star-QAM, in order to maximize the MED between two adjacent points, as denoted by [9], the following measures are to be taken in consideration:

A degree of $\theta = 22.5°$ is set between the internal ring and the external ring.

The ratio between the internal radius and the external one is set to 0.63, $\beta = \frac{1}{0.63}$.

In order to maximize both the sum distance with l, in equation (3), and the MED with r, in equation (4), of the codewords to ensure a good performance of the SCMA system, we have the following normalized minimum distances [2].

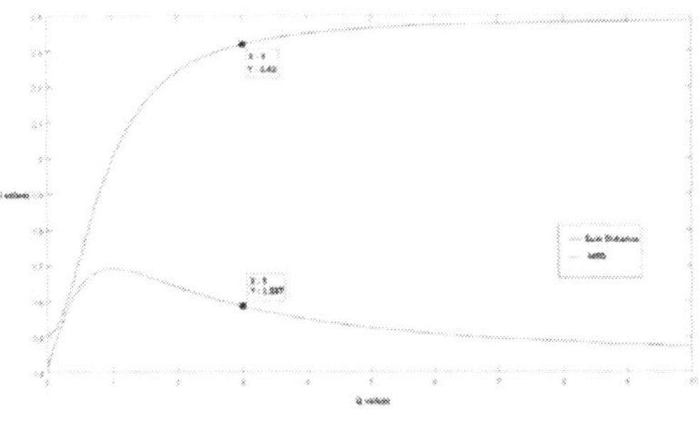

Fig. 4. Performance of Equation (3) and Equation (4).

$$r = \frac{d_{min}}{\sqrt{E}} = \sqrt{2\left(1 + 2\left(\frac{\beta^2-1}{1+\beta^2}\right)\cdot\left(\frac{1}{\frac{1}{\alpha}+\alpha}\right)\right)} \quad (3)$$

$$l = \frac{d_{min}}{\sqrt{E}} = \sqrt{8\left(\frac{1}{1+\frac{1}{\beta^2}}\right)\cdot\left(1 - \frac{1-\frac{1}{\beta^2}}{1+\alpha^2}\right)} \quad (4)$$

The average signal energy is defined as $E = \frac{1}{M}\sum_{i=1}^{M}\|x_i\|^2$; so $E = 0.25\,(R_1^2 + R_2^2 + R_3^2 + R_4^2)$.

The variation of α for equation (3) and equation (4) is represented by Fig.4. Therefore, the maximization of both r and l is achieved by increasing alpha, but at one point r becomes opposite to l where r increases and l decreases, therefore, for a range of alpha variation we found that the perfect value is α=3.

4.3 Sub-constellation generation with MED maximization

With the massive user's connection supported by the SCMA system, the decoder complexity is increased. At the receiver side, to avoid decoding confusion $d_f > U$ is a condition to satisfy [4] with d_f the number of sub-constellations and U the actual number of users on a single subcarrier.

The idea behind the MC segmentation is to extract sub-constellations, with the best and maximum MED, from the multiple choices existing from the MC divisions. We start the divisions of the MC into d_f sub-constellations as shown in Fig. 3.

For a 16-QAM MC there are 16 points to be distributed on d_f sets, with the condition of maximizing the point-to-point MED within the same sub-constellation. The first division gives two sets each with 8 points. The selection of the set's points is accomplished with help of the combination operator C_{16}^{8} (Mc) which generate several 8 points sets with different points from the MC, and then the loop for go through all the generated sets, noted as Z, searching for the sets with the largest MED. With Z a 36 × 8 matrix, select Z1=Z(i) as a candidate set and Z2 the complementary points to Z1 from the MC, with i = [1,36].

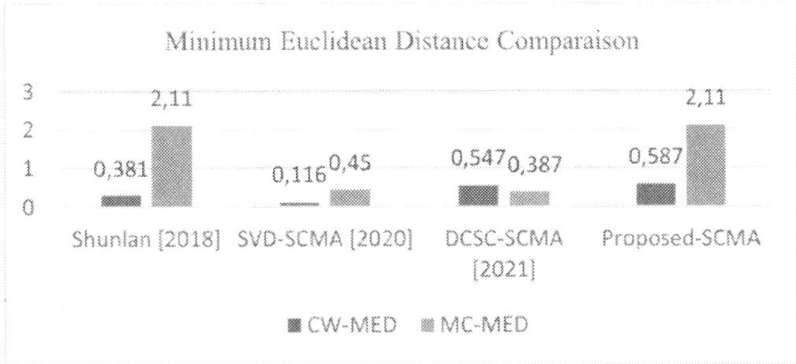

Fig. 5. Comparison of the mother constellation MED (MC-MED) and the superimposed codewords MED (CW-MED).

Operate another division to both sets Z1 and Z2 with the same pervious procedures: Generate all possible combinations of 4 points with C_8^4 from Z1 and Z2, each of the two sets (Z1, Z2) generate another two sets each of 4 points. Arriving at the end of the divisions with 4 sets, we calculate the MED of every set, after we decide whether it's the best 4 sets with the largest MED or not.

4.4 SCMA codebook generation by sub-constellation mapping with MED maximization

We have d_f sub-constellations generated from the MC $\{C_1, C_2 \dots . C_{df}\}$, such as each subset contains Q/d_f points and Q is the number of the MC QAM constellation points. These sub-constellations are the base to the codebook, we map the sub-constellations to the user's codebook as:

$$CB(i,j) = \begin{cases} C'_{i(n)} & \text{if } f_{i\,j} = 1 \\ 0 & \text{if } f_{i\,j} = 0 \end{cases} \quad (5)$$

With $C'_{i(n)}$ is the n-th value of a specific reordering of the set $\{C_1, C_2 \dots . C_{df}\}$. Using the mapping matrix F, we replace the elements "1" with $C'_{i(n)}$.

In addition to maximize MED of the sub-constellations point-to-point within the same codeword, we aim to maximize the MED between superimposed codewords of conflicting users on each subcarrier.

Therefore, we project the sub-constellations on the overlapping subcarrier, after we calculate the distances between the superimposed codewords. The best sets achieving the largest MED are to be taking in consideration to form our desired codebook. The codebook of the six users is represented in Table. 1.

5 Numerical Results and Discussions

Fig. 5. shows the comparison of MC-MED and superimposed CW-MED of the codebooks. The comparison shows the better performance of the proposed codebooks within the two different MEDs. The better performance of the proposed codebooks returns to the main focus on the MED maximization criterion. However, these papers investigated further other factors such as: with DCSC-SCMA [8], authors took the codebook design under power constraints, such as balancing the MC power, in addition to the maximization of both MEDs. Also, with SVD-SCMA [7] authors first construct the MC based on singular value decomposition (SVD) to guarantee a higher MED, then applying a rotation operator to generate different user's codebooks in which MED of the different users are equal. With [4], the main focus of the authors was on maximizing the MC MED but no optimization was on the MED of superimposed codewords. Finally, here comes the necessity of BER comparisons to be taken in consideration to more value the codebook performance.

	Codebook
1	$\begin{bmatrix} 1 & 2.781+1.123i & -1 & -2.781-1.123i \\ 1.587 & 4.414+1.783i & -1.587 & -4.414-1.783i \\ 0 & 0 & 0 & 0 \\ 0 & 0 & 0 & 0 \end{bmatrix}$
2	$\begin{bmatrix} 0 & 0 & 0 & 0 \\ 0 & 0 & 0 & 0 \\ 1.587 & 4.414+1.783i & -1.587 & -4.414-1.783i \\ 1 & 2.781+1.123i & -1 & -2.781-1.123i \end{bmatrix}$
3	$\begin{bmatrix} 1.587 & 4.414+1.783i & -1.587 & -4.414-1.783i \\ 0 & 0 & 0 & 0 \\ 1.471+0.594i & 4.761 & -1.471-0.594i & -4.761 \\ 0 & 0 & 0 & 0 \end{bmatrix}$
4	$\begin{bmatrix} 0 & 0 & 0 & 0 \\ 0.927+0.374i & 3 & -0.927-0.374i & -3 \\ 0 & 0 & 0 & 0 \\ 1.471+0.594i & 4.761 & -1.471-0.594i & -4.761 \end{bmatrix}$
5	$\begin{bmatrix} 1.471+0.594i & 4.761 & -1.471-0.594i & -4.761 \\ 0 & 0 & 0 & 0 \\ 0 & 0 & 0 & 0 \\ 1.587 & 4.414+1.783i & -1.587 & -4.414-1.783i \end{bmatrix}$
6	$\begin{bmatrix} 0 & 0 & 0 & 0 \\ 1.471+0.594i & 4.761 & -1.471-0.594i & -4.761 \\ 1 & 2.781+1.123i & -1 & -2.781-1.123i \\ 0 & 0 & 0 & 0 \end{bmatrix}$

Table 1. SCMA codebook of an optimized round 16-QAM modulation after division with (J, M, N, K) = (6, 4, 2, 4).

6 Conclusion

For the codebook design we covered the construction of the mapping matrix F with PEG algorithm, then constructed the MC from a 16 Star-QAM constellation. After generated the user's codebooks with segmentation of the MC while keeping the MED maximization as the main criterion. Finally, we had acceptable results compared to the other studies at the MC-MED and CW-MED, keeping in mind the better performances of these studies in other fields.

References

[1] S. Chaturvedi, Z. Liu, V. A. Bohara, A. Srivastava, et P. Xiao, « A Tutorial on Decoding Techniques of Sparse Code Multiple Access », *IEEE Access*, vol. 10, p. 58503-58524, 2022, doi: 10.1109/ACCESS.2022.3178127.

[2] L. Yu, X. Lei, P. Fan, et D. Chen, « An optimized design of SCMA codebook based on star-QAM signaling constellations », in *2015 International Conference on Wireless Communications & Signal Processing (WCSP)*, Nanjing, China, oct. 2015, p. 1-5. doi: 10.1109/WCSP.2015.7341311.

[3] L. Yu, P. Fan, D. Cai, et Z. Ma, « Design and Analysis of SCMA Codebook Based on Star-QAM Signaling Constellations », *IEEE Trans. Veh. Technol.*, vol. 67, n° 11, p. 10543-10553, nov. 2018, doi: 10.1109/TVT.2018.2865920.

[4] S. Liu, J. Wang, J. Bao, et C. Liu, « Optimized SCMA Codebook Design by QAM Constellation Segmentation with Maximized MED », *IEEE Access*, p. 1-1, 2018, doi: 10.1109/ACCESS.2018.2876030.

[5] Q. Wang, T. Li, R. Feng, et C. Yang, « An Efficient Large Resource-User Scale SCMA Codebook Design Method », *IEEE Commun. Lett.*, vol. 23, n° 10, p. 1787-1790, oct. 2019, doi: 10.1109/LCOMM.2019.2929766.

[6] C. Huang, B. Su, T. Lin, et Y. Huang, « Downlink SCMA Codebook Design with Low Error Rate by Maximizing Minimum Euclidean Distance of Superimposed Codewords », *ArXiv210103355 Cs Eess Math*, nov. 2021, Consulté le: 26 février 2022. [En ligne]. Disponible sur: http://arxiv.org/abs/2101.03355

[7] S. Vidal Beltran, R. Carreño Aguilera, et J. L. Lopez Bonilla, « SPARSE CODE MULTIPLE ACCESS CODEBOOK DESIGN USING SINGULAR VALUE DECOMPOSITION », *Fractals*, vol. 28, n° 07, p. 2150021, nov. 2020, doi: 10.1142/S0218348X21500213.

[8] Z. Hou, Z. Xiang, P. Ren, et B. Cao, « SCMA Codebook Design Based on Decomposition of the Superposed Constellation for AWGN Channel », *Electronics*, vol. 10, n° 17, p. 2112, août 2021, doi: 10.3390/electronics10172112.

[9] M. Celidonio et D. Di Zenobio, « An independent sharing of two 16-star QAM broadcast channels », in *Proceedings of 6th International Symposium on Personal, Indoor and Mobile Radio Communications*, Toronto, Ont., Canada, 1995, vol. 1, p. 115-119. doi: 10.1109/PIMRC.1995.476415.

Denoising Esophageal Speech using Combination of Complex and Discrete Wavelet Transform with Wiener filter and Time Dilated Fourier Cepstra

Madiha Amarjouf[1*], *Fadoua* Bahja[2], *Joseph* Di Martino[3], *Mouhcine* Chami[1] and *El Hassan* Ibn Elhaj[1]

[1]Research laboratory in Telecommunications Systems: Networks and Services (STRS), Research team: Multimedia, Signal and Communications Systems (MUSICS), National Institute of Posts and Telecommunications (INPT), Av. Allal Al Fassi, Rabat, Morocco
[2]Laboratory of Innovation in Management and Engineering for Enterprise (LIMIE), Institut Supérieur d'Ingénierie et des Affaires (ISGA Rabat), 27 Avenue Oqba, Agdal, Rabat, Morocco
[3]Loria - Laboratoire Lorrain de Recherche en Informatique et ses Applications, B.P. 239 54506 Vandœuvre-lès-Nancy, France

Abstract. Esophageal speech is one of the pathological voices, which is known to be weak in intelligibility and hard to understand. Our approach's main idea is to reduce the esophageal speech noises using two-hybrid methods. This paper aims to merge the advantages of wavelet-based methods such as DWT and DTCWT, along with the standard methods such as the Wiener filter and the time dilated Fourier. The first hybrid method applies the filters on the vocal tract cepstrum, while the second one applies them at the synthesis stage. Two experiments were conducted as well to evaluate the results by objective analysis. The results obtained by the proposed hybrid methods gave good performances.

1 Introduction

Pathological voices are sounds produced by people suffering from a dysfunction or some voice trouble [1, 2]. This voice deterioration could be temporary or permanent [1]. Besides being hoarse, this voice undergoes changes in one or multiple acoustics parameters, e.g., intensity, pitch, frequency and timbre [1, 2, 3]. Vocal pathologies are generally grouped into three main categories of origins: Functional pathologies, organic and cancerous [1, 2]. Esophageal speech (ES) is the pathological voice that our work is about to study. It belongs to pathologies of cancerous origin. This voice disorder is due to total laryngectomy, which is a surgical operation consisting of complete removal of the larynx [1]. ES is a substitution voice that facilitates communication without the need of using a device or even the hand, which makes it the most used solution to communicate [1, 2]. Otherwise, the major problem with this voice is the fact that it is weak in intelligibility and difficult to understand [3]. Improving this voice through denoising methods is the aim of this study. The methods used are DWT, DTCWT, Wiener filter and the time dilated Fourier cepstra. In this study, we combined these techniques to attenuate the noise of esophageal speech and the results were noted in order to make an objective evaluation.

Many researchers have conducted speech denoising. [4] Have proposed a time-domain speech denoising method to reduce the hoarseness of dysphonic voices. This technique separates the signal and the noise subspaces based on singular value decomposition (SVD) of appropriate data matrices. Even the fact that the original signal's spectral characteristics were preserved while reconstructing the signal, the SNR was higher, especially for the high-frequency area. The same method was applied in [5] in order to implement a real-time hoarse denoising device. It was tested on speech signals degraded by white noise and applied to dysphonic voices. The same problem of high SNR was also found. A preprocessing technique using wavelet-based denoising of non-linear dynamic analysis for laryngeal paralysis patients has been studied in [6]; the work proved that nonlinear dynamic analysis under noisy conditions was improved efficiently using the proposed method. In [7] an improving technique for speech enhancement was introduced. This method consists of decomposing the noisy signal using wavelet packet decomposition, then using adaptive thresholding and cepstral subtraction for more denoising. It is based on principal component analysis (PCA) for speech denoising. [8] have introduced a preprocessing method for denoising speech. The procedure eliminates non-speech segments with a slight cost using speech segment detection before starting the denoising process. Also, this method has demonstrated that the denoising time is lower compared to the Wavenet-based denoising method.

The rest of this article is organized as follows: The second part introduces the processes and methods used in our work. Section three describes the denoising experiments. The results and discussion are given in the

* Corresponding author: amarjouf.madiha@inpt.ac.ma

fourth section. Finally, the fifth section gives a conclusion.

2 Methods

2.1 Extraction of the vocal tract cepstrum

Firstly, the short-time speech signal was multiplied by the normalised Hann window introduced by [9], to make the discontinuities in the speech signal less noticeable. Normalised Hann Window provided by formula (1) [9]:

$$W(n) = \begin{cases} \frac{2\frac{\sqrt{L}}{\sqrt{N}}}{\sqrt{4a^2+2b^2}} * \left(\left(a - b\cos\left(\pi\frac{(2n+1)}{N}\right)\right)\right), & 0 \le n \le N \\ 0\ otherwise \end{cases} \quad (1)$$

Where:

- N: The length of the analysis window,
- L: The analysis step size
- a= b= 0.5

Then, a Fast Fourier Transform was applied to the windowed signal. After that, a log moderator was applied to the speech signal spectrum to separate the vocal tract spectrum and the excitation spectrum. The last step to obtain the cepstrum was applying the Inverse Fourier Transform (IFFT) to the logarithmic spectrum. Figure 1 resumes how the cepstrum is obtained.

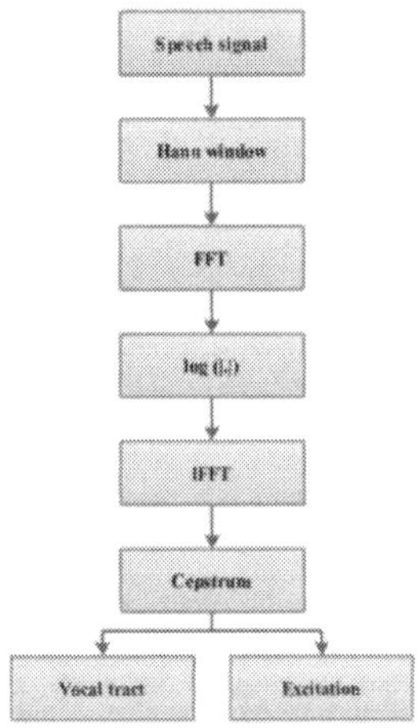

Fig. 1. The process of extracting the vocal tract and excitation cepstrum from the esophageal speech signal.

After calculating the cepstrum, it is easy to calculate the excitation data and the vocal tract data after separating them. Thus, the linguistic features of the speech signal are:

- C_0: The first coefficient
- $[C_1 \ldots C_p]$: The vocal tract cepstral vector
- $[C_{p+1} \ldots C_{N/2}]$: The excitation cepstral vector
- $[Ph_0 \ldots Ph_{N/2}]$: The phase coefficients.

Where N denotes the analysis window size, P represents the number of the vocal tract cepstrum, and $C_0 = 0$.

2.2 Denoising methods

2.2.1 Wavelet-based methods

- DWT

The Discrete Wavelet Transform gives a good localization in time and frequency [10]. It's used in several domains, especially in denoising images and speech signals [3, 11], and this is due to the multi-resolution analysis, which is the main characteristic of the wavelets. According to this analysis, the signal could be decomposed through a cascade of filters, associating a peer of filters to every resolution level: a low pass filter corresponding to low frequencies giving the approximation A_j, and a high pass filter corresponding to high frequencies giving the details D_j. This process is repeated to decompose the approximations [3, 10].

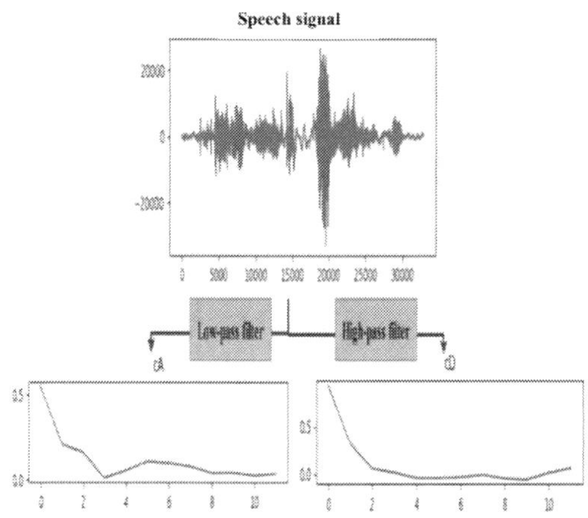

Fig. 2. The decomposition of the signal of an esophageal file sound in approximation A_j and details D_j by the DWT.

- DTCWT

The Dual-Tree Complex Wavelet Transform; often confused with continuous wavelets (CWT); could give more suitable results than the DWT [12]. This method uses two real trees of DWT, the first DWT tree generates the real part and its coefficients, while the second one generates the imaginary part of the DTCWT. Two different sets of real filters are used in the two real trees of the DWT. So, in the final, the result is approximately an analytic wavelet [12, 13]. H_0 and L_0 are respectively the high-pass filter and the low-pass filter of the first tree, and H_1 and L_1 are respectively the low-pass filter and the high-pass filter of the second tree.

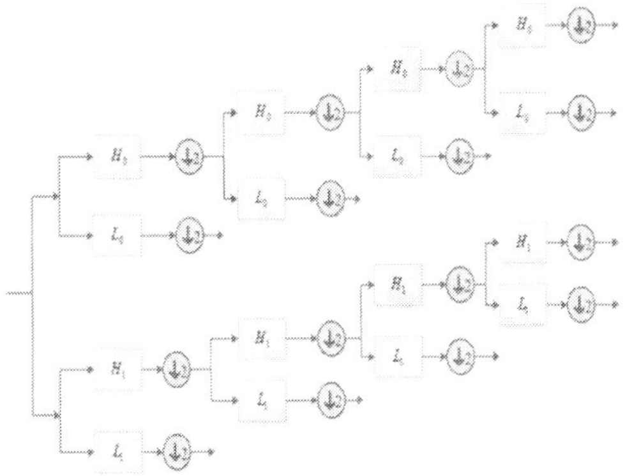

Fig. 3. Representation of the functioning of the DTCWT tree.

2.2.2 Wiener method

The Wiener method is a linear filter, belonging to the frequency filters. It is used to estimate the value of a noisy signal, by minimizing the mean squared error (MMSE) between the random estimated process and the desired process. It adapts the signal noise ratio for each treated part [14, 15].

2.2.3 The time Dilated Fourier Cepstra method

The dilation is a morphological operation, that aims to repair the interrupted signals by expanding them and making them clearer [2, 16, 17].

The time dilated Fourier Cepstra, as defined and detailed in [18], is used to enhance the esophageal speech in the frequency domain, by dilating the frequency axis of ratio $1/\alpha$. Thus, the frequency components will be changed, without corrupting the speech signal. The operator related to this operation is defined by formula (2) [18]:

$$D_\alpha[s](x) = s\left(\frac{x}{\alpha}\right), \alpha \in R^{+*} \quad (2)$$

Where α is called the α-rate homothety of a function $s \in L^2(R)$.

$D_\alpha^f = \alpha D_{1/\alpha}$ is the frequency dilated signal, it proceeds on the frequency support of a signal in the transposition direction:

- Towards the high frequencies for $\alpha > 1$
- Towards the low frequencies for $\alpha < 1$

The dilation algorithm used is detailed in [18].

2.2.4 Proposed hybrid methods

In this work, we propose two hybrid methods which aim to combine the advantages of the wavelet-based methods (DWT and DTCWT), with the standard methods (Wiener and Dilation). We proposed two models:

- Hybrid method 1

This proposed hybrid method denoised the extracted vocal tract cepstrum, using a combination of the methods described above. Figure 4 presents the schema of the first method.

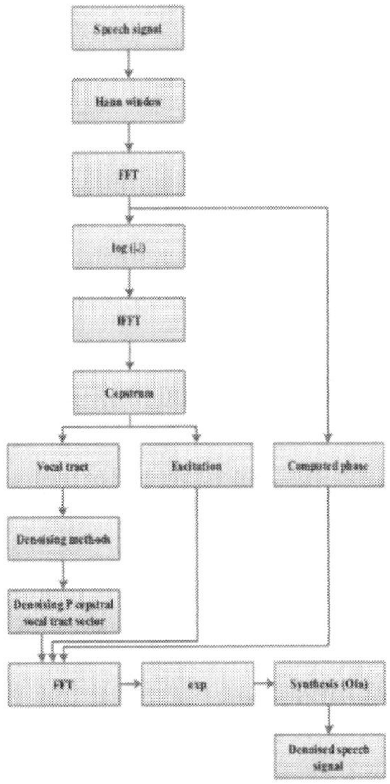

Fig. 4. Schema of hybrid method 1.

- Hybrid method 2

For hybrid method 2, the denoising methods were applied in the synthesis stage, on the Overlap-add method (Ola) without changing the vocal tract cepstrum. Figure 5 shows the schema of the given method.

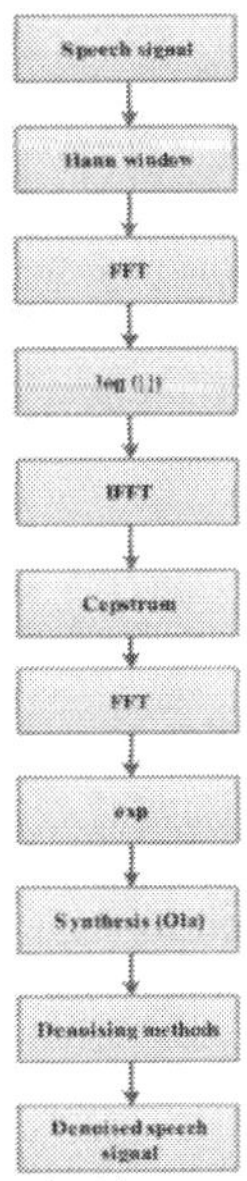

Fig. 5. Schema of hybrid method 2.

2.3 Synthesis stage

To reconstruct the speech signal and to return to the time domain [19], the IFFT operator was applied to the complex spectra, which were computed by multiplying the magnitude and the phase spectra. Then, the Overlap-add method (OLA) was used, it enables resynthesizing the original speech signal precisely from the windowed overlapping frames with low computation levels [19, 20, 21].

3 Experiments

3.1 Experiments

In this study, we conducted two experiments, the first one aimed to denoise esophageal speech signal using the proposed hybrid methods. This experiment compared the results of three cases, case 1 was about denoising the 25 vocal tract cepstrum and case 2 was about denoising the 33 vocal tract cepstrum (hybrid method 1). Case 3, used the second hybrid method. One esophageal speech file was used in experiment 1. The second experiment used the whole dataset. This experiment contained two cases; the first one was applying the first hybrid method on the 25 cepstral vocal tract vectors of each esophageal speech sound of the dataset. The choice of the number of cepstral vocal tract coefficients was made based on the first experiment. In the second case used hybrid method 2.

For experiments 1 and 2, the first cepstral coefficient C_0 was discarded. The $[C_{p+1} \ldots C_{256}]$ cepstral coefficients left were used as the excitation features. Also, the 257 phase coefficients were used as well.

3.2 The dataset

The dataset used in this work is a French database, containing three parallel corpora, spoken by three French laryngectomees male speakers: PC, MH, and GM. Each corpus contains 289 phonetically balanced phrases. Table 1 presents the experimental settings of the experiments.

Table 1. The experimental settings.

Sampling rate	16 kHz
Step size	64ms
Analysis window size	512
P1	25
P2	33

4 Results and discussion

In this work, an objective evaluation has been done in order to evaluate the results of denoising the esophageal speech sounds. For experiment 1, the Signal-to-Error Ratio (SER) has been calculated. And for experiment 2, the SER has been calculated for each file sound of the three corpora, and then the average SER of each corpus has been computed. The Signal-to-Error-Ratio is given by formula (3) [2]:

$$SER = 10\, log_{10} \frac{\sum_k \|x_k - \hat{x}_k\|^2}{\sum_k \|x_k\|^2} \tag{3}$$

Where x_k and $\hat{x}_k$ are respectively the spectrum of the initial signal and the spectrum of the synthesized signal.

4.1 Results of Experiment 1

Table 2 resumes the results of experiment 1. For denoising the Ola method, either using 25 or 33 cepstra (which correspond to the number of vocal tract cepstral coefficients) have given the same SER results, thus the choice of cepstra does not affect the results since we are working on the whole signal.

Table 2. The SER Results of the two hybrid methods for the operations performed on one sound file

Parameters / Operations	SER for 25 cepstra	SER for 33 cepstra	SER For Ola Function
Synthesis of the ES sound	<u>16,94</u>	<u>16,94</u>	<u>16,94</u>
Dilation	7,63	7,83	3,35
Wiener	10,42	9,81	<u>16,94</u>
DWT+Dilation (of cA)	11,59	10,01	5,83
DWT+Dilation (of cA and cD)	10,34	9,28	2,89
DWT+Wiener (of cA And cD=0)	10,36	10,20	7,23
DWT+Wiener (of cA And cD≠0)	<u>13,01</u>	<u>13,27</u>	<u>14,81</u>
DWT+Wiener (of cA and cD)	9,73	9,81	12,52
DTCWT+Dilation level 1	7,66	7,81	5,43
DTCWT+Dilation level 2	6,86	6,62	8,91
DTCWT+Dilation level 3	<u>12,12</u>	<u>11,66</u>	<u>13,76</u>
DTCWT+Dilation level 4	8,36	7,89	10,09
DTCWT+Dilation level 5	5,70	5,93	6,86
DTCWT+Wiener level 1	7,26	7,21	13,25
DTCWT+Wiener level 2	10,17	10,03	<u>15,33</u>
DTCWT+Wiener level 3	<u>14,19</u>	<u>14,30</u>	<u>16,13</u>
DTCWT+Wiener level 4	10,77	10,66	<u>15,31</u>
DTCWT+Wiener level 5	7,08	7,13	<u>12,74</u>
DTCWT+Wiener level 6	10,14	5,43	<u>14,07</u>
DTCWT+Wiener level 7	<u>15,20</u>	11,36	<u>15,34</u>

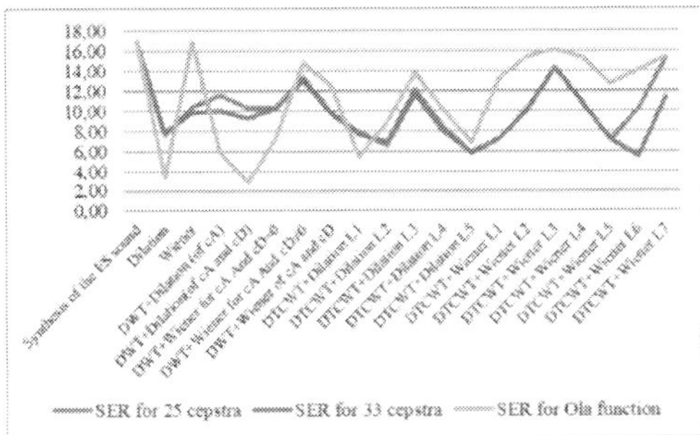

Fig. 6. Variation curves of the SER of hybrid method 1 (25/33 cepstra) and hybrid method 2 (Ola function) according to the operations performed in experiment 1.

Figure 6 shows that the Signal-to-Error Ratio curves for 25 and 33 cepstra are almost identical for all operations performed except for a few operations where the SER of the 25 cepstra exceeds the curve of the 33 cepstra. For both 25 and 33 cepstra, the SER of the combination of the 3rd level of the DTCWT with the Wiener filter gives the highest value compared to the other operations. Even though the last operation that designates the combination of the 7th level of DTCWT with the Wiener filter shows the highest peak of the results for the 25 cepstra, the successive division of the 25 cepstra by the DTCWT to obtain the 7th level decreases the number of cepstra to 1 for each tree, which means that the number of cepstra that underwent a change is almost negligible.

Also, the following operations have the highest SER values: Combining DWT with the Wiener filter applied on the approximation cA and the details $cD \neq 0$, as well as combining the 3rd level of the DTCWT with the dilation.

Concerning the operations performed at the level of the Ola function, its Signal-to-Error Ratio curve shows several significant peaks than the SER curves of 25 and 33 cepstra, precisely for the combination of the DTCWT (from level 1 to 7) with the Wiener filter, and the combination of the 3rd level of the DTCWT with the dilation as well as the combination of the DWT with the Wiener filter applied to the cA approximation and the $cD \neq 0$ details.

Furthermore, we notice that the application of the Wiener filter alone in the Ola function, without combination with other methods gives the highest peak, which value is identical to the value of the synthesis operation of the original speech sound, which means that the Wiener filter alone could not make any changes to the speech signal due to the nature of the esophageal noise. The combination of the 3rd level of the DTCWT with the Wiener filter shows the highest value and the closest to the synthesis operation value. Similarly, the two combinations of level 2 and level 4 of the DTCWT with the Wiener filter give almost the same high result.

4.2 Results of Experiment 2

Table 2 resumes the results of the average SER of the two hybrid methods for the 289 esophageal speech files from the three corpora GM, HM, and PC.

Table 2. The average of the SER of the operations performed on the three corpora, for the two hybrid methods

Parameters / Operations	The mean SER of 289 (GM corpus)	The mean SER of 289 (MH corpus)	The mean SER of 289 (PC corpus)
Synthesis of the esophageal speech sound	24,03	25,51	33,77
Dilation (Cepstra)	11,85	10,59	9,92
DWT+Dilatation of cA and cD=0	10,23	9,794	8,39
DWT+Dilatation of cA and cD	11,50	11,01	10,34
DWT+Wiener of cA and cD=0	15,15	16,04	12,51
DWT+Wiener (of cA and cD)	17,37	17,88	13,42
DWT+Wiener(cA andcD) (Ola)	19,02	19,02	13,85
DTCWT+Dilation Level 3 (Cepstra)	5,95	8,56	6,38
DTCWT+Dilation Level 3 (Ola)	9,92	11,10	7,99
DTCWT+Wiener Level 3 (Cepstra)	9,76	11,93	9,64
DTCWT+Wiener Level 3 (Ola)	16,82	17,93	16,87

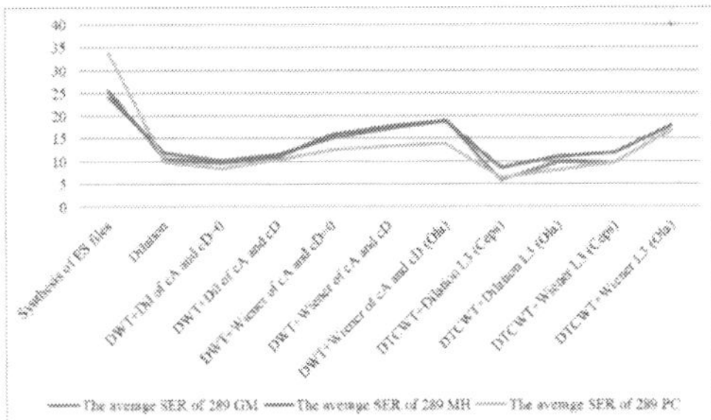

Fig. 7. The average of SER curves of the two hybrid methods for the 289 ES sound files obtained from the PC, MH, and GM corpora

Figure 7 shows that the three SER average curves of the 289 sound files obtained from the three corpora PC, MH, and GM, have the same pattern for all the operations performed. The highest mean values are those of the DWT operations combined with the Wiener filter applied on the approximation cA, and on details cD at the level of the Ola function. In addition to that, the combination of DTCWT level 3 with the Wiener filter at the level of the Ola function and the DWT combined with the Wiener filter applied on the approximation cA with cD = 0 at the level of the cepstra show remarkable results.

5 Conclusion

This study proposes two hybrid methods for reducing the noise of esophageal speech. From the first and second experiment, we can conclude that the application of the Wiener filter or the dilation without combining them with the DTCWT or the DWT (either for the 25 cepstra, the 33 cepstra, or the Ola function) do not show good results. Otherwise, the two proposed hybrid methods which consist of combining the wavelet-based methods along with the Wiener and Dilation method show better results. Also, we notice that discrete and complex wavelets (DWT and DTCWT) show significant denoising results. Furthermore, the second hybrid method shows an important improvement in the results and more precisely the combinations with the Wiener filter. The outcomes of the techniques used in this paper are still to be improved in future work, using new combinations with other denoising techniques and even using neural networks.

References

1. O. Lachhab, *Reconnaissance Statistique de la Parole Continue pour Voix Laryngée et Alaryngée*, tel.archives-ouvertes.fr, (2017) https://tel.archives-ouvertes.fr/tel-01563766/
2. I. Ben Othmane, *Conversion de la voix: Approches et applications, tel.archives-ouvertes.fr*, (2019). https://tel.archives-ouvertes.fr/tel-02276259.
3. F. Bahja, *Détection du fondamental de la parole en temps réel: application aux voix pathologiques*, tel.archives-ouvertes.fr, (2013). https://tel.archives-ouvertes.fr/tel-00927147
4. C. Manfredi, M. D'aniello, and P. Bruscaglioni, *Comparison between AR and SVD approaches for speech denoising*, (2001).
5. C. Manfredi, L. Landini, F. Faita, and V. Gemignani, *SVD-based portable device for real-time hoarse voice denoising*, IEEE Xplore, (Jul. 01, 2002)
6. Y. Zhang, J. J. Jiang, and F. A. Feroze, *Wavelet-based denoising for improving nonlinear dynamic analysis of pathological voices, ur.booksc.eu*, (2005)
7. M. Shafieian and M. Rahmanian, *An unsupervised approach for improving speech enhancement using wavelet packet transform and adaptive thresholding*, Bdigital2.ula.ve, **26**, no. 3, (2019), doi: pp 92.0200.
8. S.-J. Lee and H.-Y. Kwon, *A Preprocessing Strategy for Denoising of Speech Data Based on Speech Segment Detection*, Applied Sciences, **10**, no. 20, p. 7385, (2020), doi: 10.3390/app10207385.
9. D. W. Griffin and J. S. Lim, *Signal estimation from modified short-time Fourier transform*, IEEE Transactions on Acoustics, Speech, and Signal Processing, **32**, no. 2, pp. 236–243, (1984), doi: 10.1109/TASSP.1984.1164317.
10. Z. Khawaja, *Analyse des états de surface en science des matériaux: caractérisation multi-échelles par ondelette et détermination de l'anisotropie des surfaces*, HAL Archives Ouvertes, (2014). https://hal.archives-ouvertes.fr/tel-01081204/
11. A. Lallouani, *Débruitage d'un signal de la parole corrompu par un bruit coloré en utilisant la transformée en ondelettes et implantation sur un processeur de traitement numérique des signaux*, espace.etsmtl.ca, (2004).
12. I. W. Selesnick, R. G. Baraniuk, and N. C. Kingsbury, *The dual-tree complex wavelet transform*, IEEE Signal Processing Magazine, **22**, no. 6, pp. 123–151, (2005), doi: 10.1109/msp.2005.1550194.
13. P. Loiseau, *Ondelettes complexes pour l'analyse des lois d'échelles*, (2006).
14. K. Nabgha, M. Khannoussi, and A. Tazi, *Bruit et filtrage*, dspace.univ-adrar.edu.dz, (2018).
15. A. Jeanvoine, *Intérêt des algorithmes de réduction de bruit dans l'implant cochléaire : Application à la binauralité*, tel.archives-ouvertes.fr, (2012).
16. J. Balado, P. van Oosterom, L. Díaz-Vilariño, and M. Meijers, *Mathematical morphology directly applied to point cloud data*, ISPRS Journal of Photogrammetry and Remote Sensing, **168**, no. 168, pp. 208–220, (2020), doi: 10.1016/j.isprsjprs.2020.08.011.
17. A. Soni and A. P. Singh, *Automatic Pulmonary Cancer Detection using Prewitt & Morphological Dilation*, 2nd International Conference on Data, Engineering and Applications (IDEA), (Feb. 2020), doi: 10.1109/idea49133.2020.9170680.
18. I. Ben Othmane, J. Di Martino, and K. Ouni, *Enhancement of esophageal speech obtained by a voice conversion technique using time dilated Fourier cepstra*, International Journal of Speech Technology, **22**, no. 1, pp. 99–110, (2018), doi: 10.1007/s10772-018-09579-1.
19. W. Verhelst, *Overlap-add methods for time-scaling of speech*, Speech Communication, **30**, no. 4, pp. 207–221, (2000), doi: 10.1016/s0167-6393(99)00051-5.
20. M. Bahoura, *Efficient FPGA-Based Architecture of the Overlap-Add Method for Short-Time Fourier Analysis/Synthesis*, Electronics, **8**, no. 12, p. 1533, (2019), doi: 10.3390/electronics8121533.
21. E. B. George and M. J. T. Smith, *Speech analysis/synthesis and modification using an analysis-by-synthesis/overlap-add sinusoidal model*, IEEE Transactions on Speech and Audio Processing, **5**, no. 5, pp. 389–406, (1997), doi: 10.1109/89.622558.

Impact of Aircraft Performance and Time of the Day on Flight Arrival Delays Prediction in the United States: a Machine Learning Classification

Hajar Alla[1,2,3,*], *Lahcen* Moumoun[1,2,**], and *Youssef* Balouki[1,2,***]

[1]Laboratory of Mathematics, Computer and Engineering sciences. Mathematics and Computer Science Department
[2]Faculty of Science and Techniques Hassan First University of Settat, 26000 Settat, Morocco
[3]Corresponding Author

Abstract. The excessive growth of air traffic, with the limited airspace and airports capacity, results in a flight demand-capacity imbalance leading to air traffic delays. This paper explores the factors associated with delay in both microscopic and macroscopic ways. The aim is to develop a model which analyzes and predicts the occurrence of flight arrival delays using US domestic flight data for the year 2018. It will provide passengers, airlines and airport managers with reliable flight arrival schedules, and consequently reduce economic losses and enhance passengers trust. Beside database features, the proposed model is to the best of our knowledge the first attempt to predict flight arrival delays using three new features which are contributive factors to delays: Departure Time and Arrival Time of the day in which the flight was performed (Early morning, late morning, noon, afternoon, evening or night) and model of aircraft. Four Machine Learning classifiers namely Random Forest, Decision Trees, K-Nearest Neighbors and Naïve Bayes were used. In order to find the best parameters of each algorithm, we implemented Grid Search technique. The performance of each classifier was compared in terms of hyperparameters tuning, classification metrics and features description. The experimental results showed that the proposed system was able to predict flight arrival delays with the best Random Forest accuracy of 0.9356 and a higher number of correctly classified flights. To prove the importance of our findings, we compared our model to that of existing literature studies.

1 Introduction

Flight delays impact all aviation users such as travelers, airlines and airport authorities. It is considered as a dysfunction in the air traffic system. Bureau of Transportation Statistics (BTS) classifies flight delays in six categories: Air Carrier Delay, Aircraft Arriving Late, Security Delay, National Aviation System (NAS) Delay, Extreme Weather Delay, and Cancelled or Diverted flights. Figure 1 resumes the reasons of flight delays in all US major airports from January to December, 2017.

Among the factors influencing flight delays, we cite aircraft technical or mechanical problems, bad meteorological conditions, airport high density and capacity, late or disturbing passengers, insufficient airport infrastructure, delays in fueling, passengers embarking or disembarking, luggage loading or unloading and so on. Hence, traffic delays cost airlines million dollars per year comprising fuel costs due to fuel consumption, passenger delay costs paid to unhappy and complaining travelers and influence on next flights costs. To deal with this problem, researchers from all over the world conducted several studies to find solutions for traffic delays. In this study, a predictive classification model is proposed to estimate airline delays for United States study case using four effective Machine Learning algorithms such as Naïve Bayes, K-Nearest Neighbour, Decision Trees and Random Forest. On-time and delayed flights data for the year 2018 were extracted from BTS database.

Time of the day is an associated factor with delays. Indeed, depending on the season, airlines or travelers can pick the same part time of the day to take-in their flights (mornings, afternoons, evenings or nights). Choosing the same period can result in rush hours and an air and ground density.

According to [1], it is natural and inevitable to have differences in aircraft performances in air transportation environment. However, the differences do not only lead to congestion, density and delays but also cause economic losses and pollution of the environment through excessive gas emission and energy consumption. The delay increases in particular when slower aircraft models are followed by faster ones in a climb, cruise phase or even in the ground. For this reason and to the best of our knowledge, we proposed for the first time new features namely, Model of Aircraft, Departure Time and Arrival Time of the day. We will prove at the end of this study that the new proposed fea-

*e-mail: h.alla@uhp.ac.ma
**e-mail: lahcen.moumoun@uhp.ac.ma
***e-mail: youssef.balouki@uhp.ac.ma

https://doi.org/10.1051/itmconf/20224803005

tures were able to enhance the efficacy of our predictive model, accompanied with other features such as airport of origin and air carriers which are also contributive to flight delays.

Hyperparameters optimization is used in most studies to control the learning process, to get optimal and better values of the parameters and improve accuracy. In this paper, we chose Grid Search technique in order to find the best parameters for K-Nearest Neighbors, Random Forest and Decision Trees. We evaluated the performance of the proposed method based on hyperparameters tuning, classification metrics and features description. To prove the importance of our findings, we compared our model to that of existing literature studies.

The remainder of this paper is structured as follows: Section 2 describes existing literature on flight delay prediction. Section 3 presents in details the research methodology used in order to predict flight arrival delays. In Section 4, the performance of all classifiers is compared based on hyperparameters tuning, classification metrics and features description. Section 5 concludes the paper by summarizing the contributions of this study and recommending perspectives and further works.

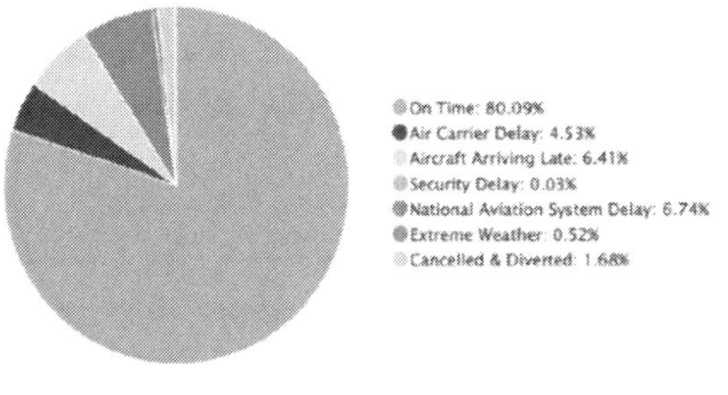

Fig. 1. Flight Delays by Cause - All Major Airports (January - December, 2017)

2 Literature Review

Due to its vital importance, the study of flight delays has taken more attention recently. Existing literature that studies flight arrival delay problems focuses the most on statistical and probabilistic techniques. Authors in [2] predicted flight delays by estimating taxi-out time using queuing model. [3] developed a probabilistic model based on delay distribution to estimate flights delays probability. [4] experimented several scenarios to estimate runway capacity of airports with Monte Carlo simulations. Authors in [5] focused on taxi-time reduction using Ant algorithm in order to optimize aircraft taxi movements. They proposed a model taking the shortest path of aircraft into account to reduce delays. Authors in [6] developed a system that allows a continuous taxiing to minimize taxi-time delays. Researchers in [7] attempted to create a statistical method to analyze departure and arrival flight delays. They developed a model for flight delays analysis using density functions. The authors claimed that Normal distribution performed better with departure delays while Poisson distribution worked better with en route and arrival delays.

Statistical methods are focused on probabilities and approximations which may lead to unsure conclusions and results. To cope with this issue, intelligent systems based on machine learning; which is a rapidly evolving field [8]; are hence needed. Authors in [9] focused on predicting flight arrival time at the block-off moment instead of airborne moment with linear regression and gradient boosting machines. Authors in [10] developed a model to forecast delays of US domestic flights based on multiple linear regression, decision trees and random forest algorithms. It has been concluded that Random forest outperformed the other models. To identify flight delay in advance, Authors in [11] described a predictive model using decision tree classifier which proved to be efficient. Decision Trees, Random Forest, AdaBoost and KNN were used as cost-sensitive classifiers to predict individual flight delays in [12]. [13] combined classification and regression. In the classification stage, Gradient Boosting Classifier performed the best and in the regression stage, Extra-Trees Regressor was the best performing classifier. Authors in [14] applied SVM, Gradient Boosting, Random Forest, AdaBoost and Decision Tree classifiers to predict on-time arrival flights. Random Forest was the best classifier with an accuracy of 77%. [15] established an improved Support Vector Machine (SVM) model to predict flight departure delay. The experimental results show that the proposed model is able to predict flight delay time with a high effectiveness and accuracy than other algorithms. In order to analyze and predict flight arrival delays, authors in [16] applied four supervised machine learning classifiers namely random forest, Support Vector Machine (SVM), Gradient Boosting Classifier (GBC) and k-nearest neighbour algorithm. The best classifier proved to be gradient boosting with an accuracy of 79.7%. Authors in [17] developed a model to estimate and predict flight arrival delay using decision tree, logistic regression and neural networks algorithms. The performance was approximately equal to 91% for all three classifiers.

In the existing studies, researchers have addressed flight delays from different perspectives. In this paper, we developed a predictive model that studies several factors influencing flight delays in the macroscopic level. Moreover, to the best of our knowledge, new microscopic features were proposed and considered in order to enhance the accuracy of the flight prediction system. The model is trained using various supervised machine learning classifiers that have demonstrated to be efficient in previous studies. The proposed system guarantees parameters tuning for a better performance.

3 Proposed Methodology

3.1 Research Motivation

Delay is the second most significant factor to consider in aviation environment, after safety. It annoys passengers and airport managers and increase airlines economic

losses. Since it is still an important ongoing research area, multiple studies have been conducted to cope with this issue. A system, which predicts flight arrival delays and provides airport personnel, companies, passengers and aviation users with the delay before its occurrence, is needed. Several macroscopic factors impact flight delays such as bad weather conditions, severe crosswinds [18], technical problems, late and disturbing passengers, airport crowdedness, runway queues, lack in airport infrastructure, aircraft maintenance delays, flight check-list delays, etc.
Our objective in this study is to identify and use new microscopic factors which participate in flight delays to develop a predictive model based on good performing machine learning classifiers. Figure 2 describes the process followed to create and generate the proposed method. Flights data are extracted from BTS database. ICAO[1] aircraft models, manufacturers and registrations are imported from Github. In order to remove redundant, duplicated, noisy and unnecessary information, preprocessing and cleaning techniques are applied. Only relevant features that are contributive to flight delays need to be selected. Hence, a features selection is necessary. In order to improve the efficacy of the proposed system, other microscopic factors of delays are considered. For that and to the best of our knowledge, three new features; namely Departure Period of the day, Arrival Period of the day and Model of aircraft; are created and inserted into the model. A data split of 70:30 ratio is applied generating training and testing data. To generate the best parameters of all four classifiers, Grid Search technique is utilized for parameters tuning. The performance of the proposed predictive model is finally evaluated using classification metrics.

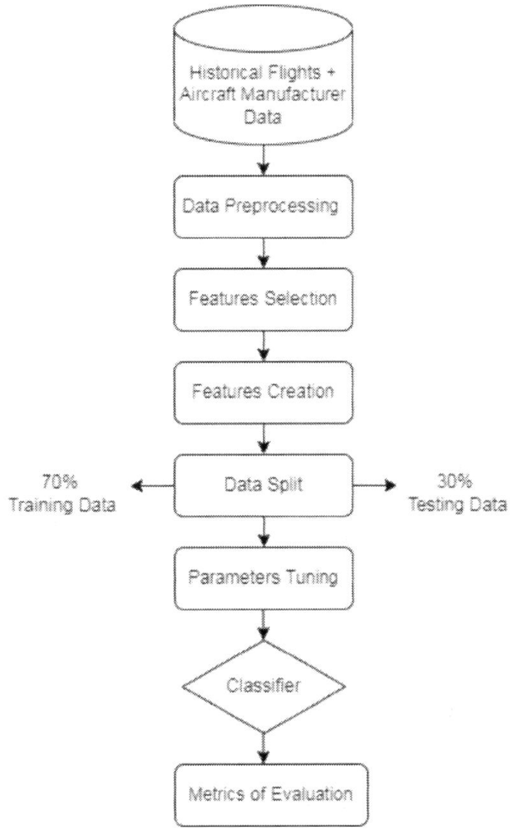

Fig. 2. Flowchart of the proposed method

3.2 Data Collection

United States on-time performance data from 1st of January to 31st of December 2018 are achieved from the Bureau of Transportation Statistics database (BTS) located in the Department of Transportation in The United States. The dataset contains about 94810 recordings of domestic flights. The accessibility of data saved since 1987, reliability, transparency and independence from political influence [19] encouraged us to choose BTS as the source of our data collection. According to BTS, all flights whose difference between scheduled and actual arrival times is 15 min or higher are considered delayed. Aircraft manufacturer, models and registration were extracted from Github[2], saved as aircraft performance and merged with BTS flight movements to create a unique dataset. The file extracted contained about 353043 instances of aircraft models and manufacturers.

3.3 Data Preprocessing

Real-world data are often incomplete, incorrect and mistaken. It may be also noisy and inconsistent [20]. Since machine learning algorithms work better with numerical variables, data cleaning need to be accomplished. Duplicate data, missing and null values need to be erased. Categorical data are converted to numeric.

3.4 Features Analysis

The data extracted contained several attributes and data fields. Only features which have an impact on flight delays were selected and unnecessary attributes were removed.

Carrier and Company Delay BTS defines a carrier or company delay as a delay or cancellation caused due to circumstances within the airline's control (e.g. maintenance or crew problems, slower aircraft cleaning, baggage loading, fueling, etc.). Due to repeated delays, passengers may change an airline for another that is less subject to delays. Furthermore, a failure in turnaround process caused by push-back, fueling/refueling, catering or mechanical check operations may engender delays. Personnel stress and fatigue [19] or personnel labor strike [21] are considered contributive to carrier traffic delays.

Origin and Destination Airports The distance between the origin and destination airports determine the duration of the flight. If it is higher, the flight is likely to be late. [21] concluded that the longer is the distance, the higher is the delay.

[1]ICAO: International Civil Aviation Organization
[2]https://github.com/RobAltenburg

Departure Delay Departure delays are contributive to arrival delays. [22] claimed that congestion at destination airport is mostly originated at the departure airport. A traffic experiencing a delay on departure may be also delayed on arrival [21]. Hence, departure delay is a relevant feature to our study.
To improve the efficiency of our system, we created three other features,namely Aircraft Model, Departure Period and Arrival Period of the day in which the flight was performed.

Aircraft Model Some aircraft can be more flexible in arriving on-time depending on the performance [23]. According to [24], aircraft performance differences cause flight delays. Lower performance aircraft determine airspace capacity. If a low performance aircraft precedes a higher performance one, it may cause a delay by retarding the following flight since the speed of the constraining aircraft is lower. The high performance aircraft is then forced to maintain the same speed or lower which causes traffic delays. In Table 1, we present, from our dataset, the remarkable difference in delays percentage depending on aircraft model and manufacturer. As we can notice, 38% from the flights performed by BOEING 7378FH were delayed while only 11% were not on-time in the case of MCDONNELL DOUGLAS AIRCRAFT MD-88. Hence, the difference in performance has an impact on traffic delays and the model of aircraft is a necessary feature in our study.

Period of the Day As we know, the period of a day is an important factor to consider when traveling. In fact, every passenger has a preferable period to take in a flight. Some prefer mornings, others choose to travel by night, etc. Also, there is some peak or rush period of the day in which traffic tends to be dense leading to flight delays. In general, we account more flights in mornings and afternoons than nights due to the visibility. For this purpose, we decided, in our work, to create two new features, namely Departure Period and Arrival Period of the day which are respectively, the time of the day in which a flight has departed and the time of the day in which a flight has arrived to its destination. According to the world's number one website on time database[3], we decided to take 06:05 US Local Time as the minimum sunrise time and 21:06 US Local Time as the maximum sunset time to cover up all the year 2018. Table 2 displays the period of the day depending on the interval of hours. Each interval is affected a label.

3.5 Supervised Learning Classifiers

In order to predict whether a flight will be delayed or not, four supervised classifiers were applied namely KNearest-Neighbors, Naïve Bayes, Decision Trees and Random Forest.
K-Nearest-Neighbors is a classifier which take into consideration k adjacent points as neighbors in order to classify a point x. The process followed by k-NN classifier is described as follows [25]:

- Calculate the number of adjacent neighbors k
- Determine the distance from the point x to all the neighboring samples
- Sort the distance and find the nearest neighbors by measuring the shortest Euclidean distance
- Collect the label of each nearest neighbor. The label with the majority vote are the one assigned for the prediction

Naïve Bayes is a simple Bayesian network with multiple children (observed nodes) and a single parent (unobserved node) based on an assumption that all the variables are independent of each other. In a study carried out by [26], Naïve Bayes has proven to be more tolerant to noise, missing values, overfitting, irrelevant, redundant and interdependent attributes such as parity problems. It has a high speed of learning with respect to number of instances and attributes. Naïve Bayes is faster than k-NN at classification.
In Decision Trees algorithm, instances are classified and sorted based on the value of each feature which represents a node. The branches contain all possible values of the chosen node. All data are stored in the root node which is the very top of the tree. Each node asks a yes or no question and as an answer the data is subdivided into two subsets. Internal nodes have arrows pointing to them and they have arrows pointing away from them. Leaf nodes have arrows pointing to them but there are no arrows pointing away from them. Each internal node designates a test on an attribute, each branch designates an outcome of the test, and each leaf node (terminal node) holds a class label. Decision tree has the advantage of adaptation with data collected, a high speed of classification, a good accuracy and being able to deal with discrete, binary and continuous attributes [26]. Authors in [27] used decision tree algorithm on crime data to predict the probability of low, medium or high violent crimes in a country.
Random Forest is an algorithm capable of performing both regression and classification tasks. It consists of an ensemble of several individual decision trees. The more trees are in the forest, the more robust is the prediction. The trees are created randomly from input features of the training dataset. The required prediction is then generated by using Bagging technique. [28] considered Random Forest as one of the most successful ensemble learning techniques to be very popular and powerful. [26] rated Random Forest among classification algorithms with a high percentage of correctly classified parameters.

3.6 Hyperparameters Tuning

To find optimal hyperparameters for Random Forest, Decision Trees and K-Nearest Neighbors and improve accuracy, Grid Search technique was chosen over Manual Search with 10-fold cross validation. For the choice of hyperparameters, we were inspired from a study conducted by [29]. We evaluate all the possible combinations of parameters and keep the best combination. The best parameters of Random Forest were: 'max_features:

[3]United States, USA – Sunrise, Sunset, and Daylength, 2018, https://www.timeanddate.com/

Table 1. Delay percentage of the most common aircraft types in the dataset

Model	Manufacturer	Percentage of delay
737-8FH	BOEING	38%
777-323ER	BOEING	33%
H101 "SALTO"	START AND FLUG GMBH	30%
DOMINATOR	BENHAM JOHN	28%
R66	ROBINSON HELICOPTER CO	28%
340A	CESSNA	27%
SR22	CIRRUS	27%
DA 42	DIAMOND AIRCRAFT IND GMBH	26%
FALCON 900 EX	DASSAULT	26%
A321-231	AIRBUS	25.5%
GVI (G650)	GULFSTREAM AEROSPACE CORP	25%
A320-214	AIRBUS	24%
ERJ 190-100 IGW	EMBRAER	24%
AT-602	AIR TRACTOR INC	22%
182R	CESSNA	19%
B739	BOEING	13%
MD-88	MCDONNELL DOUGLAS AIRCRAFT CO	11%

Table 2. US Part-Time of the Day by Local Time in 2018

Interval of hours	Time of the Day Label
From 00:00 to 06:05	Early Morning
From 06:06 to 11:59	Late Morning
At 12:00	Noon
From 12:01 to 17:59	Afternoon
From 18:00 to 21:06	Evening
From 21:07 to 23:59	Night

sqrt' and 'n_estimators: 700' with the highest accuracy of 0.936344. In case of Decision Trees, 'criterion': 'entropy', 'max_depth': 20 and 'max_leaf_nodes': 150 were the best parameters with an accuracy of 0.923033. In K-Nearest Neighbors algorithm, 'metric': 'minkowski', 'n_neighbors': 15 and 'p': 1 were the best parameters with an accuracy of 0.811090.

4 Results and Discussions

The experiments were conducted on a dataset of 94810 flights and 353043 models of aircraft. After the files concatenation and data cleaning, a final dataset of 58971 recordings was utilized. Training and testing data were generated using a 70:30 ratio split. In this study, three new features namely Departure Period of the day, Arrival Period of the day and Model of aircraft were created in order to analyze new factors contributing in flight delays and achieve a good prediction. To create the model, K-Nearest Neighbors, Naives Bayes, Random Forest and Decision Trees were applied. To maximize the accuracy and obtain a better classification, parameters were tuned using Grid Search technique.

4.1 Features Discussion

4.1.1 Period of the Day

We extracted some of the features from the database but created other relevant ones: Departure period of the day and Arrival Period of the day duly explained in the methodology section. By analyzing our data, we found that the most delayed flights on departure were performed on late mornings, afternoons and evenings. Morning is perfect for travelling since it is the least crowded time in the day with generally a calm weather that passengers prefer to take a flight in. Afternoons and evenings are subject to rush hour, bad weather conditions and traffic density which cause delays. At arrival, the flights are also delayed at night maybe because the delay propagation in afternoons and evenings affect also the night period. Finally, we notice that the less affected by delays were the flights performed at noon which are quasily non-existent according to Figure 3 and Figure 4.

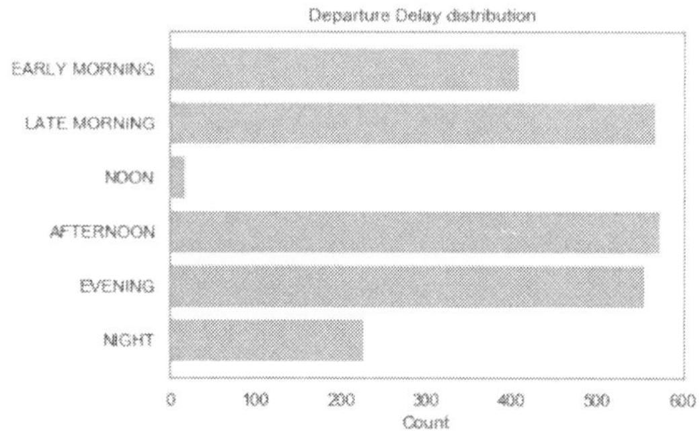

Fig. 3. Departure delay by part of the day

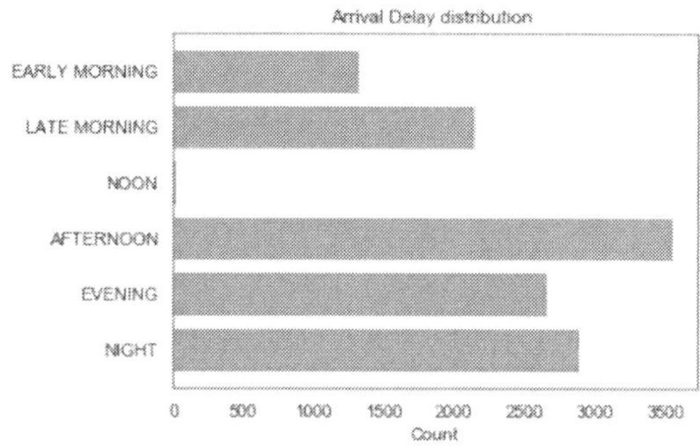

Fig. 4. Arrival delay by part of the day

4.1.2 Model of Aircraft

In Figure 5, we plot the percentage of the 5-most delayed aircraft model in our database. We notice that Airbus A321-231, A320-232 and EMBRAER ERJ 190-100 IGW were the most affected by arrival delays. The possible reason of the delay may be not being able to handle bad weather conditions.

4.1.3 Carrier Delay

We have considered also the company delay among our features. According to Figure 6, JetBlue (B6) is the most delayed company at arrivals while Alaska Airlines (AS) is the most arriving on-time company. Airlines delay may be caused by carrier issues such a slow push-back operations, mechanical check lists, fuel service, loading luggage, personnel labour strike, etc.

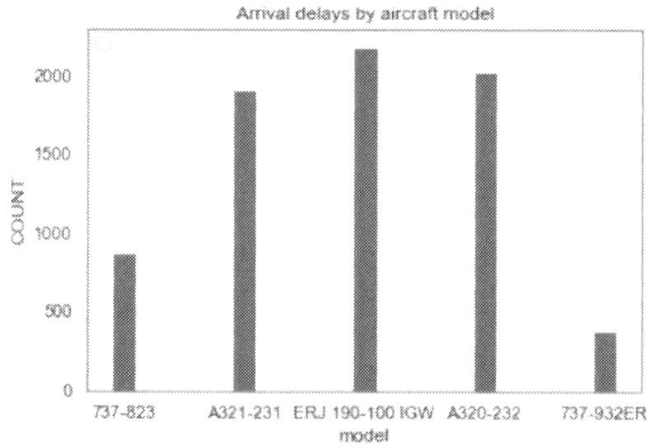

Fig. 5. Arrival delay by model of aircraft

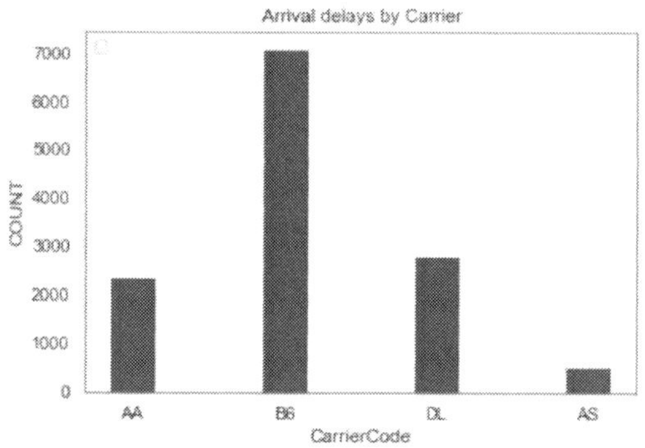

Fig. 6. Arrival delay by Carrier

4.1.4 Airport of Origin

From 63 origin airports used in this study, we plot only the 10-most delayed ones in Figure 6. We notice that Los Angeles International Airport (LAX), San Francisco International Airport (SFO) and Orlando International Airport (MCO) are respectively the 3-most delayed airports in US 2018 based on our features analysis.

4.2 Impact of Hyperparameters Tuning

In order to improve the accuracy and find optimal parameters, this study conducted tuning parameters of Random Forest, Decision Trees and K-Nearest Neighbors with Grid Search technique. We compare in Table 3 and Table 4 the performance of the algorithms with and without tuning, respectively.
We notice that the evaluation metrics of the three algorithms were improved when the tuning was applied. The experimental results show that parameters tuning has successfully generated the best classifier which is Random Forest with an accuracy of 0.9356 and a higher number of correctly classified flights.

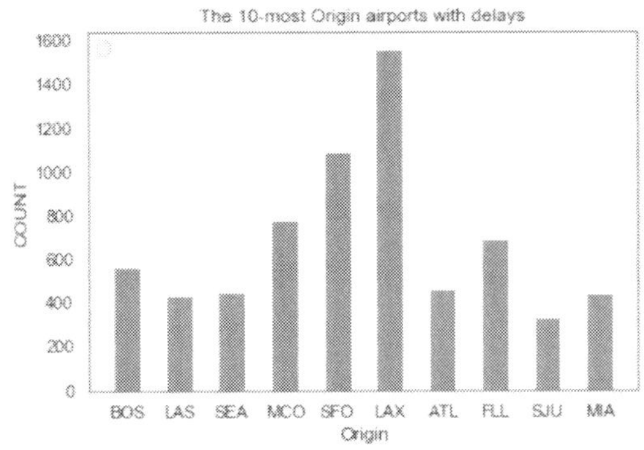

Fig. 7. The10-most delayed airports of departure

Table 3. Evaluation metrics with tuning

Algorithm	Accuracy	Precision		Recall		F1-score	
		0	1	0	1	0	1
Knn	0.81	0.75	0.90	0.92	0.69	0.83	0.78
DT	0.89	0.88	0.91	0.91	0.88	0.90	0.89
RF	0.93	0.92	0.95	0.95	0.92	0.93	0.93

Table 4. Evaluation metrics without tuning

Algorithm	Accuracy	Precision		Recall		F1-score	
		0	1	0	1	0	1
Knn	0.78	0.77	0.80	0.81	0.76	0.79	0.78
DT	0.87	0.86	0.87	0.87	0.86	0.87	0.87
RF	0.90	0.88	0.93	0.93	0.88	0.90	0.90

4.3 Models Benchmark

To prove that the experimental results of our model are fair and significant, we compare, in Table 5, the accuracy of our proposed Random Forest model to that of other existing literature results. To predict arrival delays, Jun Chen [30] chose an optimal feature selection in a Random Forest model than directly using all the features of available datasets. Jiage Huo et al. [31] developed a Random Forest system to predict flight delays for airlines in Hong Kong based on features selection. Chakrabarty et al. [16] utilized Random Forest data mining to analyze and estimate arrival delay of the flights. Gui et al. [32] implemented a random forest-based model to predict individual flight delay.

We notice that, using new features namely Departure Time of the Day, Arrival Time of the Day and Model of Aircraft accompanied with other features such as Airport of Origin, and Air Carrier Delays, generated a high accuracy model compared with existing studies.

Table 5. Accuracy benchmark of our proposed model and that of existing studies

	The proposed model	[30]	[31]	[16]	[32]
Accuracy	0.9356	0.8672	0.6639	0.7870	0.9020

5 Conclusion

Flight delay has been an important and challenging research subject as air demand grows year after year. In response to growth concerns, researchers analyzed flight delays from different perspectives. This study focused on flight arrival delays prediction using four Machine Learning classification algorithms namely Random Forest, Decision Trees, KNN and Naïve Bayes. In order to enhance the performance of the proposed prediction model, we created and added three new features: Departure period of the day, Arrival period of the day and model of aircraft. Parameters tuning using Grid Search Technique was performed in order to obtain an improved system. The model has achieved the highest accuracy of 0.9356 with Random Forest.

The proposed model can be used by airports managers, airlines and travelers to predict flight arrival delays of United States airports. It can be utilized also as a decision support tool for air traffic controllers. In fact, if they are informed of flight arrival times and possible delays in advance, they can be aware of peak hours of flights and whether it is necessary or not to add more controllers to the shift team. As a limitation, the dataset contains only on-time data for non-stop domestic flights. In future works, we consider extending the dataset to include also stop-over and international flights. Furthermore, it will be interesting to add more features and use deep learning in order to enhance the accuracy of the predictive model.

References

[1] Cavcar, Aydan, and Mustafa Cavcar. "Impact of aircraft performance differences on fuel consumption of aircraft in air traffic management environment." Aircraft Engineering and Aerospace Technology (2004)

[2] Idris, Husni, John-Paul Clarke, Rani Bhuva, and Laura Kang. Queuing model for taxi-out time estimation. (2001).

[3] Tu, Yufeng, Michael O. Ball, and Wolfgang S. Jank. "Estimating flight departure delay distributions—a statistical approach with long-term trend and short-term pattern." Journal of the American Statistical Association 103, no. 481, 112-125, (2008).

[4] Zhong, Z. W., Dhir Varun, and Y. J. Lin. "Studies for air traffic management R&D in the ASEAN-region context." Journal of Air Transport Management 64, 15-20, (2017).

[5] Nogueira, Kamila B., Paulo HC Aguiar, and Li Weigang. "Using ant algorithm to arrange taxiway sequencing in airport." International Journal of Com-

puter Theory & Engineering 6, no. 4, 857-361, (2014).

[6] Jiang, Yu, Xinxing Xu, Honghai Zhang, and Yuxiao Luo. "Taxiing route scheduling between taxiway and runway in hub airport." Mathematical Problems in Engineering 2015 (2015).

[7] Mueller, Eric, and Gano Chatterji. "Analysis of aircraft arrival and departure delay characteristics." In AIAA's Aircraft Technology, Integration, and Operations (ATIO) 2002 Technical Forum, p. 5866. (2002).

[8] Oualid, Adil, Abderrahim Hansali, Youssef Balouki, and Lahcen Moumoun. "Application of Machine Learning Techniques for Credit Risk Management: A Survey." In The International Conference on Information, Communication & Cybersecurity, pp. 180-191. Springer, Cham, (2021).

[9] Achenbach, Anna, and Stefan Spinler. "Prescriptive analytics in airline operations: Arrival time prediction and cost index optimization for short-haul flights." Operations Research Perspectives 5, 265-279, (2018).

[10] Kalliguddi, Anish M., and Aera K. Leboulluec. "Predictive modeling of aircraft flight delay." Universal Journal of Management 5, no. 10, 485-491, (2017).

[11] Katpadi, Vellore, and Tamil Nadu. "FLIGHT DELAY PREDICTION USING SUPERVISED MACHINE LEARNING."

[12] Choi, Sun, Young Jin Kim, Simon Briceno, and Dimitri Mavris. "Cost-sensitive prediction of airline delays using machine learning." In 2017 IEEE/AIAA 36th Digital Avionics Systems Conference (DASC), pp. 1-8. IEEE, (2017).

[13] Thiagarajan, Balasubramanian, Lakshminarasimhan Srinivasan, Aditya Vikram Sharma, Dinesh Sreekanthan, and Vineeth Vijayaraghavan. "A machine learning approach for prediction of on-time performance of flights." In 2017 IEEE/AIAA 36th Digital Avionics Systems Conference (DASC), pp. 1-6. IEEE, (2017).

[14] Etani, Noriko. "Development of a predictive model for on-time arrival flight of airliner by discovering correlation between flight and weather data." Journal of Big Data 6, no. 1, 1-17, (2019).

[15] Wu, Weinan, Kaiquan Cai, Yongjie Yan, and Yue Li. "An improved svm model for flight delay prediction." In 2019 IEEE/AIAA 38th Digital Avionics Systems Conference (DASC), pp. 1-6. IEEE, (2019).

[16] Chakrabarty, Navoneel, Tuhin Kundu, Sudipta Dandapat, Apurba Sarkar, and Dipak Kumar Kole. "Flight arrival delay prediction using gradient boosting classifier." In Emerging technologies in data mining and information security, pp. 651-659. Springer, Singapore, (2019).

[17] Kuhn, Nathalie, and Navaneeth Jamadagni. "Application of machine learning algorithms to predict flight arrival delays." CS229 (2017).

[18] Jarrar, Abdessamad, Abderrahim Ait Wakrime, and Youssef Balouki. "Formal approach to model complex adaptive computing systems." Complex Adaptive Systems Modeling 8, no. 1 (2020): 1-35.

[19] Alla, Hajar, Lahcen Moumoun, and Youssef Balouki. "Flight arrival delay prediction using supervised machine learning algorithms." In Intelligent Systems in Big Data, Semantic Web and Machine Learning, pp. 231-246. Springer, Cham, (2021).

[20] Moreira, Leonardo, Christofer Dantas, Leonardo Oliveira, Jorge Soares, and Eduardo Ogasawara. "On evaluating data preprocessing methods for machine learning models for flight delays." In 2018 International Joint Conference on Neural Networks (IJCNN), pp. 1-8. IEEE, (2018).

[21] Alla, Hajar, Lahcen Moumoun, and Youssef Balouki. "A multilayer perceptron neural network with selective-data training for flight arrival delay prediction." Scientific Programming 2021 (2021).

[22] Oza, Sruti, Somya Sharma, Hetal Sangoi, Rutuja Raut, and V. C. Kotak. "Flight delay prediction system using weighted multiple linear regression." International journal of engineering and computer science 4, no. 4, 11668-11677, (2015).

[23] Tielrooij, Maarten, Clark Borst, Marinus M. Van Paassen, and Max Mulder. "Predicting arrival time uncertainty from actual flight information." In Proceedings of the 11th USA/Europe Air Traffic Management Research and Development Seminar, pp. 577-586. (2015).

[24] Cavcar, Aydan, and Mustafa Cavcar. "Impact of Aircraft Performance Characteristics on Air Traffic Delays." Turkish Journal of Engineering and Environmental Sciences 28, no. 1, 23-24, (2004).

[25] Priyanka, G. "Prediction of airline delays using K-nearest neighbor algorithm." Int. J. Emerg. Technol. Innov. Eng 4, no. 5, 87-90, (2018).

[26] Osisanwo, F. Y., J. E. T. Akinsola, O. Awodele, J. O. Hinmikaiye, O. Olakanmi, and J. Akinjobi. "Supervised machine learning algorithms: classification and comparison." International Journal of Computer Trends and Technology (IJCTT) 48, no. 3, 128-138, (2017).

[27] Ivan, Niyonzima, Emmanuel Ahishakiye, Elisha Opiyo Omulo, and Danison Taremwa. "Crime Prediction Using Decision Tree (J48) Classification Algorithm." (2017).

[28] Azar, Ahmad Taher, Hanaa Ismail Elshazly, Aboul Ella Hassanien, and Abeer Mohamed Elkorany. "A random forest classifier for lymph diseases." Computer methods and programs in biomedicine 113, no. 2, 465-473, (2014).

[29] Ramadhan, Muhammad Murtadha, Imas Sukaesih Sitanggang, Fahrendi Rizky Nasution, and Abdullah Ghifari. "Parameter tuning in random forest based on grid search method for gender classification based on voice frequency." DEStech Transactions on Computer Science and Engineering 10 (2017).

[30] Chen, Jun, and Meng Li. "Chained predictions of flight delay using machine learning." In AIAA Scitech 2019 forum, p. 1661. 2019.

https://doi.org/10.1051/itmconf/20224803005

[31] Huo, Jiage, Kin Lok Keung, Carman Ka Man Lee, Kam KH Ng, and K. C. Li. "The Prediction of Flight Delay:Big Data-driven Machine Learning Approach." In 2020 IEEE International Conference on Industrial Engineering and Engineering Management (IEEM), pp. 190-194. IEEE, 2020.

[32] Gui, Guan, Fan Liu, Jinlong Sun, Jie Yang, Ziqi Zhou, and Dongxu Zhao. "Flight delay prediction based on aviation big data and machine learning."IEEE Transactions on Vehicular Technology 69, no. 1, pp. 140-150, (2019).

Non-negative Matrix Factorization for Dimensionality Reduction

Jbari Olaya
New Technology Trends (NTT)
National School of Applied Sciences
Tetuan, Morocco
olaya.jbari@etu.uae.ac.ma

Chakkor Otman
New Technology Trends (NTT)
National School of Applied Sciences
Tetuan, Morocco
otman.chakkor@uae.ac.ma

***Abstract*—What matrix factorization methods do is reduce the dimensionality of the data without losing any important information. In this work, we present the Non-negative Matrix Factorization (NMF) method, focusing on its advantages concerning other methods of matrix factorization. We discuss the main optimization algorithms, used to solve the NMF problem, and their convergence. The paper also contains a comparative study between principal component analysis (PCA), independent component analysis (ICA), and NMF for dimensionality reduction using a face image database.**

***Index Terms*—NMF, PCA, ICA, dimensionality reduction.**

I. Introduction

Dimensionality reduction is essential for extracting information from high-dimensional data. For that, PCA and ICA are the famous matrix factorization methods uses for this task. However, for many data sets such as images, text,...etc the original data matrices are non-negative. A factorization such as PCA and ICA contains negative values and is difficult to interpret for some applications. In contrast, non-negative matrix factorization restricts the elements in the data matrix to be non-negative.

The philosophy of NMF was firstly introduced, in a paper published, by Paatero and Tapper in 1994, and popularised by Lee and Seung in 1999 [12]. After that, NMF has gradually become an interesting multidimensional data processing tool to many researchers owing to its ability for giving a natural interpretation of the results (significant results) due to the constraint of the non-negativity.

NMF seeks to find two low-rank matrices $(W, H) \in \mathbb{R}_{m\times r} \times \mathbb{R}_{r\times n}$ non-negative, whose product approximates the non-negative data matrix $X \in \mathbb{R}_{m\times n}$ defined as:

$$X \approx WH \quad (1)$$

Where r is the factorization rank ($r << rank(X) << min(m, n)$) which selects how many features will be extracted from the data.

More precisely, each data point represented as a row in X can be approximated by an additive combination of non-negative vectors, which are represented as row in W (see Fig.1).

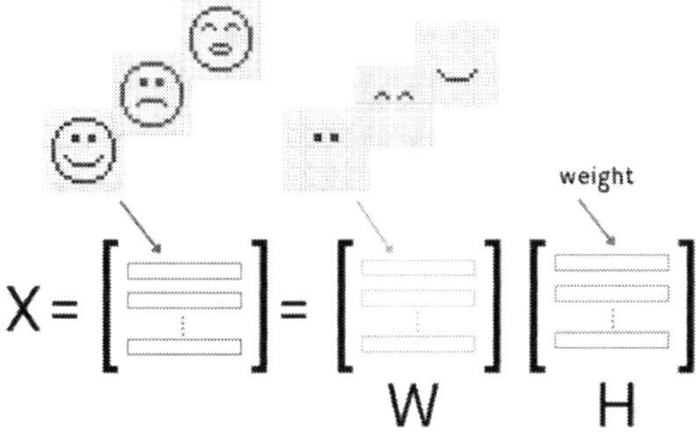

Fig. 1. Non-negative Matrix Factorization: Example of face recognition.

Different to other matrix factorization methods, NMF leads to a part-based representation (i.e, treating objects as a collection of constituent parts), because they allow only additive combinations of the original data. For example, in face recognition, It is interesting to note that the rows of the resulting W are clear parts of human faces (see Fig.1), e.g. nose, ears, and eyes, and these elements will add to each other to recreate the face (original data) [14]. On the other hand, classical factorization methods such as PCA and ICA produces both positive and negative values. Therefore, giving out components that don't offer much interpretability.

Moreover, the remarkable effectiveness of NMF in analyzing nonnegative data has attracted a significant amount of research in many others areas, such as in image processing [6],[19],[4], text mining [3],[21] and source separation problems [15],[7]. Currently, there is ongoing research on NMF to increase its efficiency and robustness.

To this end, the remaining sections of this paper are organized as follows. Section1 discusses the major challenges in solving the NMF problem, as well as the input parameters, including the initialization of the matrix W and H and the factorization rank r. In section2, several optimization algorithms are presented. In the last section, we apply the three methods, PCA, ICA, and NMF, to a face images data.

II. Non-Negative Matrix Factorization

A. Problem formulation

The non-negative matrix factorization can be mathematically formulated as a constrained optimization problem below:

$$(P)\begin{cases} \textit{Minimize:} & D(X|WH) \\ \textit{Subject to:} & W \geqslant 0 \\ & H \geqslant 0 \end{cases}$$

where $W, H \geqslant 0$ means that every element of W and H is nonnegative. $D(x|y)$ is a loss function, which is mostly chosen to be Euclidean distance (Euc) $\frac{1}{2}(x-y)^2$, Kullback Leibler divergence (KL) $xlog(\frac{x}{y}) - x + y$ or Itakura-Saito (IS) divergence $\frac{x}{y} - log(\frac{x}{y}) - 1$. The choice of the NMF cost function is made according to the type of data to be analyzed. In this article, the Euclidean distance is selected as the objective function:

$$D(X|WH) = ||X - WH||_F^2 \qquad (2)$$

D is a non-convex function in the two variables W and H. It is, therefore, difficult to find the global minima for (P). Another weakness of the NMF problem is that it is ill-posed (i.e, W and H are non-unique) [17]. For example, given a non-singular matrix A, so, $\hat{W} = WA \geq 0$ and $\hat{H} = A^{-1}H \geq 0$. Thereby $(\hat{W}, \hat{H})$ is anther solution pair. To overcome this problem usually, the recherches add prior knowledge of W and H, such as sparseness or orthogonality.

As a consequence, the optimization problem (P) cannot be solved directly. In literature, Block Coordinate Descent (BCD) is the basic framework for all NMF algorithms, based on the idea that the minimization of the cost function can be achieved by minimizing it in one direction at a time (fixes, for example, W and varies H). In this case, the optimization problem becomes convex. To sum up, we seek the solution of the sub-problems (3)(4), in order to determine the solution of the whole problem.

$$\min_H ||X - WH||_F^2 \quad \text{s.t} \quad H \geq 0 \qquad (3)$$

$$\min_W ||X - WH||_F^2 \quad \text{s.t} \quad W \geq 0 \qquad (4)$$

So, all NMF algorithms are solving (P) iteratively, and if they converge, at that time to local minima.

Initialization of W and H is needed, and a good initialization leads to an efficient local solution as illustrated in Fig.2. Due to the sensitivity of this step, researchers have used a variety of initialization methods [22][23][18], and newly, Flavia Esposito [5] has provided a taxonomy of initialization schemes appearing in the literature.

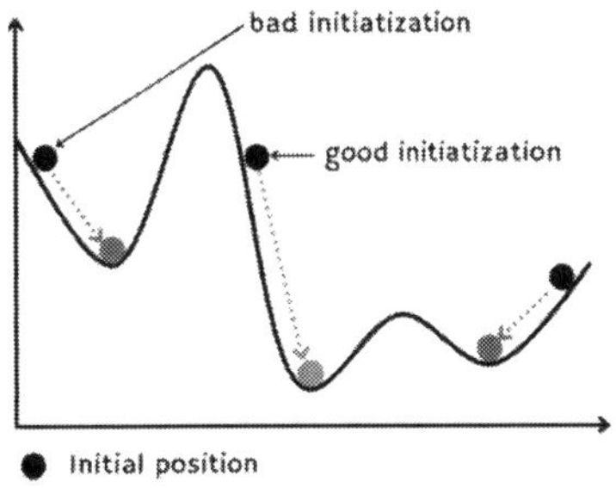

Fig. 2. NMF: bad and good initialisation: case of multiple local optima.

Furthermore, the factorization rank r is another important issue that needs to be fixed. A small value of r may lead to loss of features, and a large value of r, may be responsible for modeling noise. So, the rank r should both reduce the noise in the data and effectively model the key features. The choice of r is generally based on experiments or experience. However, several rank selection techniques have been recently proposed such as [9], [20],[16], [24].

B. Algorithms

1) Projected Gradient Descent: Gradient descent (GD) is a popular optimization algorithm used for solving NMF problem. GD find the minimum of a convex function more quickly by descending, at the current point, in the opposite direction of the gradient of the function.

In our case, the GD is used to solve each sub-problem. We consider the sub-problem (3), the steps of the GD algorithm are as the following:

1) Pick an initial point H_0
2) Loop until stopping condition is met:
   a) Descent direction: pick the descent direction as $\nabla_H D(X, WH)$
   b) Step size: pick a step size η_H
   c) Update: $H \leftarrow H - \eta_H \circ \nabla_H D(X, WH)$

With ∇D is the gradient of the cost function D. The GD update of W is similar to that of H.

The idea of the Projected Gradient Descent (PGD) is simple: if the point $H - \eta_H \circ \nabla_H D(X, WH)$ after the gradient update is leaving the nonnegative space, project it back. So, PGD has one more step which is the projection: $H \leftarrow arg\min_{H \geq 0} ||X - H||_F^2$.

2) Multiplicative Update: Multiplicative update rules (MUR) is the common NMF algorithm and the most widely used due to their simplicity. Multiplicative methods can be obtained in different ways, either by a heuristic approach, or a Majoration-Minimization (MM) approach. We present them successively below.

- **Heuristic update**

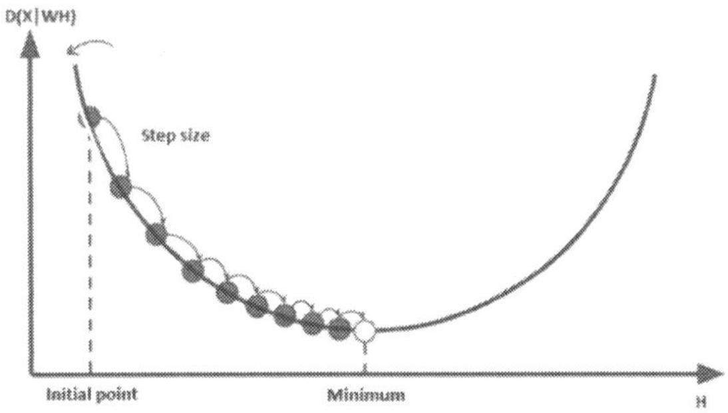

Fig. 3. Gradient Descend Procedure.

Lee and seung (1999) [12] were apparently the first to give the heuristic MU, based on the traditional GD algorithm (presented in the previous section) with an adaptive learning rate (step size). To avoid subtraction in the GD update, in [11] proposed to set the learning rate as:

$$\eta_H = \frac{H}{W^T W H} \qquad \eta_W = \frac{W}{W H H^T}$$

W and H have been derived [1] to minimize the factoriztion error between X and WH. The basic update rules are as follows:

$$H \leftarrow H \circ \frac{W^T X}{W^T W H}$$
$$W \leftarrow W \circ \frac{X H^T}{W H H^T}$$

With $\circ$ (resp. $\div$) denotes elementwise multiplication (resp.division).

- **Majorization-Minimization update**

Geting the MU by Majorization-Minimization (MM) framework means to replace the optimization problem (3 or 4) with a sequence of simpler optimization problems. In other terms, finding a surrogate upper bound of the cost function at each iteration and minimize.
Let

$$C(\theta) = D(X|WH), \ \theta = \{W, H\}$$

thus

$$\min_{H \geqslant 0} D(X|WH) = \min_{\theta} C(\theta) \tag{5}$$

The MM algorithm solving (5) by two main steps:

- 1st ***step*** : Construct a surrogate [2] function $G_k(\theta)$ of $C(\theta)$ at the current iterate θ_k.
- 2nd ***step*** : Minimize the surrogate to get the next iterate: $\theta_{k+1} = arg\,\min_\theta G_k(\theta)$.

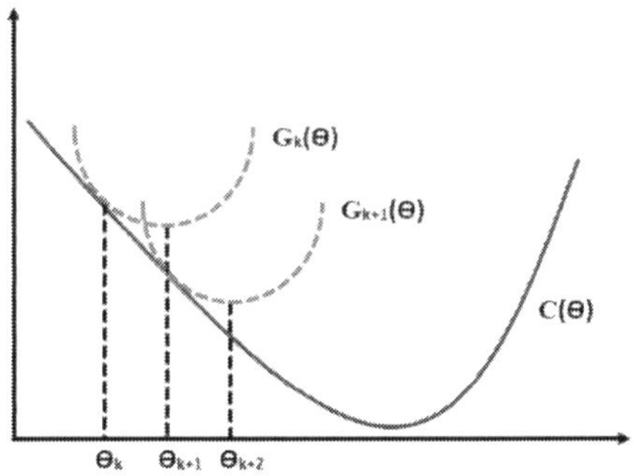

Fig. 4. Majorisation-Minimisation (MM) procedure.

The MM procedure guarantees the cost is non-increasing (monotonic descent property) at each iteration:

$$C(\theta_{k+1}) \leqslant G_k(\theta_{k+1}) \leqslant G_k(\theta_k) \leqslant C(\theta_k) \tag{6}$$

We can also observe this property from figure 4.

Building a surrogate function is a very important step for the MM algorithm, and it's not easy. However, several inequalities used in literature which helps in finding this function, including Jensen's Inequality, Convexity Inequality and Cauchy-Schwarz Inequality. In our case where,
$D(X|WH) = ||X - WH||_F^2$
$= \sum_i \sum_j (x_{ij} - \sum_k w_{ik} h_{kj})^2$
$w_{ik}, h_{kj} \geq 0$
The construction of the surrogate function is done by the convexity of the function $(x_{ij} - x)^2$

$$\left(x_{ij} - \sum_k w_{ik} h_{kj}\right)^2 \leq \sum_k \frac{a_{ikj}^{(t)}}{b_{ij}^{(t)}} \left(x_{ij} - \frac{b_{ij}^{(t)}}{a_{ikj}^{(t)}} w_{ik} h_{kj}\right)^2$$

Where

$$a_{ikj}^{(t)} = w_{ik}^{(t)} h_{kj}^{(t)}, \quad b_{ij}^{(t)} = \sum_k w_{ik}^{(t)} h_{kj}^{(t)}$$

This suggests the alternating multiplicative updates

$$w_{ik}^{(t+1)} = w_{ik}^{(t)} \frac{\sum_j x_{ij} h_{kj}^{(t)}}{\sum_j b_{ij}^{(t)} h_{kj}^{(t)}}$$

$$b_{ij}^{(t+\frac{1}{2})} = \sum_k w_{ik}^{(t+1)} h_{kj}^{(t)}$$

$$h_{kj}^{(t+1)} = h_{kj}^{(t)} \frac{\sum_i x_{ij} w_{ik}^{(t+1)}}{\sum_i b_{ij}^{(t+\frac{1}{2})} w_{ik}^{(t+1)}}$$

The update seems simpler in the equivalent matrix form,

[1]More detail on the MU derivation is given by [2].
[2]A surrogate is a function that approximates another function, it has to verify $G_k(\theta) \geqslant C(\theta)$ and $G_k(\theta_k) = C(\theta_k)$.

$$H \leftarrow H \circ \frac{W^T X}{W^T W H}$$

$$W \leftarrow W \circ \frac{X H^T}{W H H^T}$$

We notice that the MM update coincides with the heuristic update.

The main difference between the MU and PGD algorithms is that, the learning rate of PGD is flexible and that of MU is fixed. Accordingly, MU is slower in convergence than PGD.

3) Alternating Least Squares: The Alternating Least Squares (ALS) algorithm solves the problem (P) iteratively without non-negativity constraint and then projects the solution into a non-negative space. So The update rules are as follows:

$$H \leftarrow max(arg\min_{H} ||X - WH||_F^2, 0)$$

$$W \leftarrow max(arg\min_{W} ||X - WH||_F^2, 0)$$

Generally the ALS algorithm suffers from lack of convergence.

4) Alternating Nonnegative Least Squares: Unlike ALS, the Alternating Nonnegative Least Squares (ALNS) impose the nonnegativity constraints in the sub-problem (3),(4), and find their optimal solution .i.e:

$$H \leftarrow arg\min_{H \geq 0} ||X - WH||_F^2$$

$$W \leftarrow arg\min_{W \geq 0} ||X - WH||_F^2$$

Each step of the update is dedicated to finding the optimal solution to the sub-problem, there are many methods used to solve this bounded constraint problem including PGD [13], Active set [10], Projected quasi-Newton [1], and Projected Barzilai–Borwein [8]. Compared with the MU algorithm, ANLS has relatively faster convergence than MU algorithms.

III. Method

Comparisons between the three matrix factorization methods (i.e. PCA, ICA, and NMF) were made using a face image database. Before applying such a method, the data were converted into greyscale and centered (Fig.5). Also, the law rank (number of components) in the three methods is fixed as r=3. *randomized SVD*, *FastICA*, and *MU* are the used algorithms for solving PCA, ICA, and NMF respectively. Also, we use nndsvd (nonnegative double singular value decomposition) as initialization method for NMF. The source code is available online[3].

[3]https://scikit-learn.org/stable/auto_examples/decomposition/plot_faces_decomposition.html

Fig. 5. First three centered FEI faces.

Fig. 6. The first three components given by PCA.

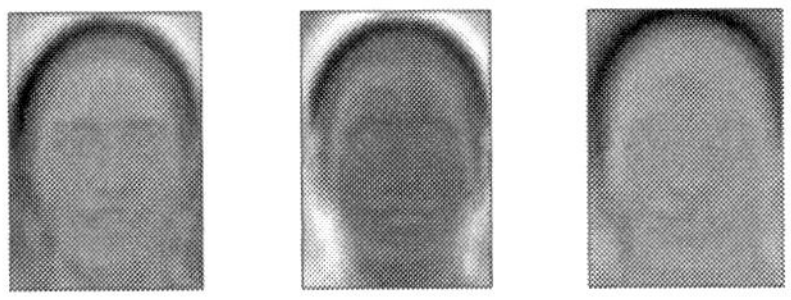

Fig. 7. The first three components given by ICA.

Fig. 8. The first three components given by NMF.

In this experiment, we have used images of a FEI face database maintained by the Department of Electrical Engineering of FEI[4], São Paulo, Brazil. The dataset contains 200 face images (100 men and 100 women), where the size of each image is 360 × 260 pixels.

IV. Result and Discussion

The first three components given by PCA, ICA, and NMF are shown in the Fig.6, Fig.7 and Fig.8 respectively. From these figures, we see that PCA was unable to reconstruct the faces because of the noisy components. ICA lost some important features in the faces and the images seem unclear. However, the images extracted from NMF represent more caractéristiques and details of the face. So, the faces are better reconstructed than PCA and ICA.

Moreover, PCA needs 2.018s for training to give the extracted components and ICA 4.032s. In contrast, NMF takes more time which is 10.609s.

V. Conclusion

In this work, we describe the NMF problem and its famous algorithms such as MU, PGD, and ANLS. We also compared

[4]This database is available on https://fei.edu.br/~cet/facedatabase.html.

NMF with PCA and ICA methods using a face image database. PCA and ICA show a significant loss of face information. On the other hand, NMF was able to extract face features and retain more information after reducing the dimensionality.

References

[1] Pierre Ablin, Dylan Fagot, Herwig Wendt, Alexandre Gramfort, and Cédric Févotte. A quasi-newton algorithm on the orthogonal manifold for nmf with transform learning. In *ICASSP 2019-2019 IEEE International Conference on Acoustics, Speech and Signal Processing (ICASSP)*, pages 700–704. IEEE, 2019.

[2] Juan José Burred. Detailed derivation of multiplicative update rules for nmf. *Paris, France*, 2014.

[3] Yong Chen, Hui Zhang, Rui Liu, Zhiwen Ye, and Jianying Lin. Experimental explorations on short text topic mining between lda and nmf based schemes. *Knowledge-Based Systems*, 163:1–13, 2019.

[4] Chen Cui, Xujun Wu, Jun Yang, and Juyan Li. A novel dibr 3d image hashing scheme based on pixel grouping and nmf. *Wireless Communications and Mobile Computing*, 2020, 2020.

[5] Flavia Esposito. A review on initialization methods for nonnegative matrix factorization: Towards omics data experiments. *Mathematics*, 9(9):1006, 2021.

[6] Yixian Fang, Huaxiang Zhang, and Yuwei Ren. Graph regularised sparse nmf factorisation for imagery denoising. *IET Computer Vision*, 12(4):466–475, 2018.

[7] Cédric Févotte, Emmanuel Vincent, and Alexey Ozerov. Single-channel audio source separation with nmf: divergences, constraints and algorithms. *Audio Source Separation*, pages 1–24, 2018.

[8] Yakui Huang, Hongwei Liu, and Shuisheng Zhou. Quadratic regularization projected barzilai–borwein method for nonnegative matrix factorization. *Data mining and knowledge discovery*, 29(6):1665–1684, 2015.

[9] Yu Ito, Shin-ichi Oeda, and Kenji Yamanishi. Rank selection for non-negative matrix factorization with normalized maximum likelihood coding. In *Proceedings of the 2016 SIAM International Conference on Data Mining*, pages 720–728. SIAM, 2016.

[10] Hyunsoo Kim and Haesun Park. Nonnegative matrix factorization based on alternating nonnegativity constrained least squares and active set method. *SIAM journal on matrix analysis and applications*, 30(2):713–730, 2008.

[11] Daniel Lee and H Sebastian Seung. Algorithms for non-negative matrix factorization. *Advances in neural information processing systems*, 13, 2000.

[12] Daniel D Lee and H Sebastian Seung. Learning the parts of objects by non-negative matrix factorization. *Nature*, 401(6755):788–791, 1999.

[13] Chih-Jen Lin. Projected gradient methods for non-negative matrix factorization. *Neural computation*, 19(10):2756–2779, 2007.

[14] Xianzhong Long, Hongtao Lu, Yong Peng, and Wenbin Li. Graph regularized discriminative non-negative matrix factorization for face recognition. *Multimedia tools and applications*, 72(3):2679–2699, 2014.

[15] Andri Mirzal. Nmf versus ica for blind source separation. *Advances in Data Analysis and Classification*, 11(1):25–48, 2017.

[16] Laura Muzzarelli, Susanne Weis, Simon B Eickhoff, and Kaustubh R Patil. Rank selection in non-negative matrix factorization: systematic comparison and a new mad metric. In *2019 International Joint Conference on Neural Networks (IJCNN)*, pages 1–8. IEEE, 2019.

[17] Weiwei Pan and Finale Doshi-Velez. A characterization of the non-uniqueness of nonnegative matrix factorizations. *arXiv preprint arXiv:1604.00653*, 2016.

[18] Hanli Qiao. New svd based initialization strategy for non-negative matrix factorization. *Pattern Recognition Letters*, 63:71–77, 2015.

[19] Yaser Esmaeili Salehani, Ehsan Arabnejad, Abderrahmane Rahiche, Athmane Bakhta, and Mohamed Cheriet. Msdb-nmf: Multispectral document image binarization framework via non-negative matrix factorization approach. *IEEE Transactions on Image Processing*, 29:9099–9112, 2020.

[20] Steven Squires, Adam Prügel-Bennett, and Mahesan Niranjan. Rank selection in nonnegative matrix factorization using minimum description length. *Neural computation*, 29(8):2164–2176, 2017.

[21] Ding Tu, Ling Chen, Mingqi Lv, Hongyu Shi, and Gencai Chen. Hierarchical online nmf for detecting and tracking topic hierarchies in a text stream. *Pattern Recognition*, 76:203–214, 2018.

[22] Stefan Wild, James Curry, and Anne Dougherty. Improving non-negative matrix factorizations through structured initialization. *Pattern recognition*, 37(11):2217–2232, 2004.

[23] Yun Xue, Chong Sze Tong, Ying Chen, and Wen-Sheng Chen. Clustering-based initialization for non-negative matrix factorization. *Applied Mathematics and Computation*, 205(2):525–536, 2008.

[24] Xiaohui Yang, Wenming Wu, Xin Xin, Limin Su, and Liugen Xue. Adaptive factorization rank selection-based nmf and its application in tumor recognition. *International Journal of Machine Learning and Cybernetics*, pages 1–19, 2021.

Parallel Hybrid 2-Opt Flower Pollination Algorithm for Real-Time UAV Trajectory Planning on GPU

Vincent Roberge[1*], and *Mohammed* Tarbouchi[1]

[1]Royal Military College of Canada, Department of Electrical and Computer Engineering, Canada

Abstract. The development of autonomous Unmanned Aerial Vehicles (UAVs) is a priority to many civilian and military organizations. An essential aspect of UAV autonomy is the ability for automatic trajectory planning. In this paper, we use a parallel Flower Pollination Algorithm (FPA) to deal with the problem's complexity and compute feasible and quasi-optimal trajectories for fixed-wing UAVs in complex 3D environments, taking into account the vehicle's flight properties. The global optimization algorithm is improved with the addition of 2-opt local search providing a significant improvement. The proposed trajectory planner in implemented and parallelized on a multicore processor (CPU) using OpenMP and a Graphics Processing Unit (GPU) using CUDA resulting in a 9.6x and a 68.5x speedup respectively compared to the sequential implementation on CPU.

Index Terms—Flower Pollination Algorithm, Graphics Processing Unit, Parallel Programming, Trajectory Planning, Unmanned Aerial Vehicle.

1 Introduction

Unmanned Aerial Vehicles (UAVs) have gained importance in military operations for a variety of reasons. Compared to manned aircraft, UAVs are small, harder to detect, inexpensive to acquire and operate, capable of reliably performing repetitive tasks, and, perhaps most importantly, do not expose aircrew to enemy threats. Although UAVs are unmanned, they still need to be operated remotely. To alleviate this task and improve the autonomy of UAVs, researchers have developed different approaches to automatic UAV trajectory planning. Many of the earlier works used simplistic model and were limited to 2D terrains. On the other hand, modern approaches strive to include the dynamic properties of the UAV, more realistic 3D terrains and multi-objective optimization functions. In the last decade, a shift from deterministic to non-deterministic methods, primarily metaheuristics, has been observed in response to this increased complexity [1]. Metaheuristics are generic optimization algorithms that depend on the iterative improvement of one or more candidate solutions in order to arrive at a near-optimal solution. Recent examples of metaheuristics used for UAV trajectory planning are the simulated annealing algorithm [2], the differential evolution algorithm [3], the genetic algorithm (GA) [4], the particle swarm optimization (PSO) [5], the ant colony optimization [6], the central force optimization [7], the predator-prey [8], the bee colony [9], and the grey wolf [10]. Despite their effectiveness, they require significant computation power which often results in an execution time that is too long for in-flight planning. To reduce the execution time, we have previously developed in [11–12] parallel implementations of the GA and the PSO for UAV trajectory planning using multicore processors (CPU), but the performance is still too slow to allow for real-time computation.

The Flower Pollination Algorithm (FPA) is another metaheuristic that has shown in many recent studies its superiority over the GA and the PSO [13–15].

The FPA algorithm is a relatively new metaheuristic that imitates the process of flower pollination to solve optimization problems and is shown to perform better than GA and PSO [13–15]. The benefit of FPA is its ability to efficiently use both global and local search for better refinement of the solutions. The local/global search traits which FPA possesses has allowed it to be used successfully in numerous fields and provide optimal solutions with minimal drawbacks [13–15]. Although various variants and applications of FPA are mentioned, very little focus was given to the applications of FPA to solve the complex problems of UAV trajectory planning in 3D environment. For that, an efficient application of FPA to successfully determine optimal trajectory for fixed-wing UAVs is needed. In this paper, we propose a parallel FPA on multicore CPU using OpenMP and on GPU using CUDA to reduce the execution time and to allow for real-time UAV trajectory planning.

The remainder of the document is organized into sections. Section 2 presents the architectures of GPUs. Section 3 covers the details of the proposed method including the solution encoding, the fitness function, the FPA and the 2-opt operator. We present in Section 4 our

* Corresponding author: vincent.roberge@rmc.ca

approach to parallelizing the FPA and the 2-opt operator. Finally, we present and discuss in section 5 the results of our numerical simulation.

2 Graphics Processing Unit

GPUs were originally built using fixed pipelines optimized for a very high throughput of operations to support graphics applications. With a constant demand from the gaming industry for new graphics features a more flexible design was needed and pipelines were replaced by programmable processors. This change has allowed GPUs to be used for scientific computing applications. Initially limited to integer calculations, the processors found on today's GPUs support single and double floating-point operations and even special functions. Equipped with thousands, and in some instances, tens of thousands of cores, GPUs have an extremely high processing capability, much more than what is typically found in standard CPUs, given that the algorithms they run can be efficiently parallelized for their architecture. The architecture of an NVIDIA GPU is shown in Fig. 1. The GPU is composed of several building blocks containing Streaming Multiprocessors (SM). These SMs have the necessary components to perform highly threaded computations such as Streaming Processors (SP), registers, control logic, and Shared Memory. The SM is interconnected to a Global Memory which receives and transfers data to and from the host computer.

On NVIDIA GPUs, a programmer can use the Compute Unified Device Architecture (CUDA) to code sequential instructions to be run on the CPU, and *kernels* which are functions that run in parallel on the GPU. When a kernel is called from the CPU, threads are launched on the GPU to execute the kernel. Threads are organized in blocks and are mapped to SMs. All threads in the same thread block can communicate together through their Shared Memory. Threads from different thread blocks can communicate through Global memory, but the consistency of the Global Memory is only guaranteed at kernel completion.

In this work, the GPU is particularly well suited to implement a parallel FPA as the algorithm exhibits a high level of parallelism and the GPU contains a very large number of compute cores which allows the parallel FPA to run efficiently.

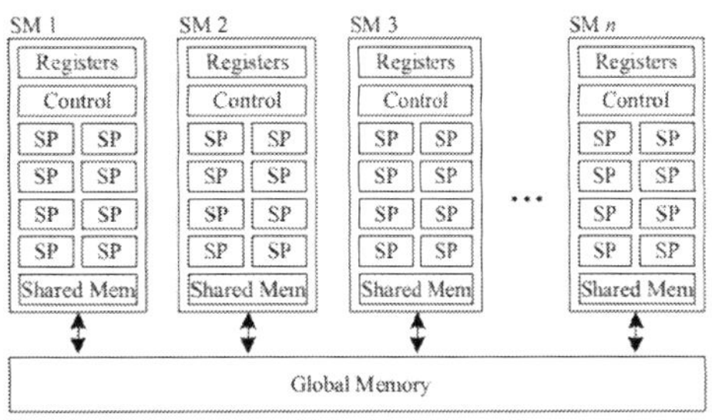

Fig. 1. GPU Architecture

3 Proposed Optimization Strategy

This section covers the optimization strategy proposed including the environment representation, the encoding of the candidate solutions, the fitness function, the FPA and the 2-opt operator.

3.1 Solution Encoding and Environment Representation

In the FPA, each candidate solution represents a path connecting the starting point to the end point. The path is defined as a series of waypoints in a 3D environment. The (x, y, z) coordinates of the waypoints are stored sequentially in a vector as in $[x_0, y_0, z_0, x_1, y_1, z_1 \ldots, x_n, y_n, z_n]$. The terrain is represented using a 2D matrix where each element of the matrix represents the elevation of the terrain at that location. Some of the terrains used in this work are fictional while others are real. Elevation maps for real terrains were taken from the Canadian Digital Elevation Model Repository [16]. The environment representation used here also allows the definition of no-fly or danger zones in the form of cylindrical areas of infinite height. These zones are inputted by identifying the (x, y) coordinates of the center of the cylinder and its radius. From these cylinders, a no-fly zone matrix is built. This matrix has the same dimensions as the elevation matrix and contains a 0 for all locations that are not in a no-fly zone and a 1 for location that falls in a no-fly zone. Using a matrix, speedup the remaining of the calculations.

By encoding the solutions as a series of waypoints, the trajectories are composed uniquely of line segments and therefore include discontinuities at every waypoint. To be flyable by fixed-wing UAVs, these trajectories, must first be smoothed. In this work, we use a technique proposed by Labonté in [17] to connect the line segments using circular arcs as shown in Fig. 2. This technique has the advantageous of generating simple trajectories. The smoothing operation is performed on every candidate solutions before its fitness is computed.

3.2 Fitness Function

Once a candidate solution has been smoothed, its fitness can be computed. This process is divided into two parts: the calculation of a penalty term and the calculation of a cost term.

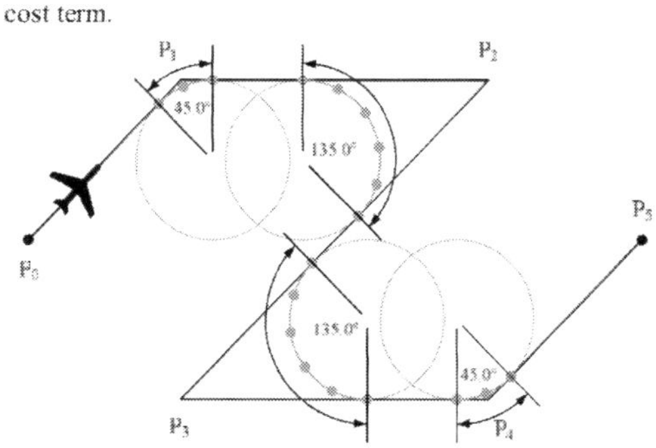

Fig. 2. Smoothing of a 6-waypoint trajectory using circular arcs

3.2.1 Penalty Term

The first part consists of computing the penalty term P which is used to penalize solutions that violate a feasibility constraint. This term is computed as follows:

$$P = d_{ut} + d_{nf} + d_{ea} + \left(\frac{N_{impossible\ arcs}}{N_{arc}} * l_{traj}\right) \quad (1)$$

where d_{ut} is the distance flown under the terrain, d_{nf} is the distance flown through a no-fly zone, d_{ea} is the distance flown at an excessive descent or climbing angle as defined by the UAV characteristics, $N_{impossible\ arcs}$ is the number of circular arcs that could not be constructed to connect two linear segments due to the incoming or outgoing segment being too short, N_{arc} is the total number of circular arcs that were attempted to be constructed and l_{traj} is the total length of the trajectory. The number of impossible arcs is normalized and multiplied by the length of the trajectory so that the last term of the equation has the same units as the other terms in the equations (i.e., meters). To calculate d_{ut} and d_{nf} we use both the elevation matrix and the no-fly zone matrix and we check at regular intervals along each segment forming the trajectory to see if the segment hits the terrain or breaches a no-fly zone.

3.2.2 Cost Term

The second part of the fitness function is calculating the cost term which assesses the quality of feasible solutions and represents the optimization objectives. In this work, the optimization objectives are to minimize the distance time altitude of the trajectory to avoid detection by enemy radars and to minimizing the fuel consumption to increase the range of the UAV. It is important to note that one cannot simply minimize the overall average altitude of a trajectory as the final path could include multiple loops at low altitude in an effort to reduce the overall altitude of the path which is obviously not wanted. These optimization objectives used here are problem specific and could be changed based on the missions. The equation for the cost term used in this work is as follows:

$$C = w_1 * \frac{a_{avg} * l_{traj}}{a_{utopia} * l_{utopia}} + w_2 * \frac{f_{req}}{f_{utopia}} \quad (2)$$

where a_{avg} is the average altitude of the trajectory, l_{traj} is the length of the trajectory, f_{req} is the quantity of fuel required to fly the trajectory calculated using [18], w_1 and w_2 are coefficients set to 0.5, but that can be adjusted to define the relative importance of the two optimization objectives. It is important to note that the altitude, length and fuel terms must be normalized before they can be added as they do not have the same units. The normalization is based on [19] and uses utopia points as references. We defined the utopia altitude a_{utopia} as the average altitude of the fictional line segment connecting the start point to the end point. Likewise, we defined the utopia length l_{utopia} and utopia fuel f_{utopia} as the length of that fictional segment and the fuel required to fly that fictional segment.

3.2.3 Combined Fitness Function

The fitness function $F(\vec{x})$ of candidate solution $\vec{x}$ is a combination of the penalty term and the cost term and is defined as follows:

$$F(\vec{x}) = \begin{cases} 0 + \dfrac{1}{1+P} & , \quad P > 0 \\ 1 + \dfrac{1}{1+C} & , \quad P = 0 \end{cases} \quad (3)$$

This function is carefully built so that infeasible solutions have a fitness ranging between 0 and 1 and feasible solutions have a fitness ranging between 1 and 2. During its operation, the FPA uses this fitness function to guide the search towards optimized solutions until an optimum is found.

3.3 Flower Pollination Algorithm

The FPA was first published by Xin-She Yang in 2012 [20] and has since been used in a wide range of fields. It is based on pollination process of flowering plants and uses global and local pollination to improve a population of candidate solutions. The global pollination step is represented by a pollinator (such as insects or wind) transporting flower pollen over a long distance, and this approach guarantees that the fittest pollen of high quality is transferred over to the next generation by following the biotic method's:

$$x_i^{t+1} = x_i^t + L(\lambda) * (x_i^t - g^*) \quad (4)$$

where x_i^t is the ith pollen or solution at iteration t, g^* is the current best solution and L(λ) is the lévy flight distribution. Lévy flight is used to imitate the movement of insects over short and long distances.

Local pollination (self-pollination) and flower constancy can be given as:

$$x_i^{t+1} = x_i^t + \epsilon * (x_j^t - x_k^t) \quad (5)$$

where x_j^t and x_k^t are pollens selected randomly from different flowers, while ϵ is a random number that follows the uniform distribution in [0, 1].

In general, the flower pollen population or solutions are randomly initialized by FPA. A new solution is generated for each generation using either global pollination or local pollination, as determined by the switch probability p [0, 1]. The FPA flowchart is given in Fig. 4.

3.4 2-Opt Operator

In the proposed FPA-based algorithms, solutions are randomly initialized and their waypoints are erratically placed on the map. This means that trajectories initially include several loops. It is hard for the FPA to remove these loops, but there exist local search algorithms that are well suited for this task. One of them is the 2-opt operator [21]. The 2-opt operator is a deterministic local search algorithm that is often used to solve the Travelling Salesman Person. Starting from the first node, it visits the next node and generates a new solution by reversing the order of the sub-trajectory and verifying if the new trajectory has a better fitness. If it does, the new trajectory replaces the solution and the 2-opt

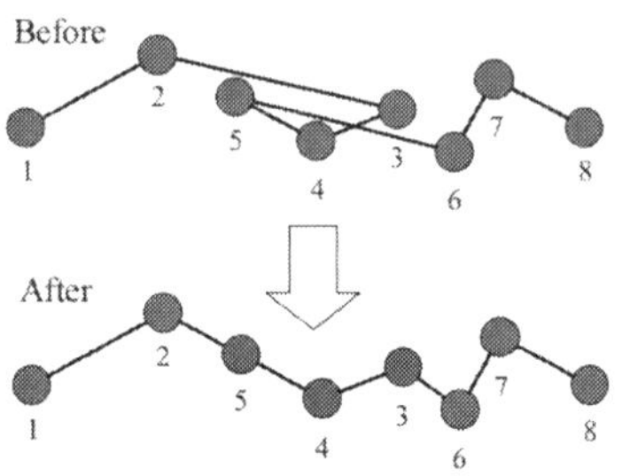

Fig. 3. 2-opt operator to remove loops

algorithm restart the search from the first node. If the new trajectory is not better, the search continues to the next node. The algorithm is implemented as two nested loops and the search restarts when a better solution is found. The details of the 2-opt algorithm is covered in [21].

An example of the 2-opt algorithm applied to the UAV trajectory planning problem is shown in Fig. 3 where it successfully removes a loop from the trajectory. Because the 2-opt operator requires the evaluation of the candidate solution after every sub-trajectory reversing, it is a very expensive function computationally. To reduce its effect on the overall runtime of the algorithm, we run the 2-opt operator every 20 iterations of the FPA as identified in Table 3. This has the benefit of using the 2-opt without affecting too much the overall runtime. Another important note regarding the 2-opt is that the number of iterations required to optimize each solution is different. This has implications for the parallelization strategy as it creates work imbalance between threads. Our parallelization strategy deals with this work imbalance.

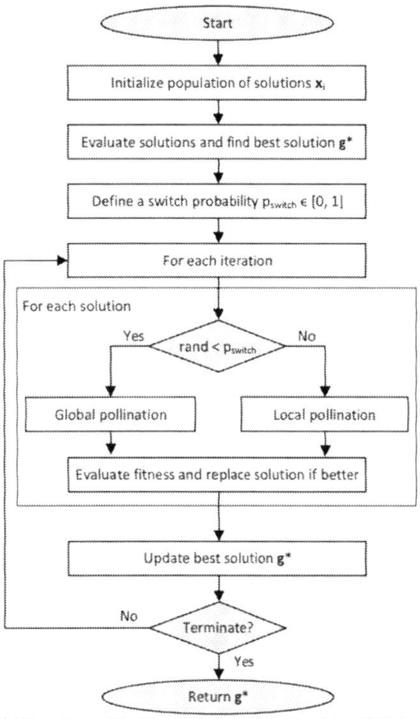

Fig. 4. Flowchart of the Flower Pollination Algorithm (FPA)

4 Parallel Implementations

In this work, two parallel implementations were developed: one for multicore CPU using OpenMP and one for GPU using CUDA. As shown in Fig. 4, at each iteration of the FPA, a for-loop is used to improve each candidate solution. The 2-opt operator is also implemented as a for-loop operation where the 2-opt operator is applied independently on each solution.

In the parallel implementation on multicore CPU, the iterations of the for-loops are distributed among multiple OpenMP threads. Because the number of solutions (i.e., the number of iterations in the for-loop) is much larger than the number of threads, work imbalance between the iterations is easily hidden resulting in a very efficient parallelization strategy.

In the parallel implementation on GPU, a much higher level of parallelism must be exploited to benefit from the massively parallel architecture of the GPU. The check for the p_{switc} is done using one thread per solution, the global and local pollination is done using one thread per dimension of each solution and the replacement of the solutions is also done using one thread per dimension of each solution.

Calculating the fitness of the candidate solutions involves several operations and one thread block is used for each candidate solution. For the smoothing of the trajectory, an initial run is performed to compute the number of points needed for each circular arc, then, each thread processes one or more points of the arc. The result of the smoothing is a trajectory with a large number of linear segments. For the calculation of d_{ut} and d_{nf}, the number of points to be verified along each linear segment from the smoothed trajectory is first calculated, then each thread verifies one or more point. To calculate d_{ea}, a_{avg} and f_{req}, each thread verifies one or more linear segment from the smoothed trajectory. To compute $N_{impossible\ arcs}$, each thread computes one or more waypoints from the original trajectory. Finally, to compute the overall fitness, the thread block performs several parallel reductions to compute the sums of all the terms found on all segments from the smoothed trajectory. The calculation of the fitness therefore uses in a very large number of CUDA threads which exploits the massively parallel architecture of the GPU.

Parallelizing the 2-opt operator on the GPU is a challenge as each solution requires a different number of iterations which creates work imbalance between the solutions. To deal with this, we use a feature from NVIDIA GPUs called *Dynamic Parallelism* which allows kernels to launch child kernels. With Dynamic Parallelism, we launch a first kernel with one thread block per solution. Each thread block contains only a single thread which runs the 2-opt operator for the solution. To evaluate the fitness of the candidate trajectories, the single thread launches a child kernel composed of a single block and 128 threads which performs the fitness calculation as explained in the previous paragraph. Using Dynamic Parallelism allows the GPU to dynamically schedule all the child kernels and handle the work imbalance between solutions. Solutions that complete the 2-opt operator early terminates and other solutions can continue longer.

5 Experimental Results

5.1 Quality of Computed Solutions

To validate the proposed trajectory planning algorithm based on the FPA and 2-opt operator, three tests were performed. The first test aims to assess the quality of the solutions computed by the proposed FPA-2-opt. Because each previous paper used their own maps and own UAV model, it is difficult to replicate and compare our work to previous works. For this reason, our assessment is qualitative and based on the feasibility of the final solutions and their quality as viewed on plotted 3D figures. For this test, we used the model for a small-scale UAV similar to the Silver Fox manufactured by BAE Systems. The specifications for the UAV are listed in Table 1. These specifications are useful to compute the fuel required to fly the trajectory. The flight properties of the UAV-autopilot system which models the dynamic behavior of the UAV are listed in Table 2. These are useful when assessing the feasibility of a trajectory and when smoothing the trajectory. To test the FPA-2-opt algorithm, four maps were used. Two are synthetic maps, the third one is from mountainous regions in North-East Afghanistan and the last one is from northern Russia, East of Norilsk. The FPA-2-opt algorithm was configured using the parameters listed in Table 3. The computed trajectories are shown in Fig. 5 to Fig. 8. All computed trajectories are feasible, do not collide with the terrain, avoid the no-fly zones, respect the maximum ascent and descent rates of the UAV. They look optimized in the sense that the minimize the average altitude and the overall fuel consumption as defined by the optimization objectives and they are smoothed using the minimum turning radius of the UAV demonstrating the good functioning of the proposed algorithm.

Table 1. Specifications of the Silver Fox-like UAV [18]

Parameter	Value
Weight (N)	113
Weight of fuel (N)	19
Wing span (m)	2.4
Wing area (m^2)	0.768
Global lift coefficient	1.26
Drag coefficient at zero lift	0.0251
Engine power (W)	1850
Specific fuel consumption (N/W*s)	*1.1569E-6
Propeller efficiency	0.8

* Modified from reference

Table 2. Flight properties of the UAV-autopilot system

Parameter	Value
Cruising speed (m/s)	23
Turning radius (m)	150
Max ascent angle (degrees)	10
Max descent angle (degrees)	30

Table 3. Flower Pollination Algorithm parameters

Parameter	Value
Number of iterations	500
Number of solutions	512
Switching probability	0.5
Initial epsilon	0.25
Final epsilon	0.02
2-opt frequency	Every 20th iteration

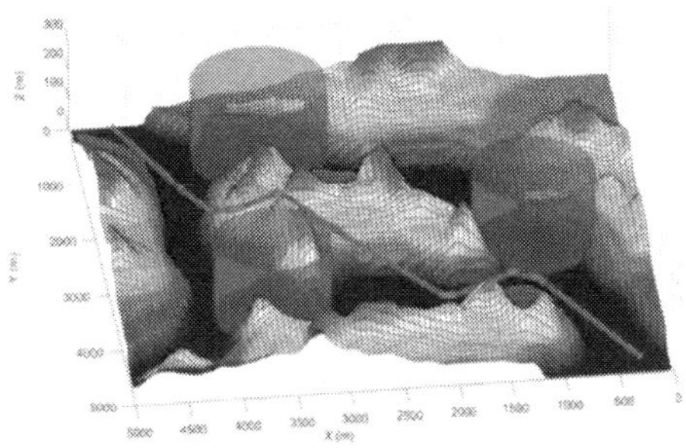

Fig. 5. UAV trajectory computed by the FPA for the canyon map (synthetic map, 25 km^2, terrain elevation from 0 to 260 m AMSL)

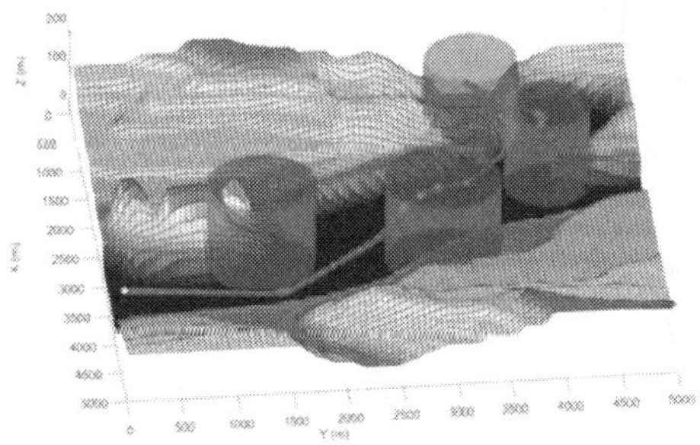

Fig. 6. UAV trajectory computed by the FPA for the river map (synthetic map, 25 km^2, terrain elevation from 0 to 217 m AMSL)

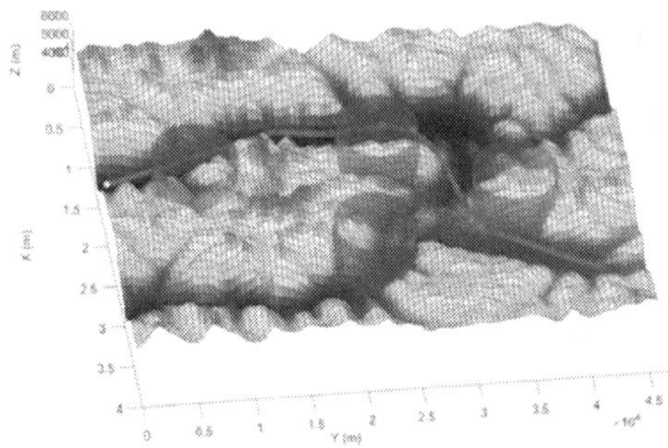

Fig. 7. UAV trajectory computed by the FPA in North-East Afghanistan (Top-left corner is 37.0705N-73.0772E, 1,672 km², terrain elevation from 3,139 to 6,082 m AMSL)

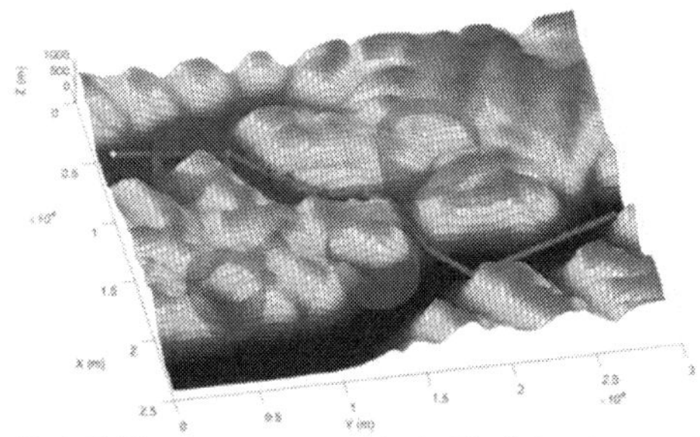

Fig. 8. UAV trajectory computed by the FPA in Russia, East of Norilsk (Top-left corner is 69.2464N 91.1934E, 716 km², terrain elevation from 40 to 1,382 m AMSL)

5.2 Benefit of the 2-Opt Operator

In the second test, we verify the advantage of the 2-opt operator. For this, we run the FPA algorithm 50 times without and with the 2-opt operator for each of the four maps and lists in Table 4 the statistics about the solutions found. To compare the fitness of the solutions computed without and with the 2-opt operator, we use a T-test and compute the p-value between the two distributions. A p-value below 0.05 means that the two means are statistically different. In the case of the data listed in Table 4, the mean fitness of the solutions found by the FPA with 2-opt is higher for all four maps and the p-value is zero for all four maps. This means that in all cases, using the 2-opt generates solutions that are statistically better proving the superiority of the hybrid FPA-2-opt compared to the FPA alone. As explained in section 3.4, the 2-opt operator is very efficient at removing loops from the candidate trajectories which are randomly initialized at the beginning of the search.

5.3 Performance of Parallel Implementation

The final test aims to evaluate the performance improvement brought by the two parallel implementations. The test is done on a Dell 5820 workstation equipped with an Intel Xeon W-2195 processors with 18 cores running at a base frequency of 2.30 GHz and a turbo frequency of 4.30 GHz. The computer is configured with an NVIDIA RTX A6000 GPU with 10,752 running at 1.41 GHz. Given the CPU frequency, the maximum theoretical speedup of a parallel application is:

$$N * \frac{F_{turbo}}{F_{base}} = 18 * \frac{2.3}{4.3} = 9.63x \quad (6)$$

In this test, we ran the FPA-2-opt algorithm 10 times using 1 to 36 threads and have measured the runtime which is plotted in Fig. 9. It can be noted that there is a performance slow done when the number of threads approaches the number of physical cores, but that a maximum speedup of 9.6x is achieved when the number of threads is equal to the number of logical cores of the CPU (here, the CPU supports hyperthreading). This maximum speedup is slightly above to the theoretical speedup which means that the proposed FPA-2-opt parallelizes very efficiently on multicore CPUs.

Next, the second parallel implementation, which is the CUDA version, is run on the RTX A6000 GPU. This version is much more complex, but benefits for a much more parallel processor. Now, the number of threads cannot be controlled, as the entire GPU is used, but we vary the number of solutions to demonstrate the point at which the GPU becomes saturated and maximum speedup is achieved. In the case of the RTX A6000 GPU, this happens with 512 solutions. Because our algorithm uses one thread block per solution, hundreds of threads are launched for each solution so saturation occurs when several thousand threads are launched. The proposed GPU implementation has a speedup of 68.5x compared to the sequential implementation on CPU and runs in just 1.14 seconds compared to 78.4 seconds for the CPU for 512 solutions. Because UAVs operate in a dynamic environment and often need to recompute their trajectory in flight, this speedup is truly an advantage.

Table 4. Mean results of 50 trials for the FPA with and without 2-opt

Scenario	FPA without 2-opt				FPA with 2-opt				T Test
	Length (m)	Average Alt (m)	Fuel (Newton)	Fitness (std dev)	Length (m)	Average Alt (m)	Fuel (Newton)	Fitness (std dev)	p-value
Canyon	8,418	119.3	0.127	1.295±0.021	7,320	86.6	0.117	1.348±0.019	0.00
River	6,559	72.2	0.099	1.363±0.025	6,126	50.0	0.090	1.420±0.018	0.00
Afghanistan	52,874	4236.2	1.003	1.451±0.024	52,737	4049.6	0.992	1.461±0.010	0.00
Russia	34,031	445.6	0.489	1.346±0.026	33,386	322.0	0.477	1.395±0.017	0.00

https://doi.org/10.1051/itmconf/20224803007

5.4 Real-Time Performance

The real-time performance of the proposed algorithm is assessed in terms of the distance travelled by the UAV during the time it takes to compute a new trajectory. This distance is computed based on the velocity of the UAV and the computation time measured in section 5.3. In this work, because the UAV flies at constant velocity, the distance can be computed using:

$$d = v * t = 23\ m/s\ * 1.14\ s\ = 26.22\ m \qquad (7)$$

This means that the Silver Fox-like UAV used in this work travels only 26.22 meters during the times it takes for a new trajectory to be computed. This distance is longer than a typical trajectory which means that the trajectory planning system will always be able to compute the next trajectory while the UAV is flying the current trajectory. Also, assuming that the onboard sensors are able to detect obstacles at a distance greater than 26.22 meters, this gives enough lead time for the trajectory planning system to replan the trajectory in case a new obstacle is detected. Based on those assumptions, we conclude that the parallel implementation on GPU allows for real-time trajectory planning which is not the case for the sequential implementation on CPU.

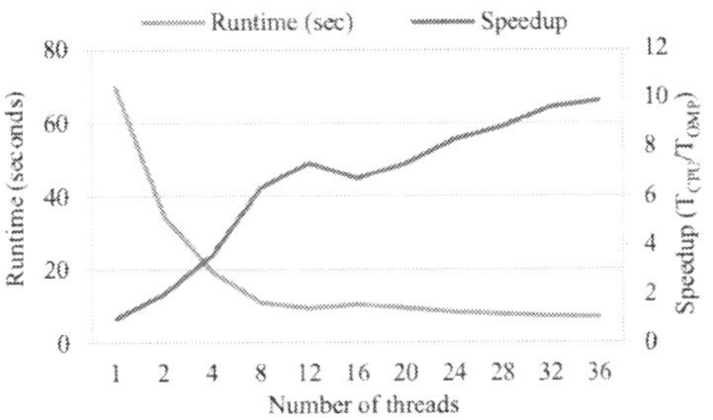

Fig. 9. Runtime and speedup for the parallel implementation on multicore CPU using OpenMP

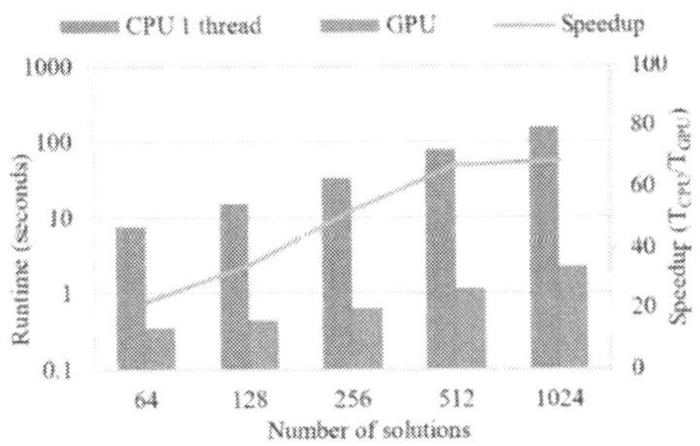

Fig. 10. Runtime and speedup for the parallel implementation on GPU using CUDA

6 Conclusion

This paper presented a parallel trajectory planning algorithm for fixed-wing UAVs based on the FPA and the 2-opt operator. The FPA is a global optimization algorithm while the 2-opt operator provides a local search strategy. The proposed approach was tested in a numerical simulation using four realistic and complex 3D scenarios and successfully computed quasi-optimal trajectories for the UAV while considering the flying characteristics of the UAV such as its minimum turning radius and maximum ascent and descent rates. To cope with the complexity of the problem and allow for real-time planning, the algorithm was parallelized on multicore CPU using OpenMP and on GPU using CUDA. A speedup of 9.6x on the CPU and 68.5x on the GPU were measured. The algorithms on GPU runs in just 1.14 seconds compared to 78.4 seconds for the sequential version on CPU. The proposed trajectory planning module contributes to the development of autonomous UAVs by allowing them to plan their route while in flight. In future works, we intent to implement and test other metaheuristics for UAV trajectory planning to compare their efficiency in this particular application.

Acknowledgement

We gratefully acknowledge the support of NVIDIA Corporation who donated the RTX A6000 GPU used for this research.

References

1. E. Masehian and D. Sedighizadeh, "Classic and Heuristic Approaches in Robot Motion Planning – A Chronological Review," *World Acad. Sci. Eng. Technol.*, vol. 1, no. 5, 2007.
2. M. Colnarič, L. P. Behnck, D. Doering, C. E. Pereira, and A. Rettberg, "A Modified Simulated Annealing Algorithm for SUAVs Path Planning," in *2nd IFAC Conference on Embedded Systems, Computer Intelligence and Telematics*, Maribor, Slovenia, Jun. 2015, vol. 48, pp. 63–68. doi: 10.1016/j.ifacol.2015.08.109.
3. X. Zhang and H. Duan, "An improved constrained differential evolution algorithm for unmanned aerial vehicle global route planning," *Appl. Soft Comput.*, vol. 26, pp. 270–284, Jan. 2015, doi: 10.1016/j.asoc.2014.09.046.
4. Y. V. Pehlivanoglu, "A new vibrational genetic algorithm enhanced with a Voronoi diagram for path planning of autonomous UAV," *Aerosp. Sci. Technol.*, vol. 16, no. 1, pp. 47–55, Jan. 2012, doi: 10.1016/j.ast.2011.02.006.
5. Y. Fu, M. Ding, and C. Zhou, "Phase Angle-Encoded and Quantum-Behaved Particle Swarm Optimization Applied to Three-Dimensional Route Planning for UAV," *IEEE Trans. Syst. Man Cybern. - Part Syst. Hum.*, vol. 42, no. 2, pp. 511–526, Mar. 2012, doi: 10.1109/TSMCA.2011.2159586.
6. U. Cekmez, M. Ozsiginan, and O. K. Sahingoz, "A UAV path planning with parallel ACO algorithm on CUDA platform," in *2014 International Conference on Unmanned Aircraft Systems (ICUAS)*, May 2014, pp. 347–354. doi: 10.1109/ICUAS.2014.6842273.

7. Y. Chen, J. Yu, Y. Mei, Y. Wang, and X. Su, "Modified central force optimization (MCFO) algorithm for 3D UAV path planning," *Neurocomputing*, vol. 171, pp. 878–888, Jan. 2016, doi: 10.1016/j.neucom.2015.07.044.
8. W. Zhu and H. Duan, "Chaotic predator–prey biogeography-based optimization approach for UCAV path planning," *Aerosp. Sci. Technol.*, vol. 32, no. 1, pp. 153–161, Jan. 2014, doi: 10.1016/j.ast.2013.11.003.
9. C. Xu, H. Duan, and F. Liu, "Chaotic artificial bee colony approach to Uninhabited Combat Air Vehicle (UCAV) path planning," *Aerosp. Sci. Technol.*, vol. 14, no. 8, pp. 535–541, Dec. 2010, doi: 10.1016/j.ast.2010.04.008.
10. S. Zhang, Y. Zhou, Z. Li, and W. Pan, "Grey wolf optimizer for unmanned combat aerial vehicle path planning," *Adv. Eng. Softw.*, vol. 99, pp. 121–136, Sep. 2016, doi: 10.1016/j.advengsoft.2016.05.015.
11. V. Roberge, M. Tarbouchi, and G. Labonte, "Comparison of Parallel Genetic Algorithm and Particle Swarm Optimization for Real-Time UAV Path Planning," *Ind. Inform. IEEE Trans. On*, vol. 9, no. 1, pp. 132–141, Feb. 2013, doi: 10.1109/TII.2012.2198665.
12. V. Roberge, M. Tarbouchi, and F. Allaire, "Parallel Hybrid Metaheuristic on Shared Memory System for Real-Time UAV Path Planning," *Int. J. Comput. Intell. Appl.*, vol. 13, no. 2, pp. 1450008-1-1450008–16, Jun. 2014.
13. Z. A. A. Alyasseri, A. T. Khader, M. A. Al-Betar, M. A. Awadallah, and X.-S. Yang, "Variants of the Flower Pollination Algorithm: A Review," in *Nature-Inspired Algorithms and Applied Optimization*, X.-S. Yang, Ed. Cham: Springer International Publishing, 2018, pp. 91–118. doi: 10.1007/978-3-319-67669-2_5.
14. S. Pant, A. Kumar, and M. Ram, "Flower pollination algorithm development: a state of art review," *Int. J. Syst. Assur. Eng. Manag.*, vol. 8, no. 2, pp. 1858–1866, Nov. 2017, doi: 10.1007/s13198-017-0623-7.
15. H. Chiroma, N. L. M. Shuib, S. A. Muaz, A. I. Abubakar, L. B. Ila, and J. Z. Maitama, "A Review of the Applications of Bio-inspired Flower Pollination Algorithm," *Procedia Comput. Sci.*, vol. 62, pp. 435–441, Jan. 2015, doi: 10.1016/j.procs.2015.08.438.
16. Government of Canada, "Canadian Digital Elevation Model." https://open.canada.ca/data/en/dataset/7f245e4d-76c2-4caa-951a-45d1d2051333 (accessed May 18, 2022).
17. G. Labonté, "Sur la construction de trajectoires dynamiquement réalisables pour les avions à partir de suites de segments de droites," Collège militaire royal du Canada, Kingston, Ontario, Canada, Nov. 2009.
18. G. Labonté, "Simple formulas for the fuel of climbing propeller driven airplanes," *Adv. Aircr. Spacecr. Sci.*, vol. 2, no. 4, pp. 367–389, 2015, doi: http://dx.doi.org/10.12989/aas.2015.2.4.367.
19. Y. Ding *et al.*, "Discussions on Normalization and Other Topics in Multi-Objective Optimization," Toronto, Aug. 2006. [Online]. Available: http://www.cas.mcmaster.ca/~romanko/fmipw-2006-moo.pdf
20. J. Holland, *Adaptation in Natural and Artificial Systems*. University of Michigan Press, 1975.
21. H. Li, S.-Y. Liu, Y.-W. Huang, Y.-Q. Chen, and Z.-H. Fu, "An Efficient 2-opt Operator for the Robotic Task Sequencing Problem," in *2019 IEEE International Conference on Real-time Computing and Robotics (RCAR)*, Aug. 2019, pp. 124–129. doi: 10.1109/RCAR47638.2019.9044008.

New control scheme of on-load tap changer for voltage regulation in active distribution systems using Fuzzy logic

Yassir Maataoui [1*], *Omar* Boutfarjoute[1], *Hamid* Chekenbah[1] and *Rafik* Lasri[1]

[1] Laboratory of Sciences and advanced Technologies (LSAT), FPL, Larache, Morocco.

Abstract. This research presents a new voltage regulation philosophy that enhances the voltage magnitude in a power distribution network where distributed generation (DG) is prevalent. Voltage regulation in power distribution networks relies heavily on the On-Load Tap-Changer (OLTC) installed in power transformers. Since it was designed for networks without DG, the OLTC's current voltage regulation scheme uses a fixed reference voltage and hence cannot regulate voltage in a distribution network. Therefore, the proposed novel system-based Fuzzy logic will compute a new set point voltage that the OLTC must follow to mitigate the undesirable DG impact on the voltage profile. The suggested system is evaluated on Morocco's MATLAB/Simulink-built 20kV distribution network.

1 Introduction

The modern world's increasing need for power, as well as the search for the best solution to environmental problems and considering the high price of fossil fuels, has prompted the use of new energy sources that are renewable, also known as distributed generation (DG). Because they are positioned closer to the consumers, renewable energy sources have the advantage of lowering transmission line losses and supporting the distribution system's voltage level **[1]**.

Solar energy (photovoltaic) and wind energy are two examples of renewable energy sources utilized in Morocco nowadays **[2]**. The conventional distribution network is passive and has a one-way power flow, but with the connection of the DG to the network, the direction of the power flow is changed, can be either from loads to the power grid or vice versa, and therefore, the network is called an active distribution network. The change direction of power flow takes place when DG output active power is bigger than the network local load.

Interference of reverse power flow with conventional voltage control components on the power distribution system produces an increase in voltage magnitude at the point of connection of distributed generation and a drop at the end of feeders without DG **[3]**. A passive electrical distribution network without distributed generation will have a voltage magnitude that decreases from the substation busbar to the end of the feeder due to transmission line losses, with the assumption that the highest voltage magnitude exists in the secondary of a power transformer in a substation **[4]**.

Because the on-load tap changer is positioned in the substation and the voltage magnitude drops from the substation busbar to the feeder end busbar, all of this will allow the OLTC to efficiently regulate the voltage. By keeping the substation busbar voltage at a little greater magnitude than the nominal voltage, the voltage throughout the distribution system can be maintained above the lowest allowable limit **[5]**.

Under extreme conditions of power flow reversal. The substation busbar may cease to be the highest voltage point in the distribution system and be replaced by the voltage at the distributed generator connection node, the OLTC would not be able to provide a better voltage control in this case because its voltage regulation scheme expects that the substation busbar with the maximum voltage level is still available **[6]**, these make the current OLTC voltage regulation scheme inappropriate for voltage control when DGs are incorporated into the power distribution network

There are already many techniques found in the literature to control the voltage in an active distribution network.

[7] Presents a fuzzy-based control of the OLTC. The OLTC takes into account the transformer's secondary voltage and the target voltage, as well as the difference between the measured and nominal transformation ratio. The tap changer is controlled by the fuzzy logic controller, which adjusts the transformer's primary voltage. nevertheless, The DG impacts on the voltage have not been taken into account in this proposed strategy.

[8] Voltage/reactive power sensitivity analysis is employed in the proposed control technique to minimise

* Corresponding author: maataoui.yassir@gmail.com

the total system deviation. A power flow analysis is carried out, and each DG acts as a local control agent to calculate its voltage deviation and the available reactive power for lowering it, as well as to monitor the load and its power consumption.

[9] Proposes an adaptive fuzzy reference that allows DGs to function at different power factors, allowing them to provide decentralized voltage control. The main drawback of this system is that it ignores the coordination between the DG and the OLTC, Furthermore, because reactive power support is constantly dependent on active power, the DG cannot regulate the voltage if the generated active power is insufficient.

[10] Another technique is presented, which divides the voltage control into three zones, with the OLTC regulating one zone and the second another zone, while the voltage regulator takes the second zone while the integrated generator with its reactive power capabilities regulates the last zone. However, the high ratio of R/X in the power distribution network does not allow the reactive power of the DG to effectively regulate the voltage.

A fuzzy voltage regulation control algorithm for the reference value of the OLTC is proposed in this study paper. With the suggested voltage regulation scheme, the standard OLTC function scheme will be modified in order to improve the voltage regulation process when DGs are present in the system. Instead of having one input from the substation busbar, the OLTC will have many measured inputs and pick between them the system's maximum and lowest voltages, rather than relying solely on the substation busbar. Then the fuzzy controller will determine the reference voltage.

The following is the structure of this paper. Section 2 explains the OLTC's voltage regulating method. Voltage regulation and distributed generators will be illustrated in section 3 via a fuzzy control system. MATLAB/SIMULINK simulation results and discussions are presented in sections 4 and 5, and a conclusion is made in section 6.

2 CONVENTIONAL VOLTAGE REGULATION SCHEME OF OLTC:

Voltage regulation in power distribution networks makes use of a wide variety of devices, among these the On-Load Tap Changer (OLTC) installed at the HV/MV transformer is the most commonly used for voltage regulation. They allow the voltage of the MV busbars to be modified as a result of variations in upstream voltage caused by changes in the loads. A typical on-load tap changer has 17 taps (8 up and 8 down) with a step size of 0.625%, i.e. ±5% of the voltage ratio can be changed with this device. The dynamic operation of the controller under load is illustrated in Figure **1 [11]**.

Due to transitory fluctuations, while connecting or detaching large loads, an additional one-minute delay $\Delta t1$ is imposed on the initial tap change if the measured voltage is beyond the permissible range threshold. If more tap modifications are required after the initial one, the delay $\Delta t2$ can be lowered to 10s.

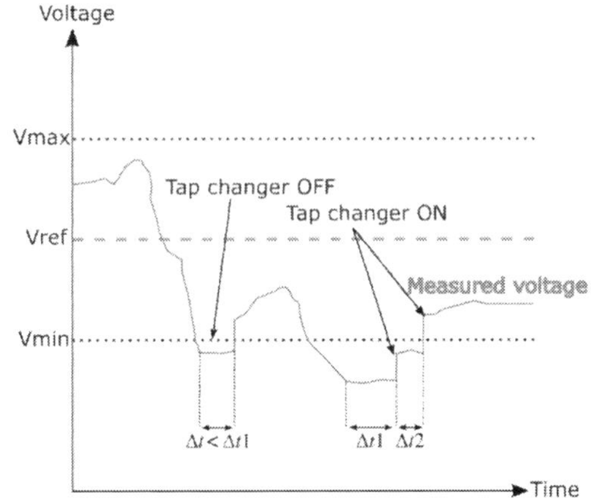

Fig. 1. OLTC under load variation.

The On-load tap changer permits the voltage to fall within a predetermined range surrounding the voltage setpoint:

$$Vmin \leq V \leq Vmax \quad (1)$$

$$Vmax = Vref + \frac{DB}{2} \quad (2)$$

$$Vmin = Vref - \frac{DB}{2} \quad (3)$$

Where the DB is the deadband and Vref voltage setpoint.

The operating principle of OLTC is shown in figure 2[12].

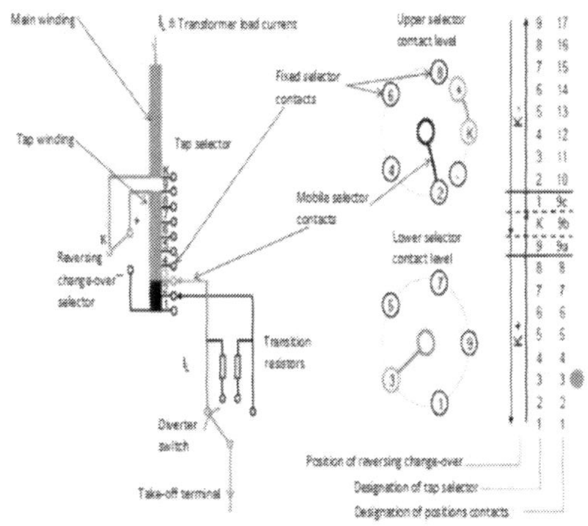

Fig. 2. The operating concept of OLTC

In most cases, the on-load tap changer's set point is determined by the voltage on the MV busbar. Figure 2 shows the operation of the OLTC while maintaining a constant voltage magnitude on the MV busbar, as illustrated. A voltage drop compensation (LDC) can also be used to connect the OLTC to the voltage at the feeder's end (LDC). Most OLTCS don't use the compounding function because it complicates things and can lead to unnecessary errors.

3 DISTRIBUTED GENERATORS AND VOLTAGE REGULATION:

A voltage drop will occur along the feeder's impedance as power is transmitted from the substation's busbar to the load busbars in a passive network, hence the MV busbar will still have the highest magnitude of voltage **[13]**. An MV distribution network is illustrated using a simplified single network line diagram shown in Figure 3, Just for understanding the distribution power network has a single load connecting to it through a feeder of Resistance R_f and Reactance X_f, the active power demand of the load is denoted by PL, and the reactive power demand of the load is denoted by QL.

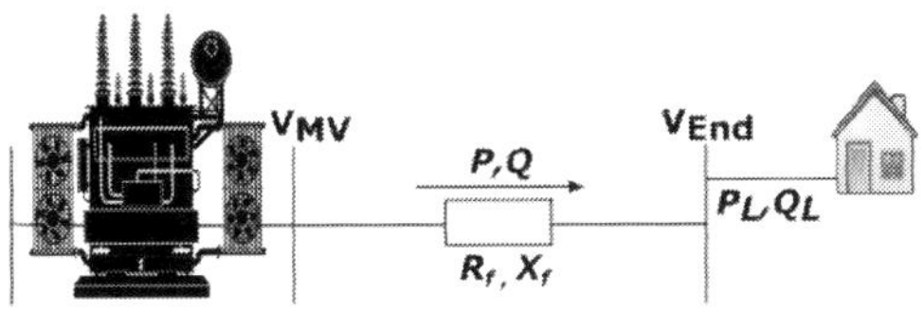

Fig. 3. Passive distribution network

The voltage drop along the line can be written by equation (1)

$$\Delta V = V_{MV} - V_{End} \quad (4)$$

$$\Delta V = \frac{R_f P_L + X_f Q_L}{V_{End}} + j\frac{X_f P_L - R_f Q_L}{V_{End}} \quad (5)$$

Equation (6) can be found by simplifying equation (5) due to the fact that the imaginary term is less significant than the real term

$$\Delta V \approx \frac{R_f P_L + X_f Q_L}{V_{End}} \quad (6)$$

In Figure 4, a distributed generator is linked to the distribution network. Voltage drop can be written in the following form:

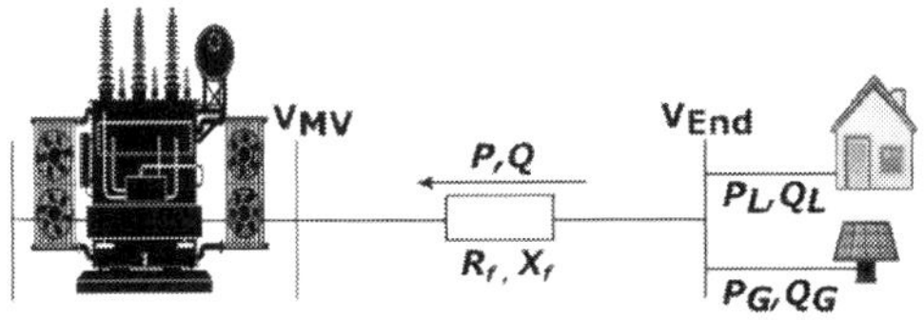

Fig. 4. Active distribution network.

$$\Delta V = V_{End} - V_{MV} \quad (7)$$

$$\Delta V \approx \frac{R_f(P_L - P_G) + X_f(Q_L - Q_G)}{V_G} \quad (8)$$

The voltage drop along the line is ΔV, in equation (8), where Rf is the line resistance and Xf is the line reactance. PG is the active power generated by the distributed generator, QG is the reactive power of the embedded generator, and PL and QL are the active and reactive powers of the load, respectively.

The reverse power becomes more aggravated when the DG active power (PG >> PL) thus the voltage rises above the statutory limits.

4 PROPOSED CONTROL SYSTEM:

The transformer's ratio is adjusted by the OLTC by adding or removing turns from zero turns in the case with the absence of voltage compensation to maximum turns in the case with the presence of maximum voltage compensation. Usually. The conventional OLTC scheme with a fixed secondary voltage magnitude setpoint is based on the assumption of voltage profile dropping from substation busbar to the load terminals; On the other hand, the incorporation of distributed generators changes this assumption and makes voltage regulation inappropriate with the conventional OLTC scheme. therefore, requires modifying the conventional OLTC scheme with a fixed setpoint to an adaptive one by using fuzzy logic inference system.

A fuzzy logic-based voltage regulation controller, which receives voltage magnitude measurements from all critical locations nodes on the power distribution network, and thus adapts the reference voltage set point of the OLTC, is proposed to overcome the current OLTC system's incapacity to manage voltage by integrating distributed generators into the electricity distribution network.

Distribution generation impacts the voltage magnitudes of the power distribution network, which means the fuzzy logic controller will adjust the OLTC's optimal voltage set point based on these changes to reduce overvoltage and undervoltage and to prevent tap hunting when the regulated voltage magnitude exceeds

 https://doi.org/10.1051/itmconf/20224804001

the allowable limit.

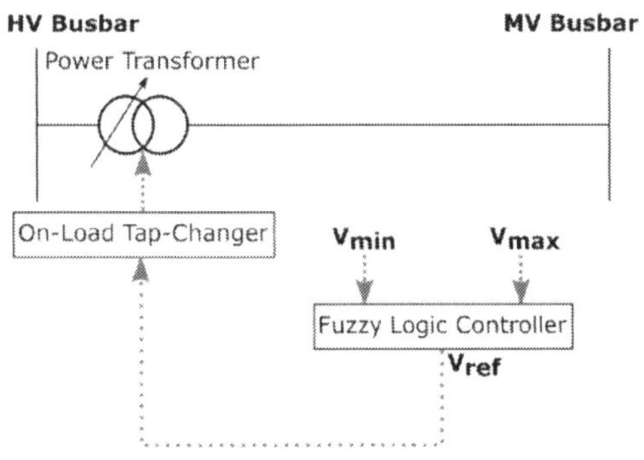

Fig. 5. New OLTC control system

For the fuzzy logic controller depicted in Figure 5 to work, two inputs are needed. Vmax and Vmin are the maximum and minimum voltage magnitudes, respectively. The controller's inputs are determined by the voltages on the secondary side of the power transformer, the voltages on the busbars to which dispersed generators are attached, and the voltages on the ends of feeders.

$$Vmax = \text{maximum}\ (V1, V2, \dots, Vi, \dots Vn) \quad (9)$$

$$Vmin = \text{minimum}\ (V1, V2, \dots, Vi, \dots Vn) \quad (10)$$

The 3 membership functions for Vmin and Vmax are shown in Figure 6, while figure 7 shows the output membership functions of the voltage step point of the OLTC.

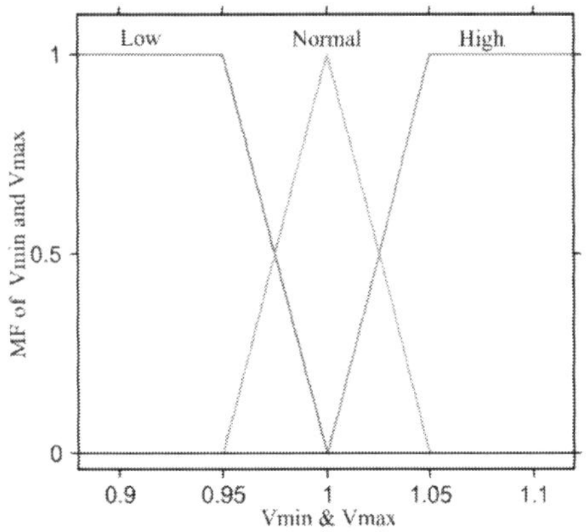

Fig. 6. Inputs membership functions of the fuzzy logic controller.

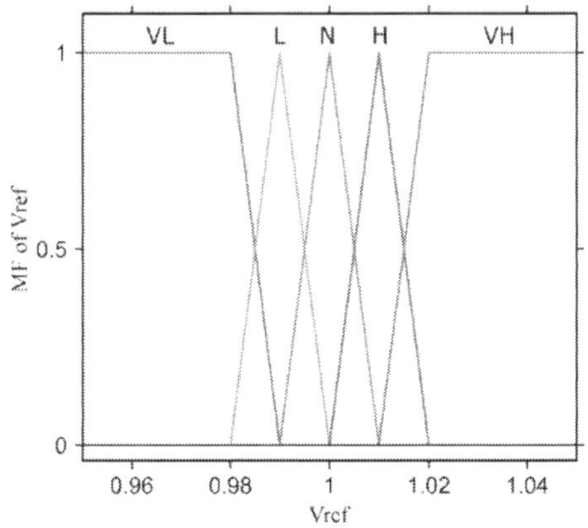

Fig. 7. Outputs membership functions of the fuzzy logic controller.

Where (VL, L, N, H, VH) are respectively very low, low, normal, high, and very high.

it can be read as follow:

- If Vmax is high and Vmin is Low then voltage reference is Low
- If Vmax is Normal and Vmin is Normal then voltage reference is Normal
- If Vmax is High and Vmin is High then step voltage reference is Very low

The fuzzy rule base is shown in **table 1**

Table 1. Presents the fuzzy rule base.

		Vmax		
		Low	**Normal**	**High**
Vmin	**Low**	VH	H	N
	Normal	H	N	L
	High	N	L	VL

5 SIMULATION RESULTS AND DISCUSSIONS:

The test network is a 13-busbar generic Moroccan MV distribution network with three radial feeders serving different loads. The single-line diagram of the network is shown in Figure 8.

The distribution feeders are connected to a 40 MVA OLTC power transformer with a nominal voltage of 63/20 kV. Feeders 1, 2 and 3 of the electrical network, as shown in the diagram, have 12 loads, respectively. It was estimated that the network's maximum load requirement would be 16 MW. The first feeder has one distributed generator linked at bus 1, and the third feeder also has also one connected at the feeder's end.

 https://doi.org/10.1051/itmconf/20224804001

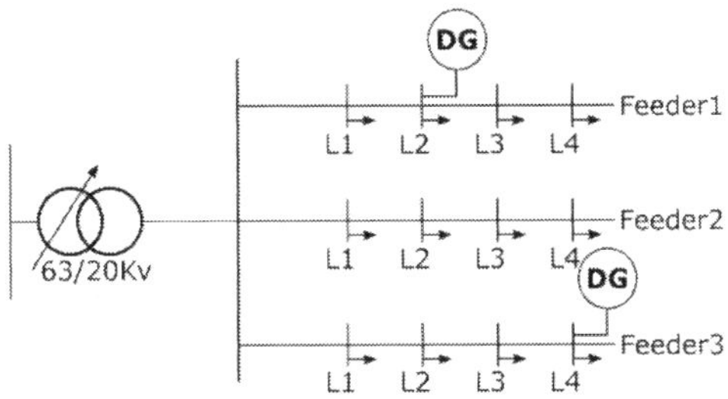

Fig. 8. The proposed Moroccan test network

Table 2 shows the genetic test system's specifications:

Table 2. Specifications of the proposed test network.

Parameter	Values
OLTC power transformer	40MVA 63/20KV 50HZ 17 taps
Network short circuit level	100MVA
Feeder	0.320+j0.119 ohm/km, feeder1=4km, feeder2=4km, feeder3=4km
Loads	Feeder1=4MW, Feeder=8MW, Feeder=4MW

5.1 First case: minimum load and maximum DG output power

In this scenario, each feeder has 4MW of load demand condition. The DG output power generated and absorbed at feeder 1 and feeder 3 is, respectively. 5MW/1.14MVAr and 10MW/2.28MVAr. The voltage of critical buses with conventional OLTC voltage regulation algorithm and after implementing the new OLTC control algorithm is shown in Fig 9 and fig 10 respectively

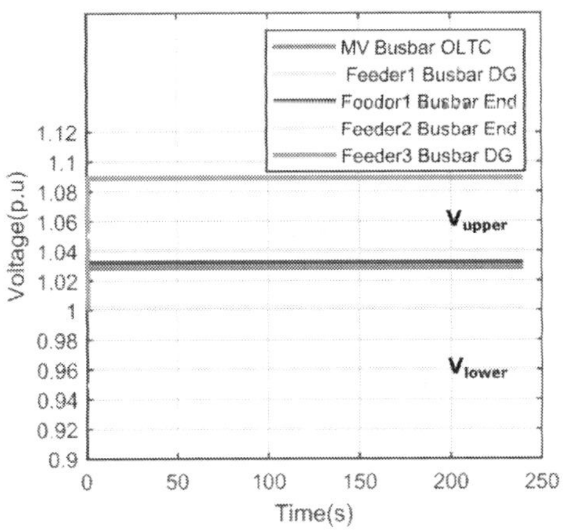

Fig. 9. voltage magnitude of the first case without the proposed controller.

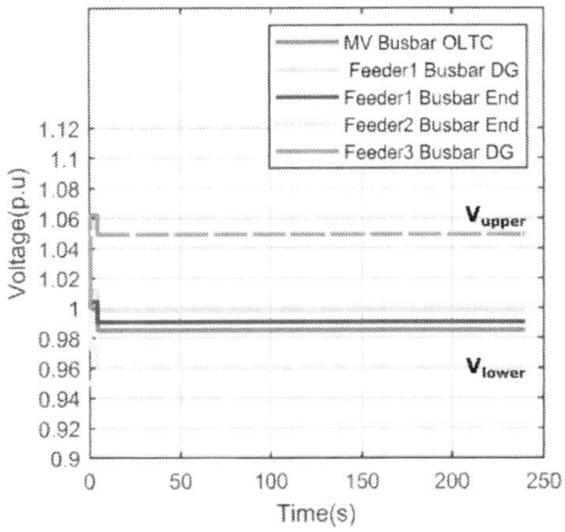

Fig. 10. voltage magnitude of the first case with the proposed controller.

From figure 9 the voltage magnitude at the end of feeder 3 is 1.09 which exceeds the upper limit. By implementing the proposed control algorithm set a new voltage reference Vref =0.985pu. Therefore, the voltage magnitude is set in a predefined limit as shown in figure 10.

5.2 Second case: Maximum load and minimum DG output power

In this scenario feeder1 and feeder3 have 4MW and the load at feeder 2 is increased from 4MW to 8MW. The DG output power generated and absorbed at feeder 1 and feeder 3 is reduced to 1MW/0.228MVAr and 1MW/0.228MVAr., respectively. The voltage of critical buses with conventional OLTC voltage regulation algorithm and after implementing the new OLTC control algorithm is shown in Figure 11 and figure 12 respectively.

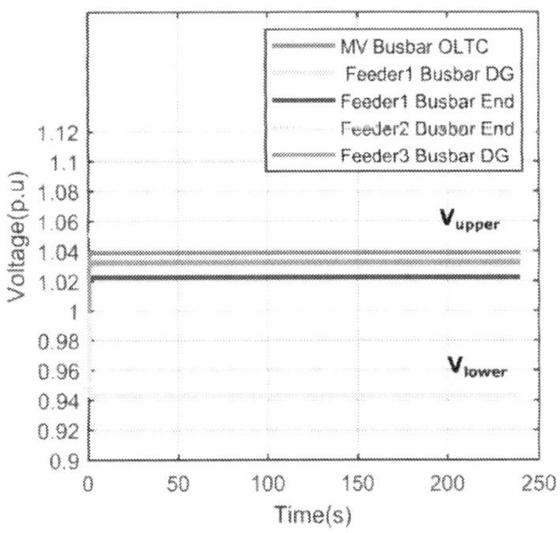

Fig. 11. voltage magnitude of the second case without the proposed controller.

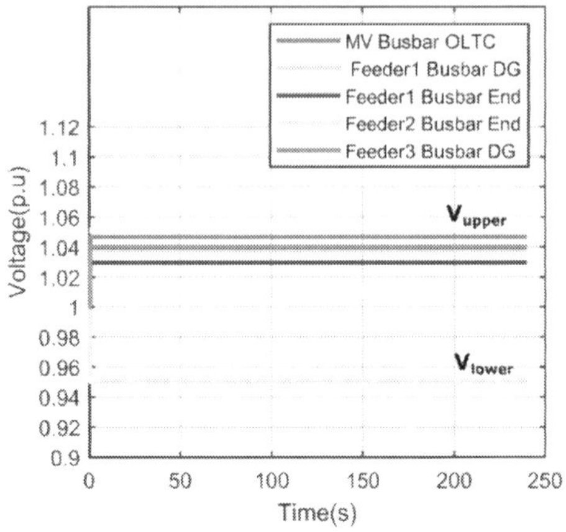

Fig. 12. voltage magnitude of the second case with the proposed controller.

From figure 11, the voltage magnitude at the end of feeder 2 is 1.04pu, which is below the lower limit. By implementing the proposed control algorithm, a new voltage reference is set Vref =1.05pu. Therefore, the voltage magnitude at the end of the feeder 2 is set in a predefined limit as shown in figure 12.

6 CONCLUSION:

Increasing the amount of dispersed generating connections necessitated the development of a fuzzy voltage control system, which is described in this study. The fuzzy logic-based control method introduced here replaces the traditional OLTC technique.

On the Moroccan distribution network, the voltage control system was put to the test. The results of the simulations show that the proposed voltage control method is able to overcome the voltage deviation caused by the substantial incorporation of distributed generators.

References

1. Sahar S. Kaddah, Magdi M. El-Saadawi & Doaa M. El-Hassanin (2015) Influence of Distributed Generation on Distribution Networks During Faults, Electric Power Components and Systems, 43:16, 1781-1792.

2. "MASEN, The Basics," Moroccan Agency for Solar Energy, http://www.masen.ma/en/lerecit/#

3. Muhammad Ahmed, Rojan Bhattarai, Sheikh Jakir Hossain, Sherif Abdelrazek, Sukumar Kamalasadan, "Coordinated Voltage Control Strategy for Voltage Regulators and Voltage Source Converters Integrated Distribution System", *IEEE Transactions on Industry Applications*, vol.55, no.4, pp.4235-4246, 2019

4. J. L. Nascimento, A. G. de Freitas Filho, C. S. Sobrinho, J. B. Almada and H. M. de Oliveira Filho, "Voltage Profile on a Feeder with the Insertion of Photovoltaic Distributed Generation," *2021 14th IEEE International Conference on Industry Applications (INDUSCON)*, 2021, pp. 1222-1227, doi: 10.1109/INDUSCON51756.2021.9529392.

5. S. S. M. Reshikeshan, S. L. Matthiesen, M. S. Illindala, A. A. Renjit and R. Roychowdhury, "Autonomous Voltage Regulation by Distributed PV Inverters with Minimal Inter-node Interference," in *IEEE Transactions on Industry Applications*, vol. 57, no. 3, pp. 2058-2066, May-June 2021, doi: 10.1109/TIA.2021.3064911.

6. N. Papazacharopoulos, M. Gibescu and P. Vaessen, "Smart voltage control in distribution networks with a large share of distributed renewable generation," *2015 IEEE Eindhoven PowerTech*, 2015, pp. 1-6, doi: 10.1109/PTC.2015.7232497.

7. G. E. Mejia-Ruiz, R. Cárdenas-Javier, M. R. Arrieta Paternina, J. R. Rodríguez-Rodríguez, J. M. Ramirez and A. Zamora-Mendez, "Coordinated Optimal Volt/Var Control for Distribution Networks via D-PMUs and EV Chargers by Exploiting the Eigensystem Realization," in *IEEE Transactions on Smart Grid*, vol. 12, no. 3, pp. 2425-2438, May 2021, doi: 10.1109/TSG.2021.3050443.

8. L. Choukri, H. Chekenbah, R. Lasri, M. Bouhorma, and Y. Maataoui, "On-load tap-changer control by a fuzzy logic controller," in Proc. 4th World Conf. Complex Syst. (WCCS), Ouarzazate, MR, USA, Apr. 2019, pp. 22–25

9. Zhenquan Yang, Yanjun Li, Ji Xiang, "Coordination Control Strategy for Power Management of Active Distribution Networks", *IEEE Transactions on Smart Grid*, vol.10, no.5, pp.5524-5535, 2019.

10. Andreea-Georgiana IANŢOC, Constantin BULAC, Ion TRIŞTIU, Irina PICIOROAGĂ, Dorian O. SIDEA, "Impact Analysis of Distributed Generation on Voltage Stability in Radial Distribution Systems", *2019 International Conference on ENERGY and ENVIRONMENT (CIEM)*, pp.510-514, 2019.

11. Qiangqiang Xie, Yongzhu Hua, Linglei Xu, "Review on voltage regulation techniques via flexible resources in power distribution systems", *2021 International Conference on Power System Technology (POWERCON)*, pp.850-855, 2021.

12. FYujun He, « Contribution au réglage de la tension sur un réseau HTA avec producteurs – Apport de la flexibilité de la demande », thèse de Doctorat, école doctorale STITS, CentraleSupélec, 2015.

13. https://www.reinhausen.com

14. Mirle Vishwanath Gururaj, Narayana Prasad Padhy, "An Improvized Coordinated Voltage Control Scheme for Better Utilization of Regulating Devices During Various Operating Conditions of a Distribution System", *IEEE Systems Journal*, vol.14, no.4, pp.4891-4900, 2020

https://doi.org/10.1051/itmconf/20224804002

Backstepping control strategy for induction motor with rotor flux and load torque observers

Mohammed El haissouf[1*], *Mustapha* Elharoussi[1], and *Abdellfattah* Ba-razzouk[1]

[1]Dept. of Applied Physics, Hassan I University, Settat, Morocco

Abstract. This paper deal with Backstepping control strategy for induction motor drive, using rotor flux and load torque observers. The proposed study uses a modified motor model with new variables, and Luenberger observers for both rotor flux and load torque. This work proposes a precise calculation method for controller parameters based on response time and damping coefficient. Simulation is performed in Simulink, results show that Backstepping works well by adding observers.

1 Introduction

Induction motor is a nonlinear system, so its control cannot be directly designed by classical methods; because it requires some special operations such as decoupling [1], switching table [2]... As a result, the traditional control strategies have complex algorithms and do not always give good results.

Backstepping is an advanced control strategy based on Lyapunov theory [3]. It is performed in several steps, depending on the order of the controlled system, and it can be easily adapted to nonlinear systems of strict feedback form, but also to nonlinear forms of systems by applying appropriate changes of variables.

This paper proposes a change of variables to establish a simple state model in the rotor flux frame for easy use of Backstepping control strategy.

Since the control law requires rotor flux and load torque, this work proposes the design of an observer allowing the calculation of these quantities using Luenberger method.

2 Induction motor modelling

In the two-phase reference frame, electromagnetic state representation of induction motor contains four variables, for this study we choose the two stator currents and two virtual currents replacing rotor flux components [4][1].

Electromagnetic state-space model is given by (1) and completed by the mechanical equation (2).

$$\frac{d}{dt}\begin{bmatrix} i_{sd} \\ i_{sq} \\ i_{\varphi d} \\ i_{\varphi q} \end{bmatrix} = \begin{bmatrix} -\frac{1}{T_c} & \frac{d\rho s}{dt} & \frac{1-\sigma}{\sigma Tr} & \frac{1-\sigma}{\sigma} p\Omega \\ -\frac{d\rho s}{dt} & -\frac{1}{T_c} & -\frac{1-\sigma}{\sigma} p\Omega & \frac{1-\sigma}{\sigma Tr} \\ \frac{1}{Tr} & 0 & -\frac{1}{Tr} & \frac{d\rho r}{dt} \\ 0 & \frac{1}{Tr} & -\frac{d\rho r}{dt} & -\frac{1}{Tr} \end{bmatrix} \begin{bmatrix} i_{sd} \\ i_{sq} \\ i_{\varphi d} \\ i_{\varphi q} \end{bmatrix} + \frac{1}{\sigma Ls}\begin{bmatrix} 1 & 0 \\ 0 & 1 \\ 0 & 0 \\ 0 & 0 \end{bmatrix}\begin{bmatrix} v_{sd} \\ v_{sq} \end{bmatrix} \quad (1)$$

$$\frac{d\Omega}{dt} = -\frac{f}{J}\Omega + \frac{T_M}{J} - \frac{T_L}{J} \quad (2)$$

With:

$$T_M = \frac{3}{2}p(1-\sigma)Ls\left(i_{\varphi d} i_{sq} - i_{\varphi q} i_{sd}\right) \quad (3)$$

$$i_{\varphi d} = \frac{\varphi_{rd}}{Msr};\ i_{\varphi q} = \frac{\varphi_{rq}}{Msr} \quad (4)$$

$$\frac{1}{T_c} = \left(\frac{1}{\sigma Ts} + \frac{1-\sigma}{\sigma Tr}\right) \quad (5)$$

$$\sigma = 1 - \frac{Msr^2}{Ls.Lr} \quad (6)$$

The reference is the same for all variables, then:

$$\frac{d\rho s}{dt} = p\Omega + \frac{d\rho r}{dt} \quad (7)$$

2.1 Induction motor modelling in rotor flux frame

Backstepping can be applied in fixed reference frame, but the equations are longer in this case as well as the resulting control law. This work proposes to apply Backstepping strategy in rotor flux frame [2][1].

In the new reference frame, electromagnetic state representation is defined by the equation (8), completed by the mechanical equation (9).

* Corresponding author: mr_elhaissouf@yahoo.fr

$$\frac{d}{dt}\begin{bmatrix} i_{sd} \\ i_{sq} \\ i_{\varphi d} \end{bmatrix} = \begin{bmatrix} -\frac{1}{T_c} & d\rho s/dt & \frac{1-\sigma}{\sigma Tr} \\ -d\rho s/dt & -\frac{1}{T_c} & -\frac{1-\sigma}{\sigma}p\Omega \\ \frac{1}{Tr} & 0 & -\frac{1}{Tr} \end{bmatrix}\begin{bmatrix} i_{sd} \\ i_{sq} \\ i_{\varphi d} \end{bmatrix} + \frac{1}{\sigma Ls}\begin{bmatrix} 1 & 0 \\ 0 & 1 \\ 0 & 0 \end{bmatrix}\begin{bmatrix} v_{sd} \\ v_{sq} \end{bmatrix} \quad (8)$$

$$\frac{d\Omega}{dt} = -\frac{f}{J}\Omega + \frac{T_M}{J} - \frac{T_L}{J} \quad (9)$$

With:

$$T_M = \frac{3}{2}p(1-\sigma)Lsi_{\varphi d}i_{sq} \quad (10)$$

It is clear that the system becomes simpler with only four variables instead of five [5].

2.2 Induction motor modelling with new variables

For simplification, another state-space representation is proposed in the equations (11), (12), (13) and (14). This representation allows to choose the adequate of variables for easy control design [6][4].

$$\frac{d}{dt}\Omega = -\frac{f}{J}\Omega - \frac{T_L}{J} + \frac{3p}{2J}(1-\sigma)Lsi_{\varphi d}i_{sq} \quad (11)$$

$$\frac{d}{dt}\left(i_{\varphi d}i_{sq}\right) = -\frac{1}{T_e}i_{\varphi d}i_{sq} - p\Omega i_{\varphi d}i_{sd} - \frac{1-\sigma}{\sigma}p\Omega i_{\varphi d}^2 + \frac{1}{\sigma Ls}i_{\varphi d}v_{sq} \quad (12)$$

$$\frac{d}{dt}i_{\varphi d}^2 = -\frac{2}{Tr}i_{\varphi d}^2 + \frac{2}{Tr}i_{\varphi d}i_{sd} \quad (13)$$

$$\frac{d}{dt}\left(i_{\varphi d}i_{sd}\right) = -\frac{1}{T_e}i_{\varphi d}i_{sd} + p\Omega i_{\varphi d}i_{sq} + \frac{1-\sigma}{\sigma Tr}i_{\varphi d}^2 + \frac{1}{Tr}i_s^2 + \frac{1}{\sigma Ls}i_{\varphi d}v_{sd} \quad (14)$$

With:

$$\frac{1}{T_e} = \frac{1}{\sigma}\left(\frac{1}{Tr} + \frac{1}{Ts}\right) \quad (15)$$

It should be noted that the modulus of the stator current is not a separate variable, as it depends on the other variables (16).

$$i_s^2 = \frac{(i_{\varphi d}i_{sq})^2 + (i_{\varphi d}i_{sq})^2}{i_{\varphi d}^2} \quad (16)$$

Proposed new state variables and input variables are defined by equations (17) and (18).

$$\begin{bmatrix} x_1 \\ x_2 \\ x_3 \\ x_4 \end{bmatrix} = \begin{bmatrix} \Omega \\ \frac{3p}{2J}(1-\sigma)Lsi_{\varphi d}i_{sq} \\ i_{\varphi d}^2 \\ \frac{2}{Tr}i_{\varphi d}i_{sd} \end{bmatrix} \quad (17)$$

$$\begin{bmatrix} u_1 \\ u_3 \end{bmatrix} = \begin{bmatrix} \frac{3p}{2J}\frac{1-\sigma}{\sigma}i_{\varphi d}v_{sq} \\ \frac{2}{Tr}\frac{1}{\sigma Ls}i_{\varphi d}v_{sd} \end{bmatrix} \quad (18)$$

The resulting state-space representation is given by the four-equations system (19), (20), (21) and (22).

$$\frac{d}{dt}x_1 = -\frac{f}{J}x_1 - \frac{T_L}{J} + x_2 \quad (19)$$

$$\frac{d}{dt}x_2 = -\frac{1}{Te}x_2 - \frac{3p^2}{4J}Tr(1-\sigma)Lsx_1x_4 - \frac{3p^2}{2J}\frac{(1-\sigma)^2Ls}{\sigma}x_1x_3 + \frac{3p}{2J}\frac{1-\sigma}{\sigma}i_{\varphi d}v_{sq} \quad (20)$$

$$\frac{d}{dt}x_3 = -\frac{2}{Tr}x_3 + x_4 \quad (21)$$

$$\frac{d}{dt}x_4 = -\frac{1}{Te}x_4 + \frac{4J}{3Tr}\frac{1}{(1-\sigma)Ls}x_1x_2 + 2\frac{1-\sigma}{\sigma Tr^2}x_3 + \frac{2}{Tr^2}i_s^2 + \frac{2}{Tr}\frac{1}{\sigma Ls}i_{\varphi d}v_{sd} \quad (22)$$

2.3 Vector modelling of induction motor

In order to simplify the synthesis of Backstepping control, this work proposes to group the state and input variables into two-dimensional vector variables [6], thus the state model is given by the vector equations (23) and (24).

$$\begin{bmatrix} \dot{x}_1 \\ \dot{x}_3 \end{bmatrix} = -\begin{bmatrix} \frac{1}{Tm}x_1 + \frac{T_L}{J} \\ \frac{2}{Tr}x_3 \end{bmatrix} + \begin{bmatrix} x_2 \\ x_4 \end{bmatrix} \quad (23)$$

$$\begin{bmatrix} \dot{x}_2 \\ \dot{x}_4 \end{bmatrix} = \begin{bmatrix} -\frac{1}{Te}x_2 - \frac{3p^2}{2J}\frac{(1-\sigma)^2Ls}{\sigma}x_1x_3 - \frac{3p^2}{4J}Tr(1-\sigma)Lsx_1x_4 \\ -\frac{1}{Te}x_4 + \frac{4J}{3Tr}\frac{1}{(1-\sigma)Ls}x_1x_2 + 2\frac{1-\sigma}{\sigma Tr^2}x_3 + \frac{2}{Tr^2}i_s^2 \end{bmatrix} + \begin{bmatrix} u_1 \\ u_3 \end{bmatrix} \quad (24)$$

3 Backstepping control design of induction motor

Vector state-space representation of induction motor can be written compactly in equations (25) and (26), where the state and input variables are two-dimensional vectors [3][7].

$$\dot{\boldsymbol{X}}_1 = \boldsymbol{F}_1(\boldsymbol{X}_1) + \boldsymbol{X}_2 \quad (25)$$

$$\dot{\boldsymbol{X}}_2 = \boldsymbol{F}_2(\boldsymbol{X}_1, \boldsymbol{X}_2) + \boldsymbol{U} \quad (26)$$

Where:

$$\boldsymbol{X}_1 = \begin{bmatrix} x_1 \\ x_3 \end{bmatrix};\ \boldsymbol{X}_2 = \begin{bmatrix} x_2 \\ x_4 \end{bmatrix};\ \boldsymbol{U} = \begin{bmatrix} u_1 \\ u_3 \end{bmatrix} \quad (27)$$

$$\boldsymbol{F}_1(\boldsymbol{X}_1) = -\begin{bmatrix} \frac{1}{Tm}x_1 + \frac{T_L}{J} \\ \frac{2}{Tr}x_3 \end{bmatrix} \quad (28)$$

$$\boldsymbol{F}_2(\boldsymbol{X}_1, \boldsymbol{X}_2) = \begin{bmatrix} -\frac{1}{Te}x_2 - \frac{3p^2}{2J}\frac{(1-\sigma)^2Ls}{\sigma}x_1x_3 - \frac{3p^2}{4J}Tr(1-\sigma)Lsx_1x_4 \\ -\frac{1}{Te}x_4 + \frac{4J}{3Tr}\frac{1}{(1-\sigma)Ls}x_1x_2 + 2\frac{1-\sigma}{\sigma Tr^2}x_3 + \frac{2}{Tr^2}i_s^2 \end{bmatrix} \quad (29)$$

It is a second-order vector system of strict feedback form which is easily amenable to Backstepping control design. Only two steps are necessary to establish the control law [8].

3.1 First step Backstepping design

In this step, variable X_1 representing flux and speed is controlled by the intermediate variable X_2. The aim is to find the desired control law X_{2d}.
We first define the first error (30), then we introduce the stabilizing control term in the error derivative (31).

$$E_1 = X_1 - X_{1ref} \quad (30)$$

$$\dot{E}_1 = F_1(X_1) + X_{2d} - \dot{X}_{1ref} = -K_1E_1 \quad (31)$$

Lyapunov candidate function is positive and its derivative is negative when the coefficients of the diagonal matrix K_1 are positive; this ensures stability in this step (32).

$$V_1 = \frac{1}{2}E_1^TE_1 > 0 \rightarrow \dot{V}_1 = E_1^T\dot{E}_1 = -E_1^TK_1E_1 < 0 \quad (32)$$

Thus, intermediate control law is given by equation (33).

$$X_{2d} = -K_1E_1 - F_1(X_1) + \dot{X}_{1ref} \quad (33)$$

3.2 Second step Backstepping design

In this step the intermediate error (34) is defined in order to calculate the main control law [9].

$$E_2 = X_2 - X_{2d} = X_2 + K_1E_1 + F_1(X_1) - \dot{X}_{1ref} \quad (34)$$

In this case we have to rewrite the error derivative of the first step considering the error on the secondary variable. This results in the equation (35).

$$\dot{E}_1 = F_1(X_1) + E_2 + X_{2d} - \dot{X}_{1ref} = -K_1E_1 + E_2 \quad (35)$$

A second positive candidate Lyapunov function must be defined on the basis of the two error vectors (36). The control law must guarantee the stability of the system, so it must be able to make the derivative (37) of this new Lyapunov function negative.

$$V_2 = \frac{1}{2}E_1^TE_1 + \frac{1}{2}E_2^TE_2 > 0 \quad (36)$$

$$\begin{aligned}\dot{V}_2 = E_1^T\dot{E}_1 + E_2^T\dot{E}_2 = E_1^T(-K_1E_1 + E_2) + E_2^T\dot{E}_2 = \\ -E_1^TK_1E_1 + E_1^TE_2 + E_2^T\dot{E}_2 = \\ -E_1^TK_1E_1 + E_2^T(E_1 + \dot{E}_2)\end{aligned} \quad (37)$$

It is necessary to introduce a stabilizing term (38) with positive coefficients control matrix K_2

$$E_1 + \dot{E}_2 = -K_2E_2 \quad (38)$$

This leads to (39).

$$\dot{V}_2 = -E_1^TK_1E_1 - E_2^TK_2E_2 < 0 \quad (39)$$

To find the main control law it is necessary to calculate the derivative of the second error (40).

$$\dot{E}_2 = \dot{X}_2 - \dot{X}_{2d} = F_2(X_1, X_2) + U - \dot{X}_{2d} \quad (40)$$

Finally, the control law is given by the equations (41), (42), (43), (44), (45), (46), and (47). The order of equations is reversed when implementing control algorithm.

$$U = \dot{E}_2 - F_2(X_1, X_2) + \dot{X}_{2d} \quad (41)$$

$$\dot{E}_2 = -E_1 - K_2E_2 \quad (42)$$

$$\dot{X}_{2d} = -K_1\dot{E}_1 - \dot{F}_1(X_1) + \ddot{X}_{1ref} \quad (43)$$

$$\dot{E}_1 = -K_1E_1 + E_2 \quad (44)$$

$$E_2 = X_2 - X_{2d} \quad (45)$$

$$X_{2d} = -K_1E_1 - F_1(X_1) + \dot{X}_{1ref} \quad (46)$$

$$E_1 = X_1 - X_{1ref} \quad (47)$$

3.3 Calculation of the Backstepping controller parameters.

According to Lyapunov method, the system is stable when parameters positive. However, it is necessary to choose values that guarantee the specific performance. First, error vectors correspond to the equations (48).

$$E_1 = \begin{bmatrix} e_1 \\ e_3 \end{bmatrix} = \begin{bmatrix} x_1 - x_{1ref} \\ x_3 - x_{3ref} \end{bmatrix}; \; E_2 = \begin{bmatrix} e_2 \\ e_4 \end{bmatrix} = \begin{bmatrix} x_2 - x_{2d} \\ x_4 - x_{4d} \end{bmatrix} \quad (48)$$

On the other hand, coefficients of the controller are given by the equations (49).

$$K_1 = \begin{bmatrix} k_1 & 0 \\ 0 & k_3 \end{bmatrix}; \; K_2 = \begin{bmatrix} k_2 & 0 \\ 0 & k_4 \end{bmatrix} \quad (49)$$

Thus, it is possible to write the error state-space equation (50) from the error vector equations. It is a fourth order system.

$$\begin{bmatrix} \dot{e}_1 \\ \dot{e}_2 \\ \dot{e}_3 \\ \dot{e}_4 \end{bmatrix} = \begin{bmatrix} -k_1 & 1 & 0 & 0 \\ -1 & -k_2 & 0 & 0 \\ 0 & 0 & -k_3 & 1 \\ 0 & 0 & -1 & -k_4 \end{bmatrix} \begin{bmatrix} e_1 \\ e_2 \\ e_3 \\ e_4 \end{bmatrix} \quad (50)$$

This system (50) is separable into two subsystems of second order each. This results in two equations (51) and (52).

$$\begin{bmatrix} \dot{e}_1 \\ \dot{e}_2 \end{bmatrix} = \begin{bmatrix} -k_1 & 1 \\ -1 & -k_2 \end{bmatrix} \begin{bmatrix} e_1 \\ e_2 \end{bmatrix} \quad (51)$$

$$\begin{bmatrix} \dot{e}_3 \\ \dot{e}_4 \end{bmatrix} = \begin{bmatrix} -k_3 & 1 \\ -1 & -k_4 \end{bmatrix} \begin{bmatrix} e_3 \\ e_4 \end{bmatrix} \quad (52)$$

From characteristic polynomial (53), coefficients are calculated to ensure a unitary damping coefficient and a response time $tr = 200ms$.

$$det \begin{bmatrix} \lambda + k_1 & -1 \\ 1 & \lambda + k_2 \end{bmatrix} = \lambda^2 + (k_1 + k_2)\lambda + k_1k_2 + 1 = \lambda^2 + 2m\omega_n\lambda + \omega_n^2 \quad (53)$$

$$m = 1 \rightarrow tr = \frac{4.75}{\omega_n} \rightarrow \omega_n = \frac{4.75}{tr} = \frac{4.75}{0.2} = 23.75 \quad (54)$$

So, controller parameters are given by (55) and (56).

$$k_3 = k_1 = \omega_n + 1 = 24.75 \quad (55)$$

$$k_4 = k_2 = \omega_n - 1 = 22.75 \quad (56)$$

3.4 Synoptic diagram of Backstepping control

Control scheme contain a Backstepping controller and observers for rotor flux and load torque.

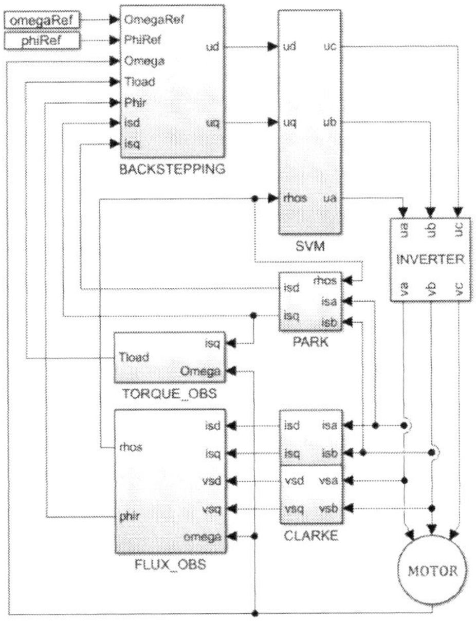

Fig. 1. Synoptic diagram of Backstepping control

4 Rotor flux and load torque observers

In this work, measured quantities are voltages, currents and speed. Rotor flux components are not measured but they are necessary for the control design, so it is necessary to estimate them using an observer [10][8][11].

The load torque also appears in control law, so it must be observed because its value is assumed to be unknown.

4.1 Rotor flux observer

This paper proposes to estimate rotor flux components in fixed reference frame, in order to calculate rotor flux modulus and its angle necessary [12][13]. The complex state-space representation of the motor is used: (57) and (58).

$$\frac{d}{dt}\underline{I}s = -\frac{1}{T_c}\underline{I}s + \frac{1-\sigma}{\sigma}\left(\frac{1}{Tr} - jp\Omega\right)\underline{I}\varphi + \frac{1}{\sigma Ls}\underline{V}s \quad (57)$$

$$\frac{d}{dt}\underline{I}\varphi = \frac{1}{Tr}\underline{I}s - \left(\frac{1}{Tr} - jp\Omega\right)\underline{I}\varphi \quad (58)$$

From this complex representation, complex equations (59) and (60) of rotor flux observer are proposed.

$$\frac{d}{dt}\underline{\tilde{I}}s = -\frac{1}{T_c}\underline{\tilde{I}}s + \frac{1-\sigma}{\sigma}\left(\frac{1}{Tr} - jp\Omega\right)\underline{\tilde{I}}\varphi + \frac{1}{\sigma Ls}\underline{V}s - C1(\underline{\tilde{I}}s - \underline{I}s) \quad (59)$$

$$\frac{d}{dt}\underline{\tilde{I}}\varphi = \frac{1}{Tr}\underline{I}s - \left(\frac{1}{Tr} - jp\Omega\right)\underline{\tilde{I}}\varphi - C2(\underline{\tilde{I}}s - \underline{I}s) \quad (60)$$

This results in the complex state-space equation of the observation errors (61).

$$\frac{d}{dt}\begin{bmatrix}\underline{\tilde{I}}s - \underline{I}s\\ \underline{\tilde{I}}\varphi - \underline{I}\varphi\end{bmatrix} = \begin{bmatrix}-C1 & \frac{1-\sigma}{\sigma}C\\ -C2 & -C\end{bmatrix}\begin{bmatrix}\underline{\tilde{I}}s - \underline{I}s\\ \underline{\tilde{I}}\varphi - \underline{I}\varphi\end{bmatrix} \quad (61)$$

Where:

$$C = \left(\frac{1}{Tr} - jp\Omega\right) \quad (62)$$

Complex coefficients $C1$ and $C2$ guarantee stability and optimal dynamics of the observer. We need to choose a response time $tr = 5ms$.

The characteristic polynomial (63) of error state-space matrix must be compared to a desired polynomial to determine complex gains.

$$det\begin{bmatrix}\lambda + C1 & -\frac{1-\sigma}{\sigma}C\\ C2 & \lambda + C\end{bmatrix} =$$

$$\lambda^2 + (C1 + C)\lambda + \left(C1 + \frac{1-\sigma}{\sigma}C2\right)C =$$

$$\lambda^2 + 2m\omega_n\lambda + \omega_n^2 \quad (63)$$

So:

$$m = 1 \rightarrow \omega_n = \frac{4.75}{tr} = \frac{4.75}{0.005} = 950 \quad (64)$$

$$C1 = 2\omega_n - C = 2\omega_n - \left(\frac{1}{Tr} - jp\Omega\right) \approx 2\omega_n + jp\Omega \quad (65)$$

$$C2 = \frac{\sigma}{1-\sigma}\left(\frac{\omega_n^2}{C} - C1\right) \approx \frac{\sigma}{1-\sigma}\omega_n^2 Tr^2\left(\frac{1}{Tr} + jp\Omega\right) \quad (66)$$

The approximate value of $C2$ is obtained for low speed; at high speed real parts of the eigenvalues of the error state-space matrix remain negative and do not change much, this guarantees stability and fast dynamics of the observer.

4.2 Load torque observer

In general, load torque is an unknown variable and independent of the system state variables. In the proposed Backstepping control strategy, this torque must be estimated and then injected into the control law.

This work proposes a Luenberger observer in the rotor flux frame from the state-space equation (67). In this equation load torque is assumed to have a zero derivative [14].

$$\frac{d}{dt}\begin{bmatrix}\Omega\\ T_L\end{bmatrix} = \begin{bmatrix}-\frac{f}{J} & -\frac{1}{J}\\ 0 & 0\end{bmatrix}\begin{bmatrix}\Omega\\ T_L\end{bmatrix} + \begin{bmatrix}\frac{3p}{2J}(1-\sigma)Ls\\ 0\end{bmatrix} i_{\varphi d} i_{sq} \quad (67)$$

State-space equation of load torque observer is given by (68).

$$\frac{d}{dt}\begin{bmatrix}\tilde{\Omega}\\ \tilde{T}_L\end{bmatrix} = \begin{bmatrix}-\frac{f}{J} & -\frac{1}{J}\\ 0 & 0\end{bmatrix}\begin{bmatrix}\tilde{\Omega}\\ \tilde{T}_L\end{bmatrix} + \begin{bmatrix}\frac{3p}{2J}(1-\sigma)Ls\\ 0\end{bmatrix} i_{\varphi d} i_{sq} - \begin{bmatrix}k_1\\ k_2\end{bmatrix}(\tilde{\Omega} - \Omega) \quad (68)$$

This gives state-space equation of observation error (69).

$$\frac{d}{dt}\begin{bmatrix}\tilde{\Omega}-\Omega\\ \tilde{T}_L-T_L\end{bmatrix}=\begin{bmatrix}-\frac{f}{J}-k_1 & -\frac{1}{J}\\ -k_2 & 0\end{bmatrix}\begin{bmatrix}\tilde{\Omega}-\Omega\\ \tilde{T}_L-T_L\end{bmatrix} \quad (69)$$

The observer must have a very large dynamic compared to the system. Since sampling period of the control algorithm is $T_{SMP} = 500\mu s$, the response time chosen for the observer is $tr = 5ms$.

From the characteristic polynomial of error state-space matrix (x), observer parameters can be determined.

$$det\begin{bmatrix}\lambda+\frac{f}{J}+k_1 & \frac{1}{J}\\ k_2 & \lambda\end{bmatrix}=\lambda^2+\left(\frac{f}{J}+k_1\right)\lambda-\frac{k_2}{J}=\lambda^2+2m\omega_n\lambda+\omega_n^2 \quad (70)$$

$$m=1\rightarrow\omega_n=\frac{4.75}{tr}=\frac{4.75}{0.005}=950 \quad (71)$$

$$k_1=2\omega_n-\frac{f}{J}\approx 1900 \quad (72)$$

$$k_2=-J\omega_n^2=-45125 \quad (73)$$

5 Validation

Proposed scheme is verified using Simulink, frequency of the inverter carrier is $f_{SVM} = 2000\ Hz$.

Thus, we validate Backstepping controller by examining the speed and rotor flux responses, and we verify the rotor flux and load torque observers by comparing the estimated values with the actual values.

The simulation results are shown in Fig.2, Fig.3, Fig.4, Fig.5 et Fig.6.

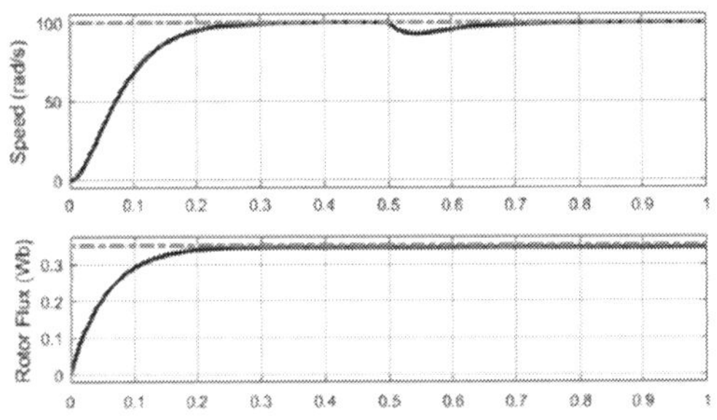

Fig. 2. Speed and rotor flux responses

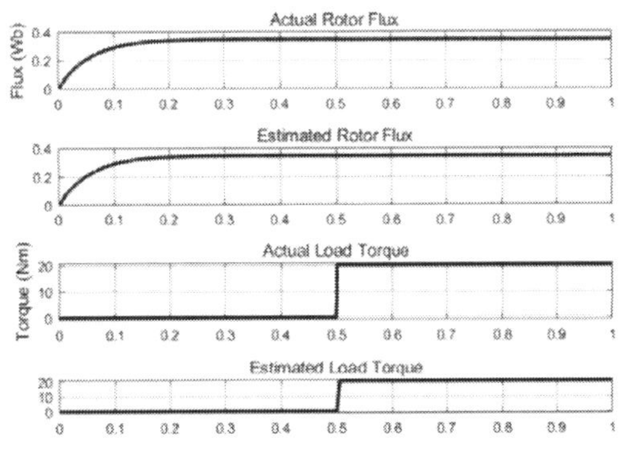

Fig. 3. Estimated rotor flux and load torque

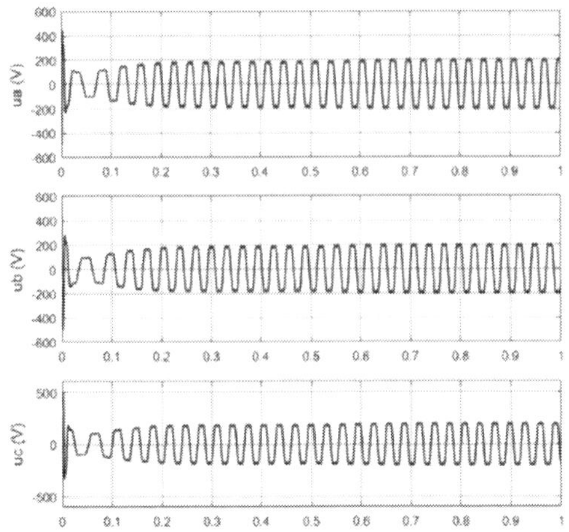

Fig. 4. Input signals for inverter

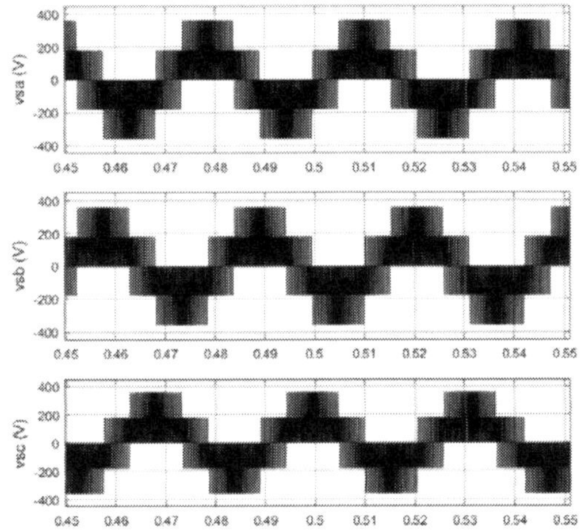

Fig. 5. Stator voltages of induction motor

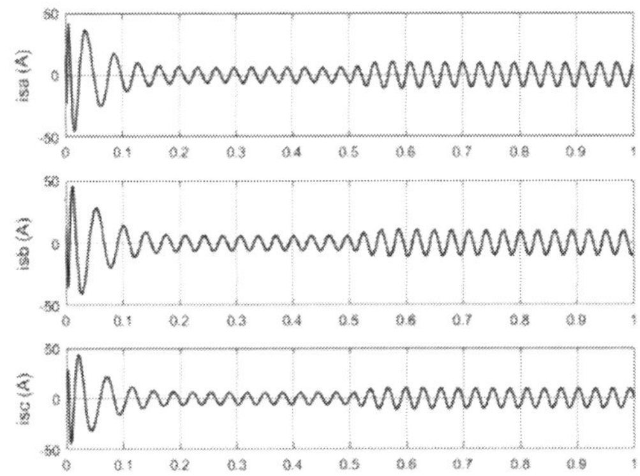

Fig. 6. Stator currents

Fig.2 shows that speed and rotor flux responses respect the dynamics fixed in the controller design. From Fig.3 the responses of the observer are satisfactory because the estimated values quickly follow the real values.

Signals in Fig.4 are not sinusoidal waveforms since they drive a space vector modulation inverter.

Waveforms in Fig.5 and Fig.6 are typical. Visually starting currents are acceptable.

6 Conclusion

Controller design by the Backstepping technique is direct and structured even when the system is non-linear. In this work, this strategy is applied to the induction motor with the use of observers of rotor flux and load torque. Simulation results are conclusive.

Compared to IRFOC and classical strategies, Backstepping is easily adapted to non-linear systems such as the induction motor.

Unlike many works, this paper proposes the numerical calculation of the controller and observer parameters according to the requirements already set.

In general, Backstepping does not require integral action, which makes it vulnerable to changes in machine and inverter parameters. Robustness of Backstepping and its improvement will be examined in next works.

References

[1] M. Elhaissouf, E. Lotfi, B. Rached, M. Elharoussi, and A. Barazzouk, "DSP Implementation In the Loop of the Indirect Rotor Field Orientation Control for the Three-Phase Asynchronous Machine," *Assoc. Comput. Mach.*, 2017.

[2] M. El Haissouf, M. El Haroussi, and A. Ba-razzouk, "Processor In the Loop Comparative Study of Indirect Rotor Field Oriented Control , Direct Self Control , Direct Torque Control and Space Vector Modulation based Direct Torque Control for Induction Motor Drives," *IREMOS*.

[3] F. Mehazzem, A. L. Nemmour, A. Reama, and H. Benalla, "Nonlinear Integral Backstepping Control for Induction Motors," *Acemp - Electromotion*, no. September, pp. 8–10, 2011.

[4] N. Mohan, *Advanced Electric Drives : Analysis, Control, and Modeling Using MATLAB/Simulink®*. WILEY, 2014.

[5] F. Wang, "Advanced Control Strategies of Induction Machine : Field Oriented Control, Direct Torque Control and Model Predictive Control," *energies*, 2018.

[6] F. Mehazzem, A. Lokmane, A. Reama, and D. Constantine, "Real time implementation of backstepping-multiscalar control to induction motor fed by voltage source inverter," *Int. J. Hydrogen Energy*, vol. 42, no. 28, pp. 17965–17975, 2017, doi: 10.1016/j.ijhydene.2017.05.035.

[7] A. Herizi, R. Balla, and S. A. Ahmani, "Backstepping Control of Induction Motors," vol. 8, pp. 132–145, 2015.

[8] B. Houcine, B. Othmane, and H. Abdelkader, "Commande de la machine asynchrone par la technique directe du couple backstepping avec onduleur à modulation vectorielle spatiale," *RRSTEE*, vol. 64, no. 2, pp. 81–86, 2019.

[9] M. Moutchou and M. Hamlich, "Robust Adaptive Backstepping Control Of The Induction Machine , Using Parallel Rotor Resistance And Stator Resistance Observer .," *Int. Conf. Smart Appl. Data Anal. Smart Cities*.

[10] C. Chen, H. Yu, F. Gong, and H. Wu, "Induction Motor Adaptive Backstepping Control and Efficiency Optimization Based on Load Observer," *energies*, 2020, doi: 10.3390/en13143712.

[11] I. Bakhti, "High performance backstepping control of induction motor with adaptive sliding mode observer," *Arch. Control Sci.*, vol. 21, no. 3, pp. 331–344, 2011.

[12] P. Martin and P. Rouchon, "Two simple flux observers for induction motors," *Int. J. Adapt. Control Signal Process.*, vol. 1115, no. March, 2000, doi: 10.1002/(SICI)1099-1115(200003/05)14.

[13] C. Kwon, S. Hui, S. Sudhoff, and S. Zak, "Rotor Flux and Speed Observers for Induction Motors," *POWER Electron. Intell. Control ENERGY Conserv.*, pp. 1–34, 2005.

[14] J. M. Dion and L. Dugard, *OBSERVER-BASED CONTROLLER FOR INDUCTION MOTORS*, vol. 35, no. 1. IFAC, 2002.

A Novel Method for DC Bus Voltage Sags Estimation Using the Switching Function Approach and ADALINE Method

Mounir Bensaid*, *Abdellfattah* Ba-Razzouk, *Mustapha* Elharoussi

Hassan First University of Settat, FST, Mathematics, Computer and Engineering Sciences Laboratory (MISI), Team: Systems Analysis and Information Processing (ASTI), Morocco.

Abstract. In this study, a functional model for diode bridge rectifiers (DB) is developed based on the concept of switching functions. The developed model is suitable for the estimation of the average voltage at the rectifier terminals from the fundamental three-phase voltage estimation based on the Adaline method and a DC bus estimator based on the switching function method, The performance of the detection estimator was validated in simulation using Matlab software under normal, unbalanced and line fault conditions. The high accuracy and efficiency of the developed estimator were demonstrated by comparison with rectifier models from SPS (SimPowerSystems), PSIM, and a reel rectifier from the pedagogical model (THREE-PHASE EDUCATIONAL INVERTER SEMITEACH 08753450BB)

1 Introduction

The conversion of energy between an alternating network and a medium or high-power electronic equipment passes more and more by a diode rectifier. However, in the context of electrical networks where it is increasingly considered, harmonic disturbances, voltage dips appear before and after the static converters [1][2]. These electrical disturbances, not very desirable, must respect standards. To analyse these electrical disturbances, models of the operation of static converters are necessary.
A voltage sag is defined as a momentary decrease in the RMS value of the nominal voltage for a duration ranging from 0.5 cycles to 1 minute [3][4]. Short circuits and the starting of large induction motors are the main causes of voltage sags. Voltage sags are the major cause of disturbances in variable speed drives, computers, and industrial process controllers [5].
A voltage dip is defined by two major parameters [15]. These parameters are as follows: the amplitude of the dip or depth, which defines the decrease in RMS voltage, and the duration, which defines the time the dip occurs [6].

The number and unit power of diode and thyristor rectifiers in electrical networks is constantly increasing. Static converters are expected to process 60% of the electrical energy produced in the future. These converters are widely used in variable speed drives, welding, electrolysis, and other applications. They are used in a wide range of applications and are required wherever AC/DC power conversion is required Variable speed drives (VSDs) appear to be more sensitive to voltage sags than data processing equipment (computers) [7].

* Bensaid mounir: bensaidmounir.eae@gmail.com

Voltage sags in (VSDs) are primarily manifested by a decrease in the voltage in the DC link [8]. In this paper, we will be interested in the variation of the average value of the rectified voltage as a function of the power supply conditions.

The difficulty in detecting voltage sags is in estimating the voltage amplitude. ADALINE, having been used to estimate the amplitude and phase of a signal's fundamental and harmonics, becomes an interesting avenue to investigate. Only the amplitude and phase of the fundamental are of interest in the case of voltage dips. The concept of function switching will be used. This concept was used to calculate the average voltage of a diode rectifier, and the accuracy of the results obtained is demonstrated.

This paper is organized as follows. The Analysis of a three-phase diode rectifier is discussed in Section 2. The algorithm of the ADALINE method and DC Bus Voltage estimation are presented in Section 3. Comparison and simulation results are presented in section 4. Finally, Section 5 concludes the paper.

2 Analysis of a three-phase diode rectifier

Consider the three-phase diode rectifier shown in Figure 1 connected to the three voltages source (v_a , v_b, v_c). The source has the internal resistance R_s and inductance

L_S. They include the values describing the connection lead. The bridge is loaded by a resistance R_d.

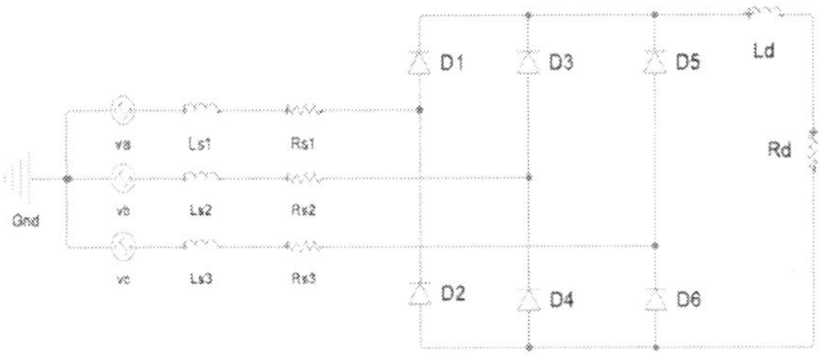

Fig. 1. Three-phase diode rectifier

Using the switching functions, we will calculate the average value of the rectified voltage. This rectifier will be analysed in both the balanced and unbalanced regimes.

The analysis hypotheses are:

- ✓ The source and power switches are perfect;
- ✓ The rectifier operates in continuous conduction mode;
- ✓ The voltages are sinusoidal in nature.
- ✓ The inductances and line resistors are identical.

2.1 Analysis of the diode rectifier in balanced regime

This is analysis under normal operating conditions in the absence of voltage dips or in the case of type A dips.

We will use the switching functions T_a, T_b, and T_c to calculate the average output value of a three-phase diode rectifier. T_a, T_b, and T_c are switching functions related to the conduction and blocking states of the diodes of each supply phase of the rectifier [9][10].

These switching functions are shown in Figure 2.

In the case of a balanced three phase source, the switching times are defined as follows:

$$\theta_1 = \frac{\pi}{6} \qquad \theta_2 = \theta_1 + \alpha$$
$$\theta_3 = \frac{\pi}{2} \qquad \theta_4 = \theta_3 + \alpha$$
$$\theta_5 = \frac{5\pi}{6} \qquad \theta_6 = \theta_5 + \alpha$$
$$\theta_7 = \frac{7\pi}{6} \qquad \theta_8 = \theta_7 + \alpha$$
$$\theta_9 = \frac{3\pi}{2} \qquad \theta_{10} = \theta_9 + \alpha$$
$$\theta_{11} = \frac{11\pi}{6} \qquad \theta_{12} = \theta_{11} + \alpha$$

α is the overlap angle

The switching functions are defined as follows:

$$T_a = \begin{cases} -1, & if \quad \theta_8 \le \theta \le \theta_{11} \\ -0{,}5, & if \quad \theta_7 \le \theta \le \theta_8 \quad and \quad \theta_{11} \le \theta \le \theta_{12} \\ 0, & if \quad 0 \le \theta \le \theta_1 \quad and \quad \theta_6 \le \theta \le \theta_7 \\ & \quad and \quad \theta_{12} \le \theta \le 2\pi \\ 0{,}5, & if \quad \theta_1 \le \theta \le \theta_2 \quad and \quad \theta_5 \le \theta \le \theta_6 \\ 1, & if \quad \theta_2 \le \theta \le \theta_5 \end{cases}$$

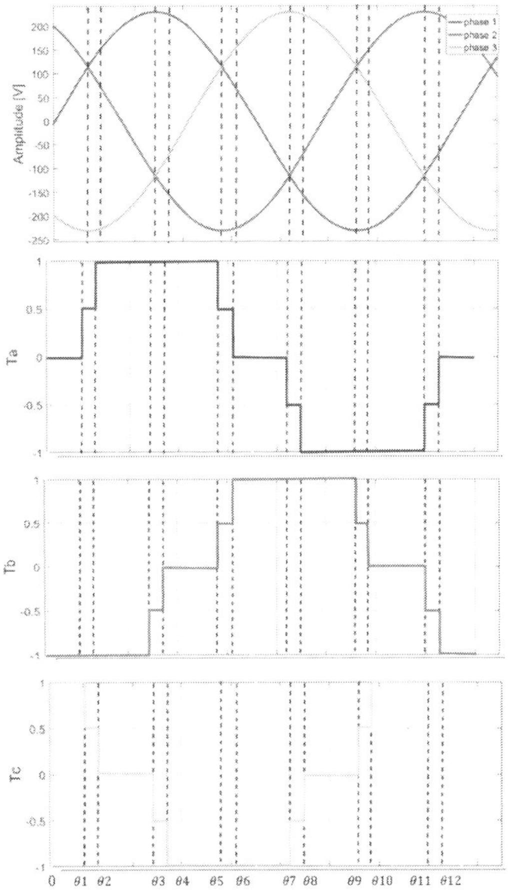

Fig. 2. Switching functions

$$T_b = \begin{cases} -1, & if \quad 0 \le \theta \le \theta_3 \quad and \quad \theta_{12} \le \theta \le 2\pi \\ -0{,}5, & if \quad \theta_3 \le \theta \le \theta_4 \quad and \quad \theta_{11} \le \theta \le \theta_{12} \\ 0, & if \quad \theta_4 \le \theta \le \theta_5 \quad and \quad \theta_{10} \le \theta \le \theta_{11} \\ 0{,}5, & if \quad \theta_5 \le \theta \le \theta_6 \quad and \quad \theta_9 \le \theta \le \theta_{10} \\ 1, & if \quad \theta_6 \le \theta \le \theta_7 \end{cases}$$

$$T_c = \begin{cases} -1, & if \quad \theta_4 \le \theta \le \theta_7 \\ -0{,}5, & if \quad \theta_3 \le \theta \le \theta_4 \quad and \quad \theta_7 \le \theta \le \theta_8 \\ 0, & if \quad \theta_2 \le \theta \le \theta_3 \quad and \quad \theta_8 \le \theta \le \theta_9 \\ 0{,}5, & if \quad \theta_1 \le \theta \le \theta_2 \quad and \quad \theta_9 \le \theta \le \theta_{10} \\ 1, & if \quad 0 \le \theta \le \theta_1 \quad and \quad \theta_{10} \le \theta \le 2\pi \end{cases}$$

The voltage at the output of the rectifier is defined as follows

$$V_d = T_a v_a + T_b v_b + T_c v_c - (T_a^2 + T_b^2 + T_c^2) I_d R_s \qquad (1)$$

Where v_a, v_b and v_c represent the three-phase voltage source and are given by:

$$v_a = V_a sin(\theta)$$
$$v_b = V_b sin(\theta - \frac{2\pi}{3}) \qquad (2)$$
$$v_c = V_c sin(\theta - \frac{4\pi}{3})$$

As an analysis assumption, we consider Rs, the resistance of the source impedance negligible, and equation (1) becomes:

$$V_d = T_a v_a + T_b v_b + T_c v_c \quad (3)$$

The Fourier series development of the rectified voltage is given by:

$$V_d = U_{d0} + \sum_{m=6}^{\infty} (A_{dm} cosm\theta + B_{dm} sinm\theta) \quad (4)$$

Where U_{d0} is the average value of the voltage V_d
And A_{dm}, B_{dm} are the Fourier coefficients of the rectified voltage's harmonic components.
We are only interested in the average value of the redressed tension in this analysis, so:

$$U_{d0} = \frac{1}{2\pi} \int_0^{2\pi} [T_a v_a + T_b v_b + T_c v_c] d\theta \quad (5)$$

equation 5 gives:

$$U_{d0} = \frac{1}{2\pi} \left[\int_0^{2\pi} [T_a v_a] d\theta + \int_0^{2\pi} [T_b v_b] d\theta + \int_0^{2\pi} [T_c v_c] d\theta \right]$$

let's put:

$$U_{d0A} = \int_0^{2\pi} [T_a v_a] d\theta$$

$$U_{d0B} = \int_0^{2\pi} [T_b v_b] d\theta$$

$$U_{d0C} = \int_0^{2\pi} [T_c v_c] d\theta$$

After development, we obtain:

$$U_{d0A} = \sqrt{3} V_a (1 + \cos(\alpha))$$
$$U_{d0B} = \sqrt{3} V_b (1 + \cos(\alpha))$$
$$U_{d0C} = \sqrt{3} V_c (1 + \cos(\alpha))$$

therefore,

$$U_{d0} = \frac{1}{2\pi} [U_{d0A} + U_{d0B} + U_{d0C}] = \frac{\sqrt{3}}{2\pi} (1 + \cos(\alpha) [V_a + V_b + V_c]$$

If $V_a = V_b = V_c = V_{LN}$ then the amplitude of the line-to-neutral voltage

$$U_{d0} = \frac{3\sqrt{3}}{2\pi} V_{LN} (1 + \cos(\alpha))$$

Or again,

$$U_{d0} = \frac{3}{2\pi} V_{LL} (1 + \cos(\alpha))$$

Where V_{LL} is the amplitude of the line-to-line voltage.
Forgetting the overlap effect, we obtain:

$$U_{d0C} = \frac{3\sqrt{3}}{\pi} V_{LN} = \frac{3}{\pi} V_{LL} \quad (6)$$

2.2 Analysis of the diode rectifier in unbalanced regime

In the presence of unbalanced voltage dips, the intervals of the switching functions change, because the terminals of these intervals represent the zero crossings of the switching voltages (v_{ac}, v_{ba}, v_{cb}, v_{ab} et v_{bc}.

For example, θ_1 and θ_7 will be found for $v_a = v_c$ (i.e. $v_{ac} = 0$); by substituting v_a and v_c by their expression we have:

$$\begin{cases} V_a sin(\theta_1) = V_c sin\left(\theta_1 + \frac{2\pi}{3}\right) \\ V_a sin(\theta_7) = V_c sin\left(\theta_7 + \frac{2\pi}{3}\right) \end{cases} \quad (7)$$

We get two angles after trigonometric development: θ_1 and θ_7:

$$\theta_1 = tan^{-1}\left(\frac{\sqrt{3}V_c}{2V_a + V_c}\right)$$
$$\theta_7 = tan^{-1}\left(\frac{\sqrt{3}V_c}{2V_a + V_c}\right) + \pi \quad (8)$$

θ_3 and θ_9 will be found for $v_c = v_b$ (i.e. $v_{bc} = 0$) and we obtain:

$$\theta_3 = tan^{-1}\left(\frac{\sqrt{3}(V_c + V_b)}{(V_c - V_b)}\right)$$
$$\theta_9 = tan^{-1}\left(\frac{\sqrt{3}(V_c + V_b)}{(V_c - V_b)}\right) + \pi \quad (9)$$

θ_5 and θ_{11} will be found for $v_a = v_b$ (i.e. $v_{ba} = 0$) and we obtain:

$$\theta_5 = tan^{-1}\left(\frac{-\sqrt{3}V_b}{2V_a + V_b}\right)$$
$$\theta_{11} = tan^{-1}\left(\frac{-\sqrt{3}V_b}{2V_a + V_b}\right) + \pi \quad (10)$$

After development, we get:

$$U_{d0A} = \frac{1}{2} V_a \{(1 + \cos(\alpha))(cos(\theta_1) - cos(\theta_5) - cos(\theta_7) + cos(\theta_{11})) - s\,in(\alpha)(sin(\theta_1) - sin(\theta_5) - sin(\theta_7) + sin(\theta_{11}))\}$$

$$U_{d0B} = \frac{-1}{2} V_b \{(1 + \cos(\alpha) - \sqrt{3}\sin(\alpha))(cos(\theta_3) + cos(\theta_5) - cos(\theta_9) - cos(\theta_{11})) - (\sqrt{3}\,(1 + \cos(\alpha)) + s\,in(\alpha))(sin(\theta_3) + sin(\theta_5) - sin(\theta_9) - sin(\theta_{11}))\}$$

$$U_{d0C} = \frac{1}{4} V_c \{(1 + \cos(\alpha) + \sqrt{3}\sin(\alpha))(cos(\theta_1) + cos(\theta_3) - cos(\theta_7) - cos(\theta_9)) + (\sqrt{3}\,(1 + \cos(\alpha)) - s\,in(\alpha))(sin(\theta_1) + sin(\theta_3) - sin(\theta_7) - sin(\theta_9))\}$$

Ignoring the overlap effect on the average voltage, we obtain:

$$U_{d0A} = V_a \{(cos(\theta_1) - cos(\theta_5) - cos(\theta_7) + cos(\theta_{11}))\}$$

$$U_{d0C} = \frac{-1}{2} V_b \{(cos(\theta_3) + cos(\theta_5) - cos(\theta_9) - cos(\theta_{11})) - \sqrt{3}(sin(\theta_3) + sin(\theta_5) - sin(\theta_9) - sin(\theta_{11}))\}$$

$$U_{doC} = \frac{1}{2}V_c\{(cos(\theta_1) + cos(\theta_3) - cos(\theta_7) - cos(\theta_9)) + \sqrt{3}(sin(\theta_1) + sin(\theta_3) - sin(\theta_7) - sin(\theta_9))\}$$

With:

$$U_{d0} = \frac{1}{2\pi}[U_{d0A} + U_{d0B} + U_{d0C}] \quad (11)$$

3 DC Bus average Voltage estimation

The proposed DC Bus voltage estimation method consists of:

- ✓ ADALINE method was used in the estimation of the amplitude and phase of the fundamental of the three-phase source, with adaptation by the least squares method (LMS)
- ✓ The estimator of the average value of the diode rectifier's DC link voltage developed on the basis of the switching function theory presented in section 2.

ADALINE (Adaptive Linear Neuron or later Adaptive Linear Element), as described in [11][12], can estimate periodic signals.

$$y(t) = \sum_{n=1}^{N}((X_n \operatorname{Cos}(n\omega t) + (Y_n \operatorname{Sin}(n\omega t)) \quad (12)$$

The structure of the ADALINE network is described in Figure 3.

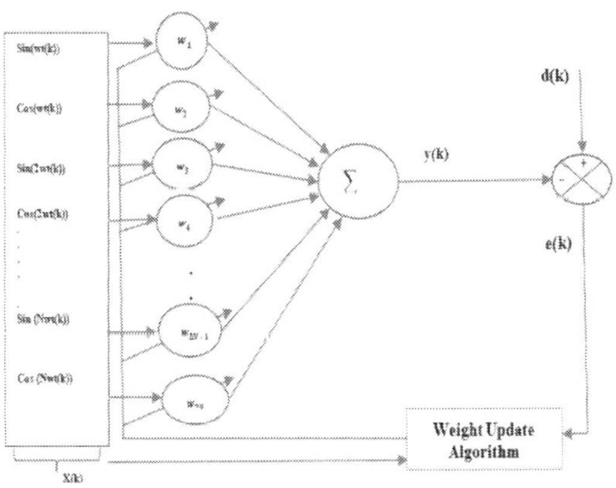

Fig. 3. ADALINE network structure

The flow chart of the ADALINE method is shown in Figure 4. This algorithm gives results with acceptable accuracies in steady-state

The three-phase detection uses three ADALINE networks to estimate the amplitude and phase of the power source voltages, and the output is used to generate the input signal of the DC bus average value estimator.

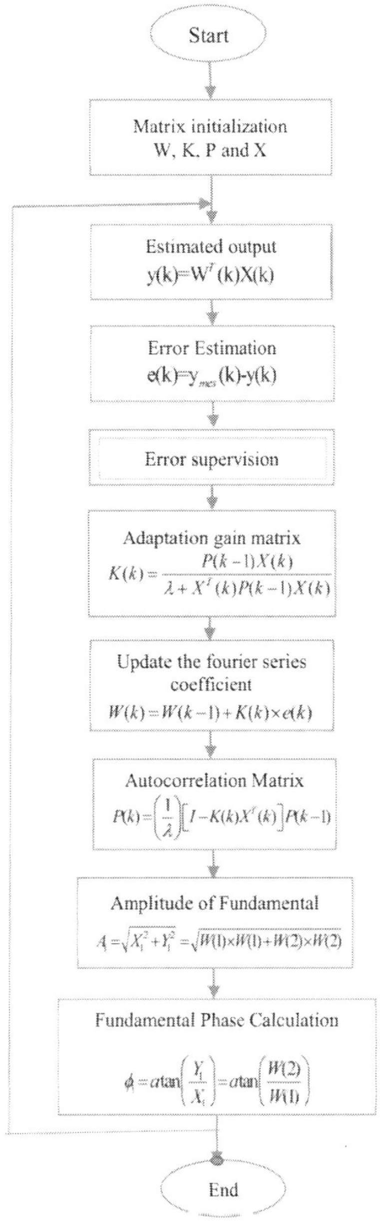

Fig. 4. Flow chart for ADALINE algorithm with Error supervision

4 Performance of DC Bus Voltage estimation

The estimator's implementation necessitates the creation of S-Functions in C. An SFunction is a programming language used to describe Simulink blocks. It can be written in the Matlab language, C, C++, Ada, or Fortran. An S-function allows you to design your own Simulink blocks. The mex command is then used to compile this program.

For the detection of the fundamental and the phase of the voltage source, three distinct estimators are implemented in the Fundamental Estimators blocks chow in figure 6. It is the same code (algorithm) with the difference is that the names assigned to them are different as well as the names of the static variables of the program

The second estimator implemented in the average DC bus voltage estimator block shown in figure 6 is that of the average value of the voltage of the dc link of the diode rectifier. This algorithm was developed on the basis of the theory of switching functions presented in section 2

The results of our estimator and the experimental results based on (THREE-PHASE EDUCATIONAL INVERTER SEMITEACH 08753450BB) shown in figure 5 available in the laboratory and the simulation results obtained using SPS (SimPowerSystems) and PSIM are shown in Table 1.

Fig. 5. ADALINE network structure

Table 1. Average values of the rectified voltage for different supply voltage conditions

		Average value [V]			
voltage type	$(V_a, V_b, V_c)[V]$	Estimator	SPS	PSIM	SEMITEACH 08753450 BB
balanced	(110,110,110)	257	255.84	257.36	263,4
	(141,141,141)	329.82	327.94	329.89	342,5
	(180,180,180)	421.04	418.65	421.14	428,3
	(200,200,200)	467.82	465.16	467.93	475,9
unbalanced	(20, 120, 120)	211,74	210,04	211,73	
	(20,20, 120)	133,67	132,07	133,80	
	(120, 60, 120)	236,48	234,88	236,58	

The accuracy of the results obtained with Estimator using the Adaline method and the switching functions is thus validated, with the small differences obtained with SPS due to differences in the switch models. For example, our analysis takes into account perfect switches, which have zero resistance for a closed switch and infinite resistance for an open switch, whereas PSIM takes into account a high resistance of 1.5 MΩ for an open switch and a low resistance of $11\mu\Omega$ for a closed switch.

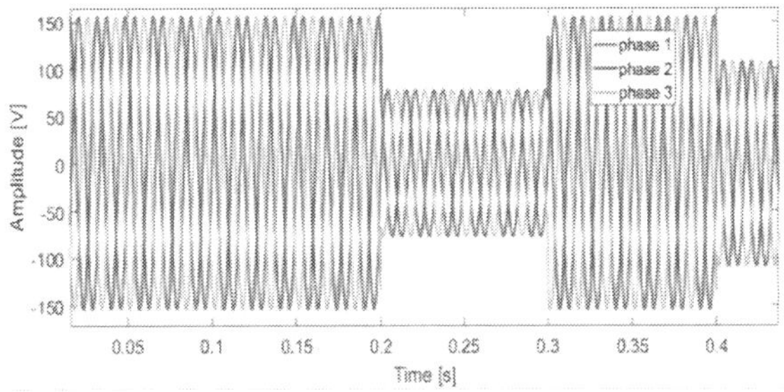

Fig. 7. Three-phase voltages for symmetrical faults, under 50% and 30%

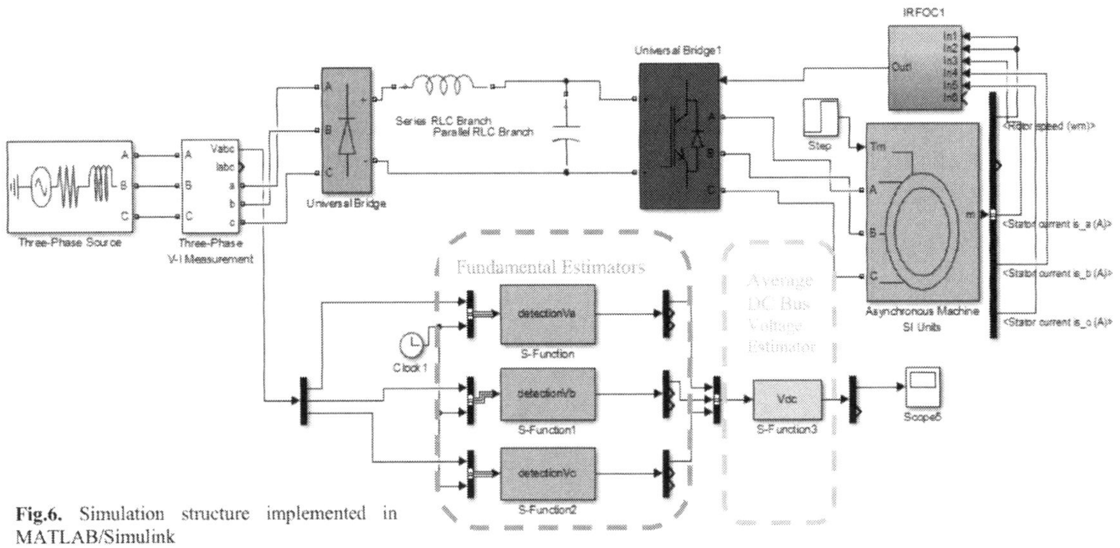

Fig.6. Simulation structure implemented in MATLAB/Simulink

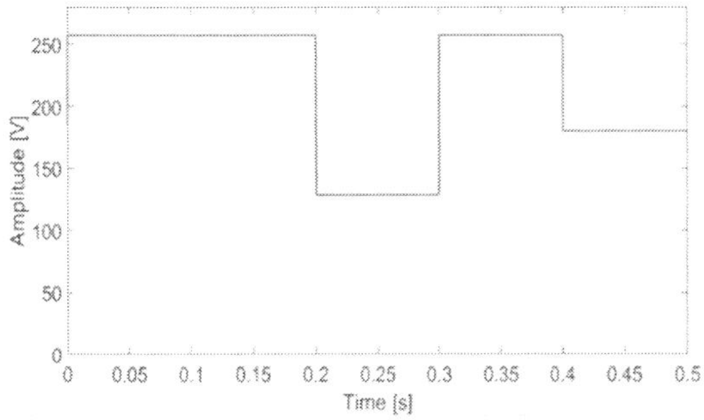

Fig. 8. Estimated DC Bus average voltage under 50% and 30%

Figure 7 represents Three-phase voltages voltage waveforms for symmetrical voltage sags from 50% between 0.2 and 0.3 s, and from 30% between 0.2 and 0.3 s respectively; Figure 8 corresponds to real data measurements estimated average voltage under symmetrical faults. The voltage sag is identified, this method can detect DC Bus average voltage in real-time, and it is very accurate in determining inception and recovery instants.

5 Conclusion

This paper presents a method to estimate the inception and recovery instants of the DC Bus average voltage sag, and the event duration, figure 8 summarizes the performance of the method. the Algorithm of the ADALINE method applied to the estimation of the fundamental and the phase of the three-phase source has been exposed, and the concept of the switching functions has allowed us to analyse and determine equations for the calculation of the average value of the rectified voltage, in the presence of balanced and unbalanced voltage dips for the different converters. Some measurements presented in Table 1 show the validity of the developed equations. These models are very useful for the development of a voltage dip management strategy for multi-motor systems.

References

1. G.Devadasu, S.Muthubalaji, S.Srinivasan, Anil Gunde " An experimental approach of DVR for relieving voltage dip and voltage swell " https://doi.org/10.1016/j.matpr.2021.01.056
2. A. Moreno-Munoz, J. J. G. de la Rosa, "Voltage Sag in Highly Automated Factories", Industry Applications Society Annual Meeting 2008. IAS '08. IEEE, pp. 1-6, 2008.
3. M. H. J. Bollen, "Understanding Power Quality Problems - Voltage Sags and Interruptions", 1st edition, IEEE Press, 2000.
4. M.Bensaid, A. Barazzouk, M. Elharoussi, and B. Rached "Effects of Symmetrical Voltage Sags on Two Induction Motors System Coupled with An Elastic Web". Proceedings of the 2020 IEEE 2nd International Conference on Electronics, Control, Optimization and Computer Science (ICECOCS), pp.1-6,2020.
5. Imdadullah , Syed Muhammad Amrr , Atif Iqbal , M.S. Jamil Asghar "Comprehensive performance analysis of flexible asynchronous AC link under various unbalanced grid voltage conditions" DOI: 10.1016/j.egyr.2021.01.028
6. Alvaro Furlani Bastos; Surya Santoso; Grazia Todeschini "Comparison of Methods for Determining Inception and Recovery Points of Voltage Variation Events" DOI: 10.1109/PESGM.2018.8585977
7. Tao Yang,Serhiy Bozhko,Greg Asher "Fast functional modelling of diode-bridge rectifier using dynamic phasors" https://doi.org/10.1049/iet-pel.2014.0414
8. Surya Hardi, R. Harahap, S. Ahmad, M. Isa"Ride through testing of variable speed drive due to voltage sag types (types I, II and III) "DOI: 10.11591/ijpeds.v10.i2.pp690-696
9. C. Marouchos, M. Darwish and M. El-Habrouk, "New mathematical model for analyzing three-phase controlled rectifier using switching function," IET Power Electron., vol. 3, pp. 95-110, Jan. 2010
10. K. L. Lian; Brian K. Perkins; P. W. Lehn "Harmonic Analysis of a Three-Phase Diode Bridge Rectifier Based on Sampled-Data Model" DOI: 10.1109/TPWRD.2008.917671
11. Mohsen Jannati , Mohammad Mohammadi « A novel fast power swing blocking strategy for distance relay basedon ADALINE and moving window averaging technique » DOI: 10.1049/gtd2.12015,2020
12. Sarawut Janpong, Kongpol Areerak and Kongpan Areerak «Harmonic Detection for Shunt Active Power Filter Using ADALINE Neural Network» doi.org/10.3390/en14144351;,2021

Design of a Particle Swarm Optimized Photovoltaic system during various partial shading conditions

Hamid Chekenbah[1*], *Yassir* El Maataoui[1], *Omar* Boutfarjoute[1], *Abdellatif* El Abderrahmani[2], and *Rafik* Lasri[1]

[1]LSTA, FPL, Abdelmalek Essaâdi University, Tetouan, Morocco

[2]CSSAC Laboratory, FSDM, University of Sidi Mohammed Ben Abdellah, Fez, Morocco

Abstract. This work discusses an enhanced photovoltaic system using the Particle Swarm Optimization algorithm. The proposed method has been evaluated under several scenarios of partial shading, and the results obtained are contrasted with the Perturb and Observe technique. Hence, it has shown its effectiveness in locating the global maximum power point whatever the meteorological conditions, especially in inhomogeneous conditions. This technique makes it possible to enhance the effectiveness of the photovoltaic system studied because it makes it possible to minimize the loss of power in the case of non-homogeneous irradiation conditions.

1 Introduction

Solar photovoltaic energy is considered the greatest reliable sustainable energy source that respects the environment. For this reason, several studies have been effectuated aiming to optimize the production of this energy. But the main drawback of this is its dependence on different parameters, such as irradiation, cell temperature, shading as well as the overall aging of the photovoltaic components.

Among the techniques for optimizing PV energy, Maximum Power Point Trackers (MPPT) plays the primary role. Indeed, MPPT algorithms enhance the efficiency of PV systems.

The usefulness of these MPPTs is thought to the non-linearity of the photovoltaic generators' characteristics, and the fact that these characteristics have a single maximum power, as illustrated in Figure 1. In other hand, we can conclude from Figure 1.a that there is only one value of the load (i.e. optimal load) for which the operating point coincides with the Point of the Maximum Power (points B), otherwise, the operating point will be further from the Point of the Maximum Power (points A, C, and D).

On the other hand, a decrease or an increase in irradiation leads to a modification of the characteristics of the photovoltaic generators, as described in Figure 1.b. This leads to the modification of the position of the MPP and thus the position of the operating points (A', B', C', and D'). Which requires the use of an MPPT to track the MPP.

These algorithms differ from each other from the simplest, like the Perturb and Observe [1] method, the Incremental Conductance technique, and the optimization technique based on the Hill Climbing (HC) method [2], [3].

In the literature, other methods can be considered complex methods, among these techniques, we find trackers based on Fuzzy Logic (FLC) [2], [4]. The FLC technique allows overcoming the restrictions observed in classical methods (P&O and I&C) concerning the choice of the incrementation step in the case of unstable meteorological conditions. Neural networks are used in particular in tracking the maximum power point [5] and make it possible to obtain significant results.

The integration of neural networks in fuzzy controllers makes it possible to adjust the internal parameters of fuzzy controllers. These techniques, used in [6], [7] for example, improve the efficiency of PV systems, but on the other hand, require significant computation time. These complex trackers based on soft computing techniques have better yields, but they necessitate some requirements when implemented in reconfigurable circuits. Therefore, the main difficulty is to build high-performance algorithms that do not require large hardware requirements.

The presented techniques present high efficiency in standard conditions, but thanks to the unstable nature of meteorological conditions, this efficiency can make remarkable decreases. In the apparition of Partial Shading Conditions (PSC), this decrease becomes important since this phenomenon influences the characteristics of the photovoltaic generator, resulting in the appearance of several maxima in these curves as shown in Figure 2.

* Corresponding author: hamid.chekenbah@gmail.com

 https://doi.org/10.1051/itmconf/20224804004

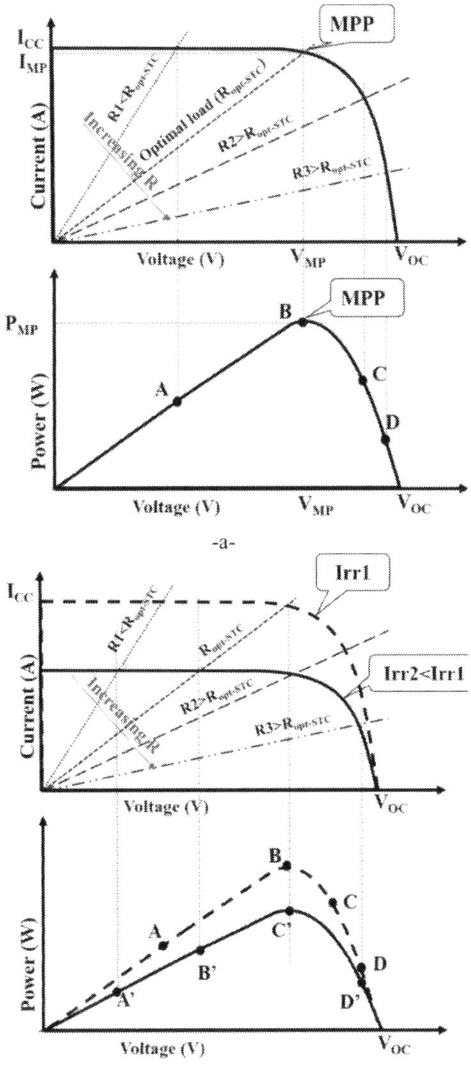

Fig. 1: Direct connection betwixt a PV model and a linear load

These conventional MPPT techniques often do not make it possible to track the true MPP (global MPP) since they cannot differentiate between local (points A, C, and D in Figure 1.a) and global MPPs (point B).

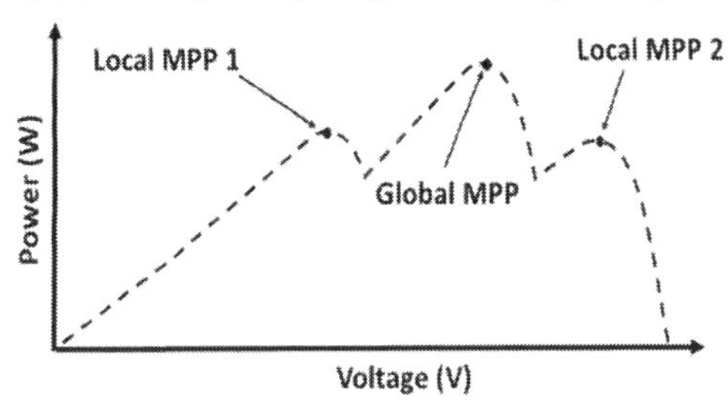

Fig. 2. Characteristics of photovoltaic generators

Among the most efficient techniques, the technique based on the β parameter demonstrates its effectiveness, especially in the detection of PSC [8]. This technique allows the detection of the existence of the PSC phenomenon, and thanks to a linear equation makes it possible to make the functional point in the proximity of the true MPP (Global-MPPT). However, an instantaneous change in weather conditions disrupts the operation of this technique.

Several other advanced Evolutionary techniques have been used to localize the global MPP, among which are the Artificial Bee Colony Optimization technique, and the Genetic Algorithm.

In this work, an MPP tracker based on the PSO technique is implemented on a PV system. This system consists of a photovoltaic array built from the storage of 4 PV modules, a Boost-type converter, and the presented MPPT-PSO. The studied photovoltaic system is summarized in Figure 3.

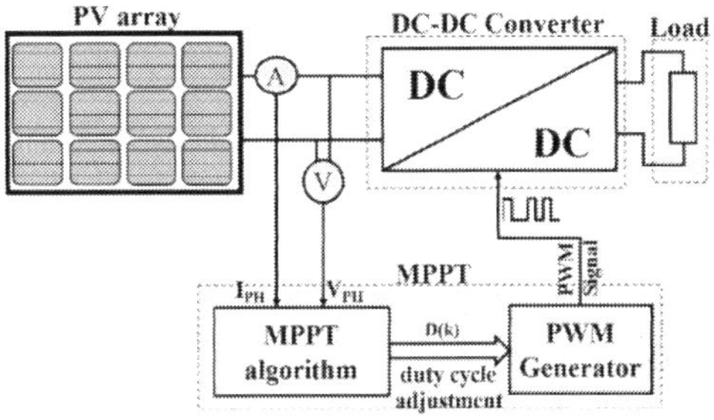

Fig. 3. PV system studied.

2 Modeling of the Studied photovoltaic System

2.1 Modeling a PV array

A PV generator is conceived by the arrangement in series and parallel of PV cells, to furnish the desired voltage (parallel connection) and the output current (cells in series) values.

➢ **Modeling of photovoltaic cells**

In the literature, we found several equivalent models of a photovoltaic cell, but the most commonly utilized is the single diode model as described in Figure 4. It includes a source of current, one diode, a series, and parallel resistance. The resistances are used to model the power dissipation through the internal resistances of the PV cell during its operation.

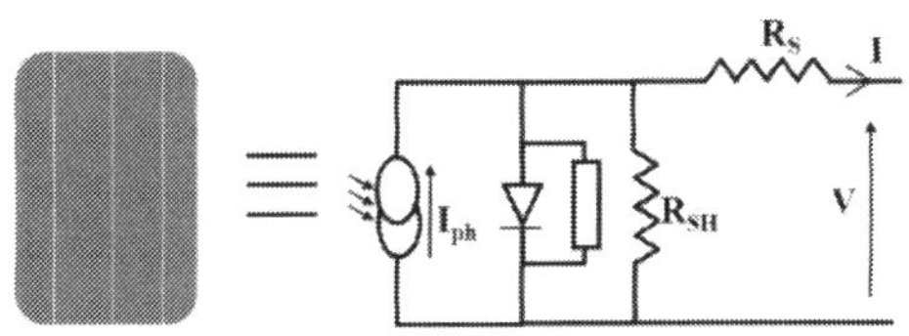

Fig.4 : Photovoltaic cell model.

From Figure 4, the generated current by the photovoltaic cell can be presented by (1) [9].

$$I = I_{ph} - I_{S0}.(e^{\frac{V+IR_S}{\alpha V_T}} - 1) - \frac{V + IR_S}{R_P} \quad (1)$$

With,

- I_{ph}: the photo-current of the cell
- Is0: the inverse PV cell diode saturation current
- A: The factor of the ideality of the diode
- T: Cell temperature
- V: The PV cell voltage
- I: The generated current of the cell

➢ **Modeling of a Photovoltaic Array:**

A photovoltaic generator consists of several photovoltaic cells arranged in series and/or in parallel. In this paper, the BP365TS module with 65.25 W of maximum power, which contains NS=18 cells mounted in series, has been selected to ratify the system.

The I(V) characteristic of such generator PV (made up of NS ×NP cells) is represented by equation (2).

$$I = N_P(I_{ph} - I_{SO}(e^{\frac{V+IR_S}{\alpha V_T N_S}} - 1) - (\frac{V + IR_S}{N_S.R_P})) \quad (2)$$

To meet the needs of consumers, PV arrays are designed to generate remarkable voltages. These arrays are made by the arrangement of PV generators in series and/or in parallel. In the present work, a PV array designed from the series arrangement of four PV modules (BP365TS) is used, Figure 5 illustrates the diagram of the PV used array.

As shown in Figure 5, anti-return diodes are used when making the PV generator. These diodes contain bypass diodes in parallel and blocking diodes in series. Blocking diodes are used in photovoltaic generators to prevent reverse currents, and bypass diodes are used to eliminate the creation of hot spot phenomena that can damage PV cells and allow PV generators to operate with high reliability throughout their lifespan.

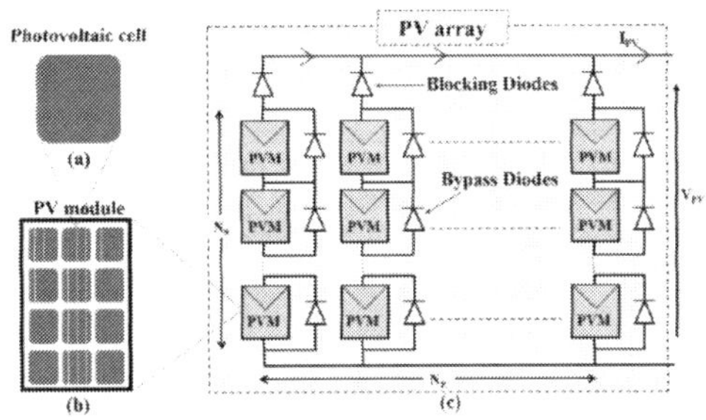

Fig. 5. Photovoltaic array.

2.2 Modeling of a boost converter

The DC-DC converter is an important component in the electronic domain, its part is to transfer a DC voltage into a same type voltage but with a different level [10]. Figure 6 presents the equivalent circuit of a boost converter[11].

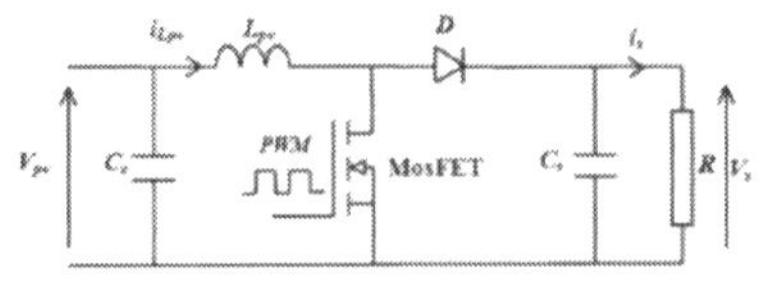

Fig. 6. Boost converter

The values of the parameters constituting the Boost converter must verify the conditions of the relations (3) and (4).

$$L \geq \frac{(1 - D)^2.D.R}{2fc} \quad (3)$$

$$C \geq \frac{V_0.D.R}{\Delta V_0.f.R_{min}} \quad (4)$$

Where ΔV0 is the ripple voltage.

By respecting the conditions mentioned in (3) and (4), the values used in the DC-DC converter are described in Table1.

Table 1. Parameter of the boost DC-DC converter used

Parameter	Value
Switching-frequency	50 KHz
Inductance, L	1.148 mH
Input Capacity, C_{in}	10 μF
Output Capacity, C_{out}	468 μF
Load, R	23 Ω

2.3 Algorithm Perturb and Observe

The Perturbation and Observation method consists of perturbing (increasing/decreasing) the PV voltage with a predetermined voltage value V_{ref} to reach MPP. The PV power $P_s(k)$ is then evaluated for each iteration, this value is then compared to P_s (k-1). If the output P_s (k) is superior to the output P_s (k-1), the variation of the next perturbation will be maintained in the previous orientation, otherwise, it will go in the observed orientation. The diagram of this technique is shown in Figure 7.

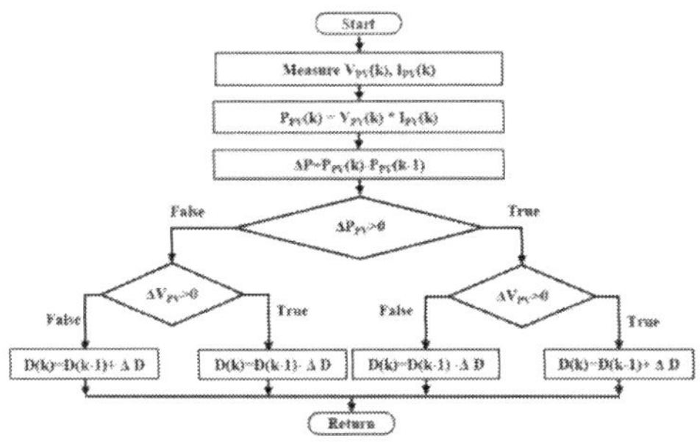

Fig. 7: Perturb and Observe flowchart

2.4 Modeling of the Particle Swarm Optimization method

The PSO method is a metaheuristic approche commonly utilized in the optimization of complex systems. This technique is inspired by the living world

https://doi.org/10.1051/itmconf/20224804004

[12], [13], it is an evolutionary algorithm that uses an ensemble of candidate solutions to locate the optimal solution to the problem under study.

Mathematically, this method can be modeled according to two main factors, the velocity defined by equation (5) and the position of the particle defined by (6) [13]. The principle of this method is described in Figure 8.

$$\vec{v_i}(k+1) = \omega.\vec{V_i} + C_1 r_1(\overrightarrow{p_{best}} - \vec{d_i}) + C_2 r_2(\overrightarrow{g_{best}} - \vec{d_i}) \quad (5)$$

$$d(k+1) = d_i(k) + v_i(k+1) \quad (6)$$

With;

- $v_i(k)$: The ith instantaneous particle velocity.
- $d_i(k)$: The ith particle.
- $\overrightarrow{P_{best}}$: Best position through the particle (i).
- $\overrightarrow{G_{best}}$: Best position (global position).
- ω : The inertia coefficient.
- $c_{1,2}$: The acceleration parameters.
- $r_{1,2}$: Random parameters on the interval [0,1] applied to the ith particle.

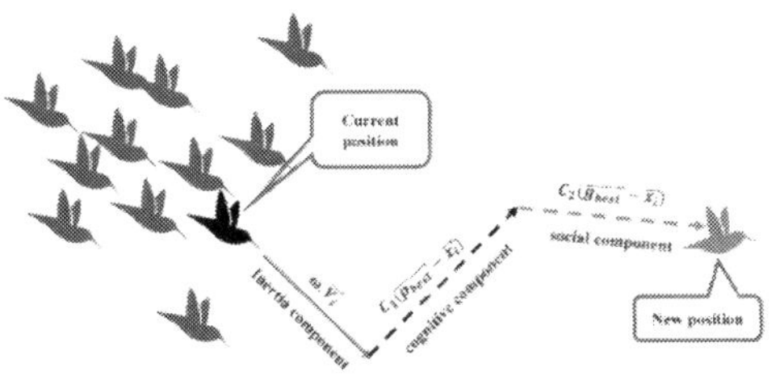

Fig. 8. Principle of the PSO technique.

In the present work, the PSO algorithm is utilized to regulate the duty cycle value used to control the PWM generator. Equation (7) shows the relationship to determine this relationship.

$$d(k) = G_{best}(k) \quad (7)$$

The parameters used during the implementation of the studied PSO technique are described in Table 2.

Table 2. The PSO main parameters.

PSO parameter	Value
Inertia-Weight(w)	0.4
Personal-coefficient(c1)	1.2
Global-coefficient(c2)	2
Number of particles N	10

A minimum number of particles (N=10) is used to avoid computational complexity per iteration, this number is necessary to cover the entire search space for each iteration. The number of iterations is also chosen to avoid unnecessary additional computational complexity. The inertia weight (w=0.4) is chosen in order to avoid delayed convergence (large inertia weight) and local trapping (low weight). This value is fixed after several trial and error tests. Finally, the values of the acceleration coefficients are chosen so that (c2 >> c1). This condition is used so that the particles are more strongly attracted toward the best global position.

3 Results and discussion

The photovoltaic generator is exposed to various partial shading situations for examining the efficiency of the presented method. Fig 9, 10, and 11 present the results obtained for the following scenarios:

- Scenario 1: Instant change of irradiation.
- Scenario 2: (1000 – 1000 – 1000 – 1000) W/m2.
- Scenario 3: (500 – 800 – 1000 – 1000) W/m2.

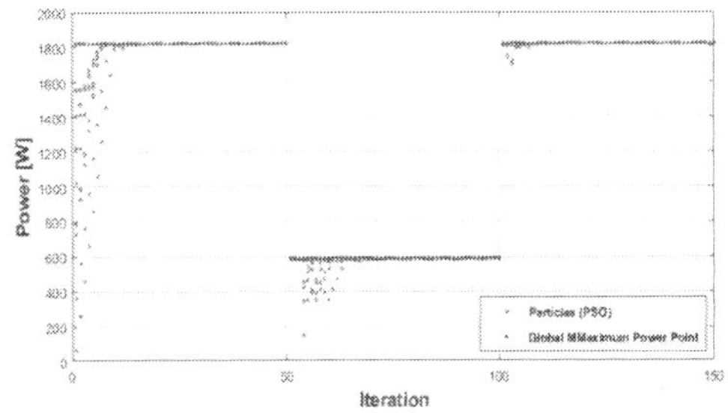

Fig. 9. Instant change of irradiation (scenario 1).

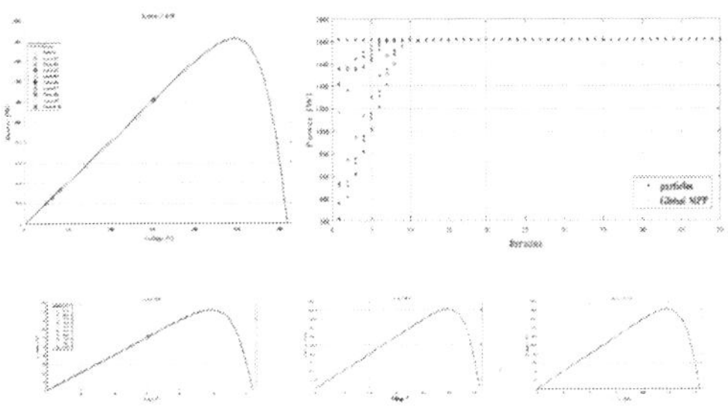

Fig. 10. Results obtained for scenario 2 (a, b, and c: Position of swarm particles, d: History of particle positions)

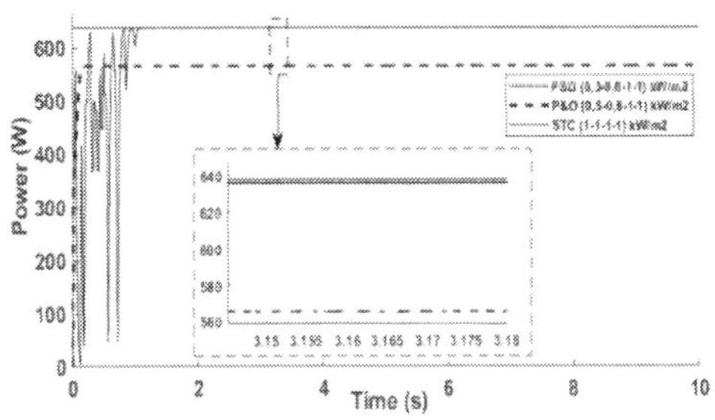

Fig. 11. Comparison between PSO and Perturb and Observe technique (scenario 3).

The obtained results prove the efficiency of the PSO approche compacted to the conventional methods since it allows to follow the true MPP with zero oscillation. In contrast to the Perturb and Observe technique, which does not differentiate betwixt maxima local and global, the PSO method allows for localization of the global MPP among a set of local maxima.

4 Conclusion

In this work, an MPPT relying on the PSO algorithm is presented and implemented in a PV system. The PSO is utilized to surmount the problem of locating the global MPP among local maxima that occurred in the presence of PSC. The obtained results prove the efficiency of the presented approach, with a 12.6% improvement in performance over the Perturb and Observe controller.

References

[1] F. A. O. Aashoor et F. V. P. Robinson, « A variable step size perturb and observe algorithm for photovoltaic maximum power point tracking », présenté à Proceedings of the Universities Power Engineering Conference, 2012. doi: 10.1109/UPEC.2012.6398612.

[2] R. Boukenoui, R. Bradai, A. Mellit, M. Ghanes, et H. Salhi, « Comparative analysis of P&O, modified hill climbing-FLC, and adaptive P&O-FLC MPPTs for microgrid standalone PV system », 2015, p. 1095-1099. doi: 10.1109/ICRERA.2015.7418579.

[3] K. S. M. Raza, H. Goto, H.-J. Guo, et O. Ichinokura, « Novel speed-sensorless adaptive hill climbing algorithm for fast and efficient maximum power point tracking of wind energy conversion systems », 2008, p. 628-633. doi: 10.1109/ICSET.2008.4747083.

[4] M. Adly et A. H. Besheer, « An optimized fuzzy maximum power point tracker for stand alone photovoltaic systems: Ant colony approach », 2012, p. 113-119. doi: 10.1109/ICIEA.2012.6360707.

[5] S. Allahabadi, H. Iman-Eini, et S. Farhangi, « Fast Artificial Neural Network Based Method for Estimation of the Global Maximum Power Point in Photovoltaic Systems », *IEEE Trans. Ind. Electron.*, vol. 69, n° 6, p. 5879-5888, juin 2022, doi: 10.1109/TIE.2021.3094463.

[6] A. Arora et P. Gaur, « Comparison of ANN and ANFIS based MPPT Controller for grid connected PV systems », présenté à 12th IEEE International Conference Electronics, Energy, Environment, Communication, Computer, Control: (E3-C3), INDICON 2015, 2016. doi: 10.1109/INDICON.2015.7443568.

[7] S. D. Al-Majidi, M. F. Abbod, et H. S. Al-Raweshidy, « Design of an efficient maximum power point tracker based on ANFIS using an experimental photovoltaic system data », *Electron. Switz.*, vol. 8, n° 8, 2019, doi: 10.3390/electronics8080858.

[8] X. Li, H. Wen, Y. Hu, et L. Jiang, « A novel beta parameter based fuzzy-logic controller for photovoltaic MPPT application », *Renew. Energy*, vol. 130, p. 416-427, janv. 2019, doi: 10.1016/j.renene.2018.06.071.

[9] S. Rustemli et F. Dincer, « Modeling of Photovoltaic Panel and Examining Effects of Temperature in Matlab/Simulink », *Elektron. Ir Elektrotechnika*, vol. 109, n° 3, p. 35-40-40, 2011, doi: 10.5755/j01.eee.109.3.166.

[10] H. Chekenbah, L. Choukri, R. Lasri, Y. Maataoui, et M. Zouiten, « Implmeneting a Fuzzy Logic Controller to Improve the Performances of an Off-Grid Photovoltaic Generator », in *2019 4th World Conference on Complex Systems (WCCS)*, avr. 2019, p. 1-6. doi: 10.1109/ICoCS.2019.8930760.

[11] N. Mohan, T. M. Undeland, et W. P. Robbins, *Power electronics: converters, applications, and design*, 3rd ed. Hoboken, NJ: John Wiley & Sons, 2003.

[12] F. Marini et B. Walczak, « Particle swarm optimization (PSO). A tutorial », *Chemom. Intell. Lab. Syst.*, vol. 149, p. 153-165, déc. 2015, doi: 10.1016/j.chemolab.2015.08.020.

[13] S. Makhloufi et S. Mekhilef, « Logarithmic PSO-Based Global/Local Maximum Power Point Tracker for Partially Shaded Photovoltaic Systems », *IEEE J. Emerg. Sel. Top. Power Electron.*, vol. 10, n° 1, p. 375-386, févr. 2022, doi: 10.1109/JESTPE.2021.3073058.

AUTHOR INDEX